HOG TIES

HOG TIES

WHAT PIGS TELL US ABOUT AMERICA

Richard P. Horwitz

University of Minnesota Press
Minneapolis
London

Originally published in hardcover by St. Martin's Press, 1998. Republished by arrangement with the University of Minnesota Press. First published by the University of Minnesota Press, 2002.

Published by the University of Minnesota Press
111 Third Avenue South, Suite 290
Minneapolis, MN 55401-2520
http://www.upress.umn.edu

A Cataloging-in-Publication record for this book is available from the Library of Congress.

Printed in the United States of America on acid-free paper

12 11 10 09 08 07 06 05 04 03 02 10 9 8 7 6 5 4 3 2 1

Thanks are due for permission to reprint extracts
from the following copyrighted material:

From *Charlotte's Web: A Pig's Salvation,* by John Griffith.
Copyright 1993 by Twayne Publishers.
Reprinted with permission of Twayne Publishers,
an imprint of Simon and Schuster Macmillan.

From *In a Pig's Eye,* by Karl Shwenke.
Copyright 1985 by Karl Shwenke.
Reprinted with permission from Chelsea Green Publishing Company.

From *The Pig Poets: An Anthology of Porcine Poesy,* by Henry Hogge.
Copyright 1995 Ralph Rochester.
Reprinted by permission of HarperCollins Publishers Ltd.

Excerpts from *The Porcine Canticles,* by David Lee.
Copyright 1984 by David Lee.
Reprinted by permission of Copper Canyon Press, P.O. Box 271,
Port Townsend, Washington 98368.

From a letter by E. B. White in "Trade Winds," by Bennet Cerf.
Reprinted by permission of *The Saturday Review,*
copyright 1979, General Media Communications, Inc.

From "Dave Berry's Vacation in Iowa," by Dave Berry.
Copyright Tribune Media Services, Inc. All rights reserved.
Reprinted with permission.

From letter to Gene Dietch in *Letters of E. B. White,* edited by Dorothy Lobrano Guth.
Copyright 1976 by E. B. White, Harper and Row Publishers.
Reprinted by permission of HarperCollins Publishers.

From "Opinion" by Dale Miller in *National Hog Farmer.*
Copyright 1994. Reprinted by permission of Dale Miller.

From *The Little Town Where Time Stood Still,* by Bohumil Hrabel,
translated by James Naughton. Copyright 1993, Abacus.
Reprinted by permission of Pantheon Books.

From "Hog Slaughter," by Garrison Keillor, from original monologues, 1983.
Reprinted by permission of Garrison Keillor.

CONTENTS

PART V:
DISEASE, BODY AND SOUL

PART VI:
SAY IT WITH SWINE

ACKNOWLEDGMENTS

First, last, and greatest thanks must go to my family, especially Noni and Carl for again forbearing my *meshugaas*. In working through it, I drew inspiration from key mentors in the academy—Tony Wallace, John Caughey, and Al Stone—and on the farm—Roger Stutsman and Phil Berglund. Much of the most tedious labor—hunting down obscure references, transcribing miles of tape, copy editing and the like—was done by graduate research assistants at the University of Iowa: Jay Satterfield, Richard Sober, Allison McCracken, and Jennifer Mach.

The actual ideas for writing the book were, no doubt, cultivated by others, long before I was born. Many of them were there, ripe for the picking, when I got around to looking. In endnotes I try to give credit where credit is due, though other sources surely escape my memory. I cannot recall, for example, many of the university colleagues and students or farmers and farm families who, in the course of just being themselves, taught me a lot about American culture, agriculture, and life in general. I am sorry that I cannot thank each one of them by name, although, since they did not know what I would do with their help, they may consider neglect a blessing.

I here list people whom I told what I was up to and who therefore had reason to be suspicious but contributed nonetheless.

Ramon Abad
Jack Albright
Susanne Ammendrup
Jerry Applegate
Charlie Arnot
Alan Beck
Michael Bell
Mary Bennett
Anne Berglund
Audrey Berglund
Cathie Berglund
Emily Berglund
Jayne Berglund
Phil Berglund
Gregg Bevier
Lawrence Birchmier
Edward Bohl
Dain Borges
Joe Bosco
Birte Broberg
Norma Bruce
Eric Bush
John Butler
Jack Carlson
John Caughey
Mike Chibnik

Al Christian
Lauren Christian
Betty Chmaj
Ebbe Christensen
Kirk Clark
Ken Cmiel
Karen Coble
Robin Gross Connell
Ham Cravens
Fabian Faurholt Csaba
Stan Curtis
Osha Davidson
Jack Deeney
Jane Desmond
Anne Donadey
Kelley Donham
J. F. Donnelly
Maureen Donnelly
Mary Dudziak
Paul Durrenberger
Judith Laikin Elkin
Tim Emerick
Glenda Farrier
Walter Felker
Jeff Finlay
Jørgen Flensburg
José Flores
Staci Ford
Ken Forsythe
Sally Fry
Ian Gardner
Don Gingerich
Dan Gleason
Roy Goldblatt
Temple Grandin
Paul Greenough
Lucille Gregory
Hank Harris
Gordon Hendrickson
F. John Herbert
Thayer Hoover
Marsha Hopkins
Will Houston
Scott Hurd
Jim Julich
Imre Kacskovics
Ken Kephart
Glenn Keppy
Denise Klehm
Jim Klibenstein
Tip Kline
Jerry Kunesh
Charlotte Lackender
Dennis Lackender
Doug Lackender
Russell Lackender
Brent Ladd
Kelly Lager
Paul Lasley
Beth Lautner
Susan Lawrence
Danny Levitas
Carrie Louvar
Tom Lutz
Donnarae MacCann
Randy Madsen
Miriam Mandel
Alan Marcus
Mac Marshall
Gordon Massman
Emily Martin
Peter Matthews
Deirdre McCloskey
John McGlone
James McKeen
Kevin McNamara
John McNutt
Jay Mechling
David Meeker
Bill Mengeling

Gloria Mercado-Martin
James Merchant
John Moe
Jean Morrow
Jim Morrow
Julie Morrow-Tesch
Don Muhm
Marius Nabuurs
Steve Ohrn
Norman Owen
Prem Paul
Carl Pederson
Lauren Rabinovitz
John Raeburn
Emily Roberts
Paul Roberts
Priscilla Roberts
Wayne Roberts
Earl Rogers
Ron Rowland
Jeff Ruoff
Linda Saif
John Schacht
John Schiltz
Roy Schultz
Ken Schwartz
Sharon Schwartz
J. A. Smak
Finn Sørensen
Ole Stahlheim
Jack Stapleton
Al Stone
Grace Stone
Eldon Stutsman
Jim Stutsman
Mark Stutsman
Mike Stutsman
Roger Stutsman
Ron Stutsman
Sally Stutsman
Scott Stutsman
Hongjun Su
Bonnie Sunstein
Richard Tegner
C. Terpstra
Kendall Thu
Eldon Uhlenhopp
John Vahle
Deborah Van Dyken
A.P. van Nieuwstadt
Mike Varley
Ana Lydia Vega
Dick Vegors
Kim and Lori Verdeck
Andy Viner
Shirley Wajda
Dave Waldemeyer
Tyler Walters
Annette Weber
Joel Weinstock
Doug Weiss
Mark and Joseph Welter
Gert Wensvoort
Alan Widiss
Martha Wilding
Lauren Will
Janet Wilson
Phil Winborn
Margery Wolf
Paul Yeske
Norm Yetman
Ningping Yu
Halina Zaleski
Jeff Zimmerman

Although none of these people necessarily approves of what I did with their help, I am deeply grateful for it.

EXTRACTS

One of my favorite parts of the American classic, *Moby Dick,* is the collection of quotations with which it begins. Rather than take credit for these opening "extracts," Herman Melville blames "a mere painstaking burrower of a grub-worm of a poor devil of a Sub-Sub" librarian. Selections range from aphoristic to downright goofy. The "Sub-Sub appears to have gone through the long Vaticans and street stalls of the earth, picking up whatever random allusions to whales he could anyway find in any book whatsoever, sacred or profane." One of the very few passages of literary art that I ever committed to memory is among them:

> *The whale is a mammiferous animal without hind feet.*
>
> —Baron Cuvier

For the life of me, I am not sure which qualities of the quotation most draw me to it. On the one hand, I first really considered rather than skimmed the section (as I always had before) only because a college English professor insisted on it. "Do not let blowhard guardians of Art and Culture lead you astray," she said. "The novel welcomes a laugh. It was and remains one of the best farces ever written, and the extracts clinch the point." On the other hand, I also remember flipping through them after having given the novel the "serious" reading that a high school teacher required. Rather than seeming goofy, the quotations then encapsulated the unparaphrasable profundity of the book as a whole.

If Melville's publishers were stuck with a book about farming and swine dysentery rather than whaling and the sea, they might accept something like the following:

> *And the Lord said to Moses and Aaron, "Say to the people of Israel . . . the swine, because it parts the hoof and is cloven-footed but does not chew the cud, is unclean to you. Of their flesh you shall not eat, and their carcasses you shall not touch; they are unclean to you."*
>
> —Leviticus 11:1-2, 7-8

It's kind of an axiom: "What's good for the pigs is good for the people."

—Terry Coffey,
head of Research and Development
for Murphy Farms Inc.

During the late nineteenth century campaign in China against foreign missionaries, pictures were distributed showing pigs representing Christians being attacked by Buddhist priests. This was a fine visual depiction of a pun: The word for pig in Chinese differs only in tone from the word for Jesus.

—Emily Martin,
"Pigs as People; People as Pigs"

In America swine are said to have been introduced into Hispaniola by Columbus in 1493; into Florida by De Soto in 1538; into Nova Scotia and Newfoundland in 1553, into Canada in 1608, and into Virginia in 1609. So great was the fecundity of swine in Virginia forests, that in eighteen years after their introduction the inhabitants of Jamestown had to palisade the town to keep them out.

—J. Russell Manning,
The Illustrated Stock Doctor and Livestock Encyclopedia

And now there was this pig. Unblinking, the faded blue eyes looked into mine. . . . "Good afternoon," I croaked with a dry throat. Nothing. The pig's knowing eyes never left mine.

It is not unusual or silly for a farmer to talk to his animals. What is unusual and silly is for the farmer to expect his animals to answer back, in English. "Hot, ain't it?" I ventured. Nothing—not so much as a blink.

—Karl Schwenke,
In a Pig's Eye

Venture out in public these days, and ten bucks says you'll be drawn into a conversation about how clean, civilized, and downright intelligent pigs are. It's all lies, of course. As Ann Tanzer *discovered at this year's World Pork Expo in Iowa, few things deserve their reputations as feces-snuffling hairless blobs more than do our curly-tailed cousins—with the possible exception of the men who breed them.*

—*Spy* magazine

I could now see how beautiful she was. My pig . . . she was clean white all over, with just enough pink to be sweet as candy. "Pinky," I said. . . .

"Pinky's a fitting name," said Mamma.

"Never heard of naming a pig," Aunt Carrie said.

"But Solomon [the ox] has a name," I said, "and so do Daisy [the milch cow]."

"Let's eat," Pap said, "before we have to name every weed on the place."

—Robert Newton Peck,
A Day No Pigs Would Die

Pigs is Pigs in China
And Pigs is Pigs in Cork
But Pigs Ain't Pigs in frying pans
Because there Pigs is Pork

—Ellis Parker Butler,
1912, inscription in the copy of
Pigs is Pigs presented to the Iowa Authors Collection

A farm is a peculiar problem for a man who likes animals because the fate of most livestock is that they are murdered by their benefactors. The creatures may live serenely but they end violently, and the odor of doom hangs about them always.

I have kept several pigs, starting them in spring as weanlings and carrying trays to them all through the summer and fall. The relationship bothered me. Day by day I became better acquainted with my pig, and he with me, and the fact that the whole adventure pointed toward an eventual piece of double-dealing on my part lent an eerie quality to the thing.

—E. B. White,
author of *Charlotte's Web,*
in a letter to Bennet Cerf

They were a sprightly lot of five-week-old youngsters. When I climbed into their pen they stopped their mauling play to stare at me, and I suddenly became aware of the overall silence. Looking around, I found that every pig in the place, big and little, had stopped what he or she was doing and was looking my way.

"Makes you want to check that your fly is buttoned, doesn't it?"

—Karl Schwenke,
In a Pig's Eye

In my book, Charlotte dies. If in the screenplay, she should turn up alive at the end of the story in the interests of a happier ending, I would consider this a gross violation.

—E. B. White,
in a letter to his attorney, Alexander Lindey

Gideon's wife, Helen, died on the day that my sow gave birth to thirteen squirming babies. I know that only because I called minutes after the ambulance had come to take her body away. Embarrassed, I apologized for calling at such an awkward time, and he did his best to assure me that it was all right. "It's fittin'," he said, "for new life to come along to replace an old one that got done." Looking back on our painful conversation that day, I find it fitting that Gideon did not draw a distinction between pigs and people.

—Karl Schwenke, *In a Pig's Eye*

The hog is a mammiferous animal with *hind feet.*

—Anonymous

HOG TIES

PART ONE

CARING FOR HOGS AND THE MEANING OF LIFE

You cannot live long in our town without learning a lot about pigs and, as a consequence, also learn a lot about your neighbors. . . . I'm not sure what it is about the animal, but pigs bring out the yarn spinner in the shyest of folks. One of the first people I ever met in town was Gideon Whitman, a retired swineherd. . . . It was from Gideon that I got my first inkling of how a lifetime association with pigs inevitably affects your relationship with your fellow man. In self-defense swineherds become sociologically insightful, or they become liars. Gideon was both, and he was a neighbor.

—Karl Schwenke, *In a Pig's Eye*

Too rigidly we divide mind from body, privileging mind as the human *observer that must police the base, slovenly flesh involved in its own piggish desires. Every intellectual hides a partitioned pig whose enjoyments can . . . only [be] seen through a thick barrier and at a distance.*

—Richard Shusterman, "A House Divided"

Believing, with Max Weber, that man is an animal suspended in webs of significance he himself has spun, I take culture to be those webs, and the analysis of it to be therefore not an experimental science in search of law but an interpretive one in search of meaning.

—Clifford Geertz, *The Interpretation of Cultures*

"We have received a sign, Edith—a mysterious sign. A miracle has happened on this farm. There is a large spider's web in the doorway of the barn cellar, right over the pigpen, and when Lurvy went to

feed the pig this morning, he noticed the web because it was foggy, and you know how a spider's web looks very distinct in a fog. And right spang in the middle of the web there were the words 'Some Pig.' The words were woven right into the web. . . . It says, 'Some Pig,' just as clear as clear can be. There can be no mistake about it. A miracle has happened and a sign has occurred here on earth, right on our farm, and we have no ordinary pig."

"Well," said Mrs. Zuckerman, "it seems to me you're a little off. It seems to me we have no ordinary spider*."*

—E. B. White, *Charlotte's Web*

ONE

A BIT OF AN EXPLANATION

THIS IS A BOOK ABOUT FARMING, pigs, and disease in American culture.

The "American culture" part is probably easiest to understand. Admittedly, when you get down to cases, it gets a bit fuzzy. The expression "American" only crudely corrals "culture," itself a rough, qualitative generalization. National borders bound geopolitical differences more crisply than ways of life. For example, there are many parts of the United States that are now, culturally speaking, more "Mexican" than they were when legally subject to Mexican sovereignty. And exclusive, expatriate facilities—such as grocery aisles in the PX or those schools and clubs that are reserved for Yankees on Third World assignment—can be "more American" than anything back home. Nevertheless, having grown accustomed to almost any neighborhood within the United States, you probably will have little trouble adjusting to another. A month or two of orientation will do. But there is no mistaking Seattle or Topeka for a sector of Siberia or Tanzania. Getting along there would require learning much more than where the Wal-Mart is, which channel has *All My Children*, or how to avoid the most dangerous streets. There are ways things are done in the United States—guesses you can make about what people will do or the ways they might respond to each other—that are usually serviceable, even if occasionally woefully inadequate. You might or might not share or appreciate those ways. If you leave, you might sense a great burden being lifted or something precious slipping away. Or you might be indifferent. But you can apprehend the thing, those ways, and give it a name. "American culture" can suffice.

An interest in farming also might be easy enough to understand when you recall how important it has been for Americans. For centuries, pundits have spoken as if farmers were the quintessential "American." "Old MacDonald" is the real McCoy.[1] The movie screens portray national destiny through Sally Field, madly hoeing seedlings in a deluge, and John Wayne, battling ranchers or rustlers. The small screen makes fun of those fables through *Green Acres* and *The Beverly Hillbillies,* but it leaves in place a rural mystique. Whenever journalists ask a policymaker, "How will that play in the Heartland?" we can hear: "Come off it. What do 'real' (or 'regular') Americans think?"—even though most of them actually better know beltways than barnyards. Of course, I am also interested in farming because it occupies a lot of the actual surface of the land and its people, especially in the Heartland, where I just so happen to live. It is both the symbolic and practical importance of agriculture that draws me to it.[2]

Likewise, "disease" has myriad significance. You can make sense out of much human history by noting the ways various peoples have fared with infection. It is striking how the march of time changes when the majorette contracts syphilis or when cholera comes to town. Most recently, HIV ("the [sic] AIDS virus"), STDs (sexually transmitted diseases), and "emergent" or "tropical" pathogens like Ebola have renewed public fears that antibiotics and vaccines, developed during the Cold War, no longer allay. The era of "containment" has ended for contagion as well as Communists.[3]

Among the rewards of looking at disease in food animals rather than in humans is a bit of clarity and focus. Superfluous complexities can be culled. Popular responses to HIV (*human* immunodeficiency virus), for example, are deeply mired in homophobia and racism as well as disease lore. But just about any human infection tends to draw histrionics. For people living with disease, knee-jerk expressions of sympathy can get tiresome and begin to sound cruelly dishonest. "We'll do everything possible," the doctor says. "We'll spare no expense." Well, that pair of promises is impossible to keep. At some point, far short of infinity or even the possible for a select few, the spending and doing stop. Why not discuss openly if/when discontinuing might be desirable, anyway? For better or worse, people are much more willing at least to consider such questions (for example, the merits of accommodating individual suffering for the sake of long-term, collective good) when the victim is a member of a different species. In the case of agriculture, they are also more willing to address frankly the economic consequence. In short, illness in livestock helps provide a clearer picture of public understandings of disease and the more general, spiritual but still distinctly bodily disquiet that the word "dis-ease" connotes.

But why, for goodness' sake, pigs?

TWO

PORCINE WAHINIS

BY THE TIME YOU HAVE FINISHED WRITING A BOOK, you have loved and hated the topic, pitched it and apologized for it so many times that just about every conceivable tale of its attraction has been told. Every one of them is a bit sensible and suspect. Even the preceding passages, I regret, might make it sound as if I deduced one day that farming and food animal disease deserve a decade or so of obsession. No, it did not happen that way. Interests accumulate more haphazardly. Swine are only the most unusual among mine. The mere mention of such lowly creatures flaunts the tenuousness of pure-principle justification. Try saying to someone "I am really interested in hogs," and check out the response. If you are a liberal arts professor, someone who supposedly prepares students for the real world or provides an enlightening break from it, people are apt to assume that you are kidding.

So, it is tempting to trot out some autobiographical imperative—say, anecdotes from a hog-induced coming-of-age. Of course, given my Jewish name and liberal arts vocation, it may be a little hard to believe. But even before my birth, my parents embraced a Reform interpretation of Levitical law. On weekends we might have bacon for breakfast or BLTs for lunch, and there was nothing even remotely transgressive about it. Contrary to stereotypes of New York Yids, my father grew up on a New England farm, and I grew up in a village whose selectmen were usually the dairymen with the largest herds. The place became resolutely suburban, with tract homes, a highway to Hartford, and town-manager rather than town-meeting government, only after I left. When I was a boy in the 1950s, my father, the only dentist around, was still doing house calls for rural shut-ins. There was a hulking, "portable" drill in a plywood case that filled the trunk of his Ford sedan and that he occasionally set up in farmhouse kitchens.

Around our house, education was highly prized but only if the lessons were hand-forged. Hence, although we could afford almost any luxury, from the age of sixteen I worked blue-collar jobs and, even before that, aimed to know how to build or repair almost anything. I was ashamed of every manual skill I lacked and did my best to learn by doing and by paying close attention to pros whenever I could. From high school through graduate school and now full professorship, I have always kept a manual occupation full-time for part of the year or part-time year-round. Move me to a job in Iowa, the heart of the country's Swine Belt, and you might predict porcine epiphany around the bend.

It is not such a far-fetched possibility. Some of the most effective works of American artistry, particularly in the literature for children, have been swine-inspired. I am thinking of not only shrewd amusements like Porky or Miss Piggy but also stirring ones, porcine classics such as Robert Newton Peck's daunting *A Day No Pigs Would Die*. The book is at once a fond reminiscence of Peck's boyhood in 1930s Vermont, a testimony of Shaker faith, a eulogy for his father, a nightmarish vision of mortality and a celebration of it, all fashioned from life with hogs.[1] Or there is Ellis Parker Butler's more whimsical parable, *Pigs is Pigs*, about which I will have more to say later.[2]

My favorite example, though, is *Charlotte's Web* by the master of middlebrow style, E. B. White. Few Americans who have read anything in English survive childhood without encountering the book and being moved by it.[3] Kids' faces clearly register a realization—often their first, through their identification with a sweet pig named Wilbur—that death is a fact of life. Innocent creatures gain their vitality only because loving protectors surrender theirs. Even very small children seem to get the bittersweet message. It is all the more poignant because Charlotte, the spider protector, is so obviously a model mother. Her death in the book is affecting both because it challenges literary convention (matricide is rarely children's fare) and because it faces (or adroitly effaces) issues that Americans prefer dipped in honey. A critic explains: "Its subject matter—death, redemption, nature, and the love of animals—is both serious and playful; its tone is both mellow and light. It is a genteel book, civilized and literate." Improvising on barnyard gabble, White is able to impress graybeard guardians of literary taste no less than fidgeting tykes: "Millions of readers, both children and parents, are fond of it, and professional people in children's literature have given it the highest marks. These conditions establish *Charlotte's Web* as an immortal children's classic, and it is by no means certain that similar conditions will ever come together in and around one children's book again."[4]

Among those conditions is White's own livestock experience. He was brought up in New York, but he also lived on a farm in Maine every summer from his birth in 1899 through his early teens and then again after 1933, when he and his wife, Katherine Angell, bought their own place Down East. Among his early duties was tending the family pig that would become the family pork. Chores eventually occasioned one of those coming-of-age epiphanies. It came in the form of a bout with swine disease, something like erysipelas. He struggled to save the pig and to confront his feelings about it. When the animal died, White suffered an intense, curious remorse—remorse, because he lost a creature that he cared for; intense, because keeping it alive was both a practical and reverential responsibility; and curious, because the animal was destined for slaughter, anyway.[5] Out of that experience came a moving essay, "Death of a Pig," in *The Atlantic Monthly*:

> I wanted no interruption in the regularity of feeding, the steadiness of growth, the even succession of days. . . . I just wanted to keep on raising a pig, full meal after full meal, spring into summer into fall. . . . From the lustiness of a healthy pig a man derives a feeling of personal lustiness; the stuff that goes into the trough and is received with such enthusiasm is in earnest of some later feast of his own, and when this suddenly comes to an end and the food lies stale and untouched, souring in the sun, the pig's imbalance becomes the man's, vicariously, and life seems insecure, displaced, transitory.[6]

Immediately thereafter White began writing *Charlotte's Web*. When it was done, he warned against reading too much into it. It was just "a straight report from the barn yard," at most "a paean to life, a hymn to the barn, an acceptance of dung." But it is hard to resist finding profound life lessons in those humble pig tales. Millions of Americans have done so for generations.[7] So why should I feel foolish in following White's lead?

One reason is that I might meet too few of the requisite conditions. White's wordsmithery is probably beyond me. There is also the fact that, as rural- and manual-oriented as I may be, I did not grow up or even summer on a farm or do a single day of wage-work on one until the mid-1970s, when I moved from Maine to Iowa. Furthermore, it is only fair to say, much of the success of *Charlotte's Web* had nothing to do with White, with words, erysipelas, or barnyard experience.

A major share of the credit has to go to Garth Williams, the illustrator who invented Charlotte's and Wilbur's looks. He was born in New York and spent his boyhood in rustic New Jersey, but his parentage and most of his education were urban and British. His vocation was decidedly more attuned to fine than swine art. He attended English schools, received a scholarship to attend the British Royal College of Art, and won the 1936 Prix de Rome for sculpture. Only the dislocations of World War II occasioned his return to the United States. His collaboration with White began when their portfolios happened to cross on the desk of Ursula Nordstrom at Harper and Brothers. Bankable aptitude rather than particular life experience seems to account for Williams's success. *Charlotte's Web* is just one of nearly a hundred books he illustrated, including many of the famous Golden Books and an array of other children's classics. His formula for success would impress a Tory bureaucrat: "I'm pragmatic. It's a job. My inspiration is my deadline." When he reluctantly agreed to reillustrate Laura Ingalls Wilder's *Little House* books, he had not been west of the Hudson River. Much of the actual drawing was done in Italy. When last tracked down by a critic, just a few years ago—at the age of eighty, living with his fourth wife and their teenage daughter in their Southwest compound—there was still a "BBC tint" to his voice. So, there goes the special relationship between swine and all-American wisdom. Whatever the pedigree, it is not mine.[8]

For better or worse, though, as I finished a book on an unrelated topic, I felt pressure to settle on a rationale for porcine obsessions.[9] Such pressure is familiar, at least in kind, among academics. In the absence of bottom lines and benchmarks, people account for themselves prosaically. If a university colleague or, more perilously, a dean asks you, "How have you been doing?" the correct answer includes a paragraph on your latest, cutting-edge research. You are unlikely to qualify for tenure or those "merit" pay raises unless you are well prepared. And for many of us, the urge to prepare is also heartfelt. A compelling rationale can be considered a matter of both professional advantage and personal integrity. I resolved that the question—"Why, for goodness' sake, pigs?"—would get more than a shrugging "Why not?"

The perfect opportunity to hone an excuse arose courtesy of one of the professional societies to which I belong. I was at the time (1992-1993) president of the American Studies Association in my region (the eleven states of "Mid-America"—hence, "MAASA"). Among the very few and final duties attending that office is a "presidential address." It follows a banquet and business meeting at an annual conference, most of which is dedicated to scholarly papers. Since presidential addresses are much anticipated, free-form events, I decided to make mine my "Why swine?" high noon.[10]

Of course, the august title of the occasion increased the pressure, the mixture of good fortune and desperation that I felt. Anything "presidential" sounds self-satisfied and old boy, but I was neither. I could not shake the feeling that plagued my last presidential address, at Elizabeth Green Elementary School where I was chief executive of Mrs. Miller's second-grade science club. As best I can recall, there was the same mix of thrill and dread. Also as best I can recall, my presentation was on something like "The Boreal Owl." I stole every dreary word from the *World Book Encyclopedia*, and no one noticed or (as far as I could tell) even listened. It was not the sort of experience to amplify the thrill or mute the dread.

But I also knew that there was really little cause for concern. The way you become president of most academic associations is by volunteering for (or, more likely, failing to avoid) marathon meetings and clerical follow-through. These duties may have weighty consequence and may be recalled with satisfaction, but they are innately, at times excruciatingly unattractive in themselves. The Executive Board usually has to convince one of its own to accept nomination, and the nominee runs for election unopposed. The fancy title is reward for sucker service. So, not much stands between anyone and high office. Although MAASA presidents have been distinguished, the association has always been—in principle, anyway—equally vulnerable to takeover by intellectual giants and gnats. With such an indiscriminate qualification procedure, why should anyone expect much presidential wisdom? No pressure at all.

On the other hand, there was that need to justify my research and self-respect. I had a full year to prepare an address, and I could say whatever I wanted. The other conference participants, the majority in the audience, had to prepare proposals and pass muster with a program committee months in advance. Moreover, while your average academic prose must go through the referee reaper, a presidential address is presumed publishable in the association journal. So I had every encouragement—the freedom, the time, and the near-captive audience—to do more than a passable job. All things considered, it ought to be absolutely stellar.

I knew, too, that there were tough acts to follow, addresses so memorable that they still occupy association lore. For example, in 1989 Steve Watts delivered one, "The Idiocy of American Studies," that I found incisive and amusingly transgressive, but that a share of the audience greeted like a fart in a pup tent. A milder version found its way into a 1991 issue of the national journal and was promptly Lysoled by commentators in a following issue "Forum" (among the few times I can recall when anything in the journal stirred more than an hour of cocktail-party bluster).[11] So there was at least one presidential address that incited passions and then suit-and-tie respectability before being deloused, shredded, bagged, and hauled to the landfill. There was also the daunting precedent of Professor Wayne Wheeler, twice president of MAASA, who delivered a raucous parody of B. F. Skinner's *Beyond Freedom and Dignity* while methodically shedding his clothes. He eventually accepted the applause of the membership, clad only in undershorts (though Professor Wheeler assures me that they were bathing trunks). How could anyone beat something like that?

I found a solution for that occasion that seems appropriate to this one, as well. Judging from Wheeler's success and Watts's debacle, I figure that academic pretense is the problem. First to go, at least in spirit, has to be the suits and "sensible" shoes—flats, Rockports and Birkenstocks. Let's face it: If you are going to ply American byways, you have to be ready for off-roading. In practice this means that I must forewarn readers not to expect exhaustive citations or elegant deconstructions of woefully misunderstood text. Such print worship has to be among academia's least attractive features. Frankly, I suspect that heavy exposure to the sixty-cycle hum of library fluorescent bulbs causes brain damage, at the very least a perverse tendency to add "-ization" onto nouns that already have four syllables.

I also must cast off the usual rules of academic justification. They require embracing—in the foreword or certainly before the end of chapter one—a particular theory or two and attendant, so-called methods, package tours to understanding. What is your sampling procedure? Do you highlight typical or path breaking cases? Do you analyze institutions or ideas? Individuals or groups? Art or society? Great Books or pop schlock? Is the goal analysis or advocacy? Each of these questions has a stock set of answers, one for each discipline and school of thought, in varying states of fleeting favor. They are often called "tools," akin to that portable drill my father hauled around, a boxed contraption for drilling, cleaning, and filling operations. Hence people talk of "applying" Marxism as if it were a torque wrench or French Feminism as if it were a shop-vac. Better, I think, to treat method as a human disposition in action, a vaguely purposeful manner of paying attention and forming a response.[12]

Rather than pursue these points here—yet another post mortem on the "context of justification"—I want to direct attention to a methodological moment that is less easily dissected: the act of engaging a topic, particularly the process that occurs before anyone could know which theory or method "fits" or even if the tools are better left in the trunk.[13] I have in mind that exhilarating instant when a subject—even one as unlikely as a sickly sow and its relation to meaningful life—becomes interesting. For at least a minute (maybe months or even years), it does not depend on or imply a particular theory of its importance, at least none that can yet be named. No particular technique for exploring it is (clearly, anyway) more "appropriate" than another. It does not yet "have" a distinct logic, politic, or aesthetic but is in the midst of acquiring them. Academics seldom address that occasion.[14] Through swine, then, I am trying to explore the half-baked: What goes on when we first realize that we are onto something? We do not yet know what that something is, how we will get it, or why anyone should care, but we feel a buzz of recognition. It is just . . . something.

The observational documentary filmmaker Ross McElwee calls that something "a great Wahini." It is a term that he ripped off Haole surfers (to mean that one ocean swell that will give you the ride of your life) who in turn ripped it off Hawaiians (who use it to refer to a kind of person, namely a woman). Whether this figurative leap, like those in his films, is best considered humorous or offensive, the resemblance is striking. In making a film, you look through the viewfinder at wave after wave of imagery, like a surfer scanning the horizon. How will you ever edit all that footage so that the remainder has a shape, a look, a pace, a center that makes sense? Then that centering moment, a perfect wave, appears: In making *Silver Valley*, Michel Negroponte catches Jerry salvaging a "Peter and the Wolf" record from a Dumpster; in making *Sherman's March*, Ross McElwee finally confronts the real Burt Reynolds. These are, Negroponte says, their "great, great Wahinis."[15] It is hard to say how you find a great Wahini beyond "Shit happens." But what is going on? *How* does shit happen? I want to try to explain by exploring a single porcine case—by turning the compost, so to speak.

THREE

SHIT HAPPENS

I HAVE LONG AIMED TO FIND SOMETHING TO SAY ABOUT FARMING. By luck and pluck, I have been introduced to a corner of it unusually well. For nearly two decades I have been working part-time, one or two days each week, as a hired hand on a large family farm near my home in Iowa. Of course, I have also read a fair amount of agricultural history, rural sociology, and associated commentary, hoping to find a place marked "Contribution to the literature wanted here." But none was readily apparent. Instead, each week I donned my farm getup—boots, coveralls, seed cap, chore gloves, and holstered pliers—anticipating mundane rewards: a little exercise outdoors, some male bonding, and the simple pleasure of learning to accomplish tasks that seem novel and tangible when compared to university routines.

One January morning, after an especially frustrating week at the university, I saw my wife off to work and our son off to school, and they could tell by my upbeat mood that I was going to work on the farm. I glanced at the thermometer, put on an extra layer under my insulated coveralls, and drove off to take care of the round of hog chores that I always do before joining my friend and boss on the farm, Phil Berglund, for the more particular demands of the day.

It is actually an object of some pride that they trust me to handle such responsibilities on my own. Although your average Iowa farm kid earns that trust by the age of eleven, I had to work pretty hard to prove that I, Jewish egghead from out East, could develop decent hog sense and cope with emergencies that can be anticipated only in the abstract.

In farming you cannot know what will go wrong, but something surely will, as likely as not when most costly or dispiriting. Every farmer must find a way to cope with God's regular reminder that nature still is not (as they say of show cattle) "broke to lead." Although people who can cope and have decent hog sense may be a dime a dozen in this corner of the world, I am flattered that Phil occasionally asks me to take care of his herd when he leaves town for a few days.

Hogs are in one sense a small part of the operation. Phil, who is a few years younger than I, and the Stutsman brothers, Phil's in-laws, a little older than I, work about 2,000 acres, mainly feed corn and soybeans, and each year they fatten and market more than a 1,000 beef cattle. All of this is accomplished with the unpaid labor of backstage kin and an awesome array of gargantuan machinery but very few people on the job: just two "farmers" (Phil and Roger), one full-time hand, part-time or contract help for planting and harvest, and me. Hog work is left for hurried moments when field work or cattle allow.

In another sense, though, the hog business is very important. In bookkeeping, sidelines are not necessarily lesser ones. Hogs are miraculously efficient converters of grain (which often does not fetch its production cost on the market) to meat (which usually does). Moreover, hog checks can be written and received without going through the partnership. Hence, although occupying only a small share of farm equipment or attention, hogs are among the few reliable sources of cash flow that pays interest on loans, puts clothes on the kids and food on the table.

Phil does the farrowing, arranging the conception, birthing, nursing, weaning, and care of baby pigs (gilts and barrows) till they reach "feeder weight," 50 to 60 pounds. His breeding stock includes about eighty sows and a half-dozen boars. They can be nasty beasts, huge, stubborn, and competitive, especially when confined to small spaces. But we move them in as orderly fashion as possible in groups from one hog lot to another, then to various buildings ("confinement systems") through the breeding cycle. The produce of this cycle is a stream of feeder pigs, more than a thousand each year, that Phil sells to Roger, who "finishes" them to market weight, 260 to 280 pounds.

As in any family business, relations are generally congenial but fragile. In part to avoid conflict, Roger, who is the middle son of the farm patriarch (Eldon), and Phil, who married into the family and grew up "in town," have over the years divided the labor so as to have minimal daily contact. They may go as long as a week without seeing or speaking to each other, not necessarily because anything is wrong but because they can get along without it. From my vantage the lack of communication is remarkable—very "Euro-Midwestern" and "male"—but they seem to deal with it just fine, thank you. And they both prize the hog arrangement. Phil has a ready buyer for his feeder pigs (without having to establish their worth in a fickle open market), and Roger a dependable supply (without having to dicker on price, transport, or quality or risk introducing disease to the herd). Of course, there are periodic tensions, as when Phil's pigs are not ready on time (and Roger's expensive finishing unit idles below capacity) or when Roger's unit is full (and Phil's nursery building is bursting at the seams). Maybe more communication would help, but, then too, those rare breakdowns can be shrugged off along with other "natural" frustrations of the business and reminders of a willful God. The finished hogs consistently grade high enough to demand top dollar at the local packing plant.

The farrowing—the actual birthing and nursing—takes place inside one of two buildings reserved for that purpose. That is where I was headed that January morning, the farrowing house at Eldon's. I wish I could say that I was intently honing some American Studies tools, but I was mainly looking forward to a day of labor away from the office and fluorescent bulbs.

Appreciating the farrowing house at Eldon's place requires, let us say, a tolerant spirit. As much as the newer unit at Phil's place bespeaks the fastidiousness of 1980s agriscience, the one at Eldon's bespeaks "making do" in the Great Depression. Of course, any building that is packed full of swine is a far cry from your living room, even if your roommate is a slob. No amount of exposure can dull the sense that swine buildings are noisy and dirty. Experts recommend breathing through a respirator when you enter even the cleanest, most up-to-date unit, where manure falls through slatted floors into a pit that needs to be pumped out only three or four times per year. The best of units are architected septic tanks.

And the farrowing house at Eldon's pales before that ideal. Insofar as it was ever "designed," it was not with human comforts in mind. It is, though, quite hospitable for sows, each confined to a cagelike "crate" made of tubular steel. The crates keep them pointed straight ahead at a feeder and waterer and make them less likely to abandon, lay on, or stomp their offspring (or you). Sows are famous for the way they may at one moment maul you to protect an offspring and at the next maul that offspring to beat it to a kernel of corn.[1] With pipes and wires of various generations dangling from the ceiling to each crate, the farrowing house at Eldon's could not be redesigned or sanitized easily, even if we had the time, which we usually do not. It is a constant battle to keep the rodent population under control, and a little more than a week after fumigating the place, you must hack your way through cobwebs covered with manure dust to navigate the narrow walkways between rattling crates and screaming sows.

In fact, this was the place where my mettle was tested, back around 1980. I had been working irregularly for a few years, helping sort and feed cattle, hauling manure, running a disk or chisel plow or harrow in the spring, making hay in the summer, filling silos or catching corn off the combine in the fall. I worked just for the experience or in exchange for meat or use of the equipment on my own place. We decided it would be better all around if my position were more regular: I would be on payroll, working as long and hard as anyone else on a given day. No one should have to wonder, "Is it okay to ask Rich to do this job?"

Although Roger does not remember much of this exchange, I can well recall the gut check that was my first assignment under the new terms: Hop in the truck, drive to the farrowing house at Eldon's, and clean it from top to bottom. "If you run into trouble [in other words, if you're a wimp], you should be able to find one of us around somewhere."

In the summer heat scraping well-crusted manure out from under eighteen crates and wrestling with a pressure sprayer designed for roomier environs is a nasty day of work for two, a much longer and nastier day for one. Of course, the assignment is no worse than any Phil or Roger ordinarily expects of each other. But it is still among the most infamously unpleasant chores in a massively varied repertoire, the kind that incites a knowing smirk: "So . . . going to clean the farrowing house, eh? Have fun." "Yeah, right!"

In fact, I did not really mind. As a seasoned ethnographer, I had weathered such trials before and welcomed the chance to prove myself. Besides, I could always use the exercise, the break from the sweeter-smelling but too-familiar shit at the university, and the reward of a cleaner farrowing house to work in thereafter, at least for a couple of weeks till the sows, flies, rats, and spiders reestablish their ambiance.

In the truck on the way to Eldon's on that January morning, I could chuckle to recall that test so many years before and to anticipate the fact that the farrowing house this time had been cleaned long enough ago that the university/farm divide would be abundantly obvious once I opened the door.

I was feeling downright giddy when I did. Rather than proving myself, I was basking in self-confidence, a feeling gained through years of farm experience and bloated through contrast with the university world, a bracing wind on one side of the door and dusty heat on the other. As usual, the sows rose, rattling crates, and grunted in response to the opening door, a sign that feed was on its way. Like the hush that overcomes a lecture hall when you approach the podium, it is an odd collective "hello," disconcerting at first but comfortingly familiar once you get used to it, and I was amused that I was.

So, we began our normal routine. I flipped on the lights and the switch on the auger that pulls feed from the bulk bin outside to the southeast corner of the building. As feed drops into a five-gallon bucket, I mentally rehearse: Sows with nursing pigs get a half bucket; those without, a third. As always, watch your step! The floors are slippery with manure and tipped to drain to a central pit, which you may have to vault, if no one replaced the makeshift bridges when the place was last cleaned. When walking between the third and fourth sets of crates, remember to duck or you will hit your head (again!) on the thermostat on the ceiling. Do not try to take two buckets down the second walkway; in coveralls you will not fit. Remember that the sow in the southwest corner is a "bitch." She will stand quietly until the moment you reach her feeder (and your head is about a foot from hers); then she will leap to attack, her jaw crashing the bars, and bark loud enough to drain the blood from your face. Be ready! And by all means, distribute the feed before the bedlam gets deafening. Once the sows are preoccupied eating, you can go back, check pigs, and make sure everything else is in order. No big deal.

I am relieved that the auger is working smoothly, so I do not have to go out and pound on the bin again to loosen frozen feed and allow more time for the sows to grow restless. They are still just beginning to stir as I finish filling the second bucket, turn off the auger, and head through the cobwebs to the feeders. I am just a little dirty and sweaty and the sows a little impatient as I work my way down the first row. Everything is going smoothly, a condition in farming, especially in winter, that is remarkable in itself. This may be no big deal, but I feel great, well into Zen and the art of hog maintenance.

Teetering on the edge of the manure pit, as I lift a bucket shoulder high to fill another feeder, my eyes drop to admire the baby pigs. (They can be awfully cute.) But what I see is more appalling than anything I have ever seen or could have imagined. The crate floor is littered with tiny dead pigs, and the live ones are pathetically sprawled on all fours, racked with tremors, and soaking wet in their own milky diarrhea and vomit. Oh, my God! What the . . . ?!

Just as the horror begins to register, the farrowing house door crashes open, and I hear Phil bark, "Don't feed them!" And just as quickly, the door crashes closed. He's gone.

Oh, my God! What have I done? How could I have been so damned self-indulgent, so careless? Sure, I have seen plenty of sickness before. Deaths now and then are always troubling but hardly unexpected. In fact, sows regularly drop more than a dozen pigs, and if you can keep ten or eleven alive, you are doing well.[2] Nearly every birth entails handling at least a couple of little corpses. But this? I feel like a tourist who anticipated a Kodak moment in a quaint native village only to find a raging plague. Even worse, I am the tourist caretaker. In about thirty seconds, my emotions have tumbled from elation to shame and despair, now bordering on panic.

I race out the door to catch up with Phil. I know he will be in a peach of a mood, even less tolerant of my inquisitiveness than usual, but I have to know: What went wrong? Is there some way this could have been avoided? What the hell can we do now? And we better do something fast!

I catch up and momentarily loose my ethnographic cool. I assault him with questions, and this is not what he wants to hear. In a transparent attempt to get me to shut up, he mumbles something technical, a few letters that are supposed to suffice as an explanation, and then begs, "Rich, will you give it a rest?!"

I do; I shut up, but my heart is still pounding, and the thoughts churn. I am not sure what the letters are—did he say "EDE"?—much less what they stand for or how they are related to the dozens of other abbreviations that fill farm talk: "Rich, did you get the heifers at North Place? Make sure you give them twenty seconds of MGA. You can use the 4020, but if it doesn't start, use the 50. We've got to hit some of the new pulls off the south lot with Naxcel."[3]

Maybe what the pigs have is a special strain of E. coli? I know a little about that.

In any case, a few things are clear. We have a serious disaster on our hands. From their look, I would bet that no more than a couple of pigs will survive the night, and I do not have the slightest idea what to do. But best to keep mum for now and hope that over the course of the day I can find a way to ease into the subject while we work on other things. The cattle call.

Much to Phil's annoyance, I keep veering back to the subject, but he offers little. At one point, with great disgust, he explains: "No, Rich, you did not do anything wrong." An hour later: "It just happens." Yet another hour later: "I'm giving them some electrolytes, but I don't know if it will matter. There isn't much of anything you can do."

So, they are all just going to die this horrible death? The whole idea runs against everything I have learned to expect on the farm. Misfortune may be a normal, "natural" part of the business, but we always put up the good fight, usually the fiercest when it looks like a loser. By the time we break for dinner, I recommit: I had better hold my tongue and ponder the possibility that I still do not understand the play of fate and control, of bad luck and screw-up on the farm.

But after dinner, while working on some equipment, off on my own, Roger pulls up and steps from his pickup to chat.

God, I wonder if he knows? There go his feeder pigs.

He probably could explain all of this to me, but if he does not know what is going on, I do not want him to hear about it first from me. Relations with Phil could get touchy, and I do not want to add complications, not only because of my ignorance but also because I could too easily destroy their détente. In fact, in generous response to my eternal questioning, Roger and Phil each tell me things—such as speculations about the thinking of the other—before, if ever, they tell each other. And they each know that I have this share of the other's confidence. So I have to be ever wary to avoid inadvertent betrayals. Again, this is not because either harbors secrets, but because I am confident that they are apt to get along better without this egghead's meddling. But the plague will be a difficult topic to avoid.

Roger hints, "So, you've been down to the farrowing house at Eldon's?"

I volunteer only, "Yes."

"Terrible, huh?"

So he knows!

"Yes, it's TGE," he says, as if those three letters would clear up the whole business. When it is obvious that they do not, Roger begins to explain. "It stands for 'transmissible gastroenteritis.' You mean you don't know about that!?" He struggles to spare me his incredulity.

"I thought I told you about it, you know, about how I got TGE, in that very same building back . . . not long after I finished at Iowa State and came home to farm. I had just bought a set of gilts and got TGE their first farrow. It was awful. You have to feel for Phil—sleepless nights. It made me sick. I couldn't handle it. I mean the responsibility, all that death. That's why I stopped farrowing. I just sold them off."

As Roger speaks, I begin to understand. Evidently TGE is this angel of death that visits all hog farmers at one time or another. Apparently you avoid talking about it above a whisper, at least in public, if you are the one it has visited. But it is always there, hovering just above the farrowing house.[4]

Roger explains, "It's amazingly contagious. Once it hits a farm, you can smell it in the air, and every hog will get it. The sows will probably be okay, but there isn't much you can do for the pigs. . . . And no one is sure how it spreads. I told Phil to wash his boots when he was up at my place." (I think: "You mean, I have been feeling so bad, and Phil may have brought this on himself?") But then Roger adds, "No one can know where it came from. Supposedly birds can carry it, just flying from one hog lot to another."

After Roger leaves, I finish bolting some machinery back together and have time to think while I drive to find Phil for another assignment. I wonder, "So, here I was feeling so damned irresponsible and dumb, when the whole business could have been Phil's screw-up?" But just as important, Roger reminds me, no one can know. You have to feel for Phil and envision the elusive demon he battles.

My bookish side begins to churn: "Maybe I should write about this episode, say, from the vantage of a starling *nosferatu.* At one moment I soar, innocently scanning a pastoral landscape, and at the next, alight to spread death with a touch. . . ."

Just as the compositional difficulties of this option begin to register, I find Phil who still looks upset but maybe a little more open to questions.

"I bumped into Roger. [Painful pause.] I didn't know that he knew."

Phil, reluctantly: "Yeah."

"He said that TGE is amazingly contagious and that you can't do much about it."

"Yeah, I *told* you."

"He also said something about washing boots. Can TGE spread that way?"

"Yeah, I guess so. Roger had a lot of room in his finishing unit. So he bought a load of feeder pigs last week from someone, and I went up to help him sort and unload. But who knows?"

So I begin to think, "You mean, here I am congratulating Roger for his magnanimity, and he was the one who brought in the TGE in the first place? Phil goes up to help preempt space that could have been reserved for his own feeder pigs, maybe driving them onto a weak market, and this is his reward?"

Yet again I have to refigure the play of luck and screw-up. As gracefully as possible, I ask Phil to help me get the story straight. "Geez, Rich, who the hell knows? Can't we talk about something else?" We did and have continued to file it under "shit happens" for the several years that passed since the episode occurred.

But it has become a great Wahini for me. I am not yet sure I can say how or why in any way other than in telling the story itself. Of course, even the story must be distinguished from the moment, the happening shit that I set out to explore. In the telling, the episode acquires order and themes that bear the mark of standard tools, methods at work. In fact, "method" is a decent name for the craft of turning episode to story.

The marks are already quite deep. For example, through its intimacy—just three people around one building during one day—the story presumes intricacies that anchor actions in a larger world. As much as the story might tell us about hog sense, it calls for attention to other senses of time, space, and place and to the institutions—regional, national, and global—in which they are embedded. Likewise, key themes—such as the play of thrill and dread, fate and control, communication and resignation, the farm and the academy—say as much about here and now, us and this moment, as they do about the farrowing house when TGE hit. For example, it may be the resemblance of TGE and AIDS that makes this tale a Wahini. Pursuing those methods, themes, and links is my next order of business. "Shit happens" remains a serviceable gloss for the way that business began.

PART II

STEPPING INTO IT

Until maybe 20 years ago, a few country people still fattened a pig each year, reared piglets, kept a pigsty in the garden. And then the pig seemed to disappear with the decisiveness of an extinct species. In fact, pig farming had become an indoor, concentration-camp affair, and pigs were unseen and forgotten by most people in Britain. Now, however, in the last year, we have begun to see fields full of open-air pigs once again. This appears to be the result of some new fashion in agriculture. But their reappearance has the freshness of unfamiliarity and has a feeling good factor about it. . . . Perhaps we should designate 1996 the Year of the Return of the Pig.

—Christopher Andreae,
The Christian Science Monitor

The current pig-meat producing areas of Britain, Holland, Denmark, N.W. Germany and (maybe) Brittany, France, are just too overcrowded with humans and animals in the land area available to cope with the effluent. The crops can't cope, neither can the rivers. Pollution is wrecking our environment and starting to affect the quality of our lives. There is a steady rebellion among our meat-buying public against intensive agricultural production—often viewed as over-production—and farmers have joined politicians, lawyers (and journalists!) at the top of the unpopularity poll. They are considered arrogant, complaining and uncaring, with their palm open for any hand-out which is going. Our pork-buying public is calling for moderation in all production methods, and a reduction in farm support grants, which means fewer animals and less profit for European hog farmers. So I see livestock production declining in the traditional areas of Europe and moving East to the plains of

Russia and Ukraine and West to the former prairies of North America. Here is the food's raw material and the land to dispose of the pollution safely and profitably. How long this will take to be noticeable is an open guess. But it is starting now. I think you are seeing the start of it here in the U.S.

—John Gadd,
European correspondent to the *National Hog Farmer*

FOUR

SWINE SYMBOLISM

MOST PEOPLE, I GATHER, DO NOT KNOW MUCH ABOUT PIGS. Americans, for example, complain that they "sweat like" one—meaning to say "a lot"—when they are hot, even though pigs do not sweat at all. Or they exchange euphoric grins when "happy as a pig in shit," even though pigs are blasé about the stuff.[1] In fact, when first given the opportunity, hog herds quickly tussle out a pecking order and a tidy zoning code. Thereafter everyone seems to know the proper place for the proper pig to do the proper thing. Spaces are reserved for eating, lounging (one for sun; one for shade), sleep, play, and poop. Although people routinely lounge with a magazine atop the bathroom throne, hogs in their "dunging area" (as it is known in the facilities-design trade) usually just relieve themselves and move on. Granted, swine are awfully comfortable with excrescence, even curious about it. They welcome almost anything new to the pen, especially if it arrives below snout level. And they are fond of mud, even when it is foul. But attributing euphoria to any of this requires long-jump imagination. Nevertheless, people seem to insist on finding something powerfully human in such vaguely porcine concerns. So, what I am doing here is just pursuing ordinary connections more doggedly.[2]

A pig is both an animal and a trope *extraordinaire*. Pork may be just a food (the most popular meat on earth), but it is also a taste of the Resurrection among Christians, come Easter ham, and of the taboo among Jews and Muslims. People all over the world have made swine stand for such extremes of human joy or fear, celebration, ridicule, and repulsion. For example, for more than 400 years millions of Chinese have envisioned the qualities of vice through the comic antihero Zhu Bajie. His body may be human and his girth merely porky, but his head is absolutely pig.[3] And children in Puerto Rico hear of basic, country virtues (*jíbaro*) through ever-popular stories of "la puerca de Juan Bobo."[4] But listing only the most important of swine shrines, rituals, and allusions would occupy more than a lifetime, certainly more than the topic deserves. I was particularly daunted, for example, to discover that the literature on pigs in New Guinea alone would fill bookshelves and that several major art exhibits and another trunkful of monographs tackle what I had hoped was the short story of hog iconography in the Western world.[5]

One of the most impressive works is *The Role of Swine Symbolism in Medieval Culture* by Milo Kearney. He documents and classifies an awesome array of forms. A proper understanding of European civilization, Kearney asserts, elevates the pig to "its integral and rightful place as a medieval symbol amidst the flying buttresses and the coats of arms."[6] The dozens of icon variants that evolved from antiquity to the end of the fifteenth century prefigure those still circulating today. In literature, art, music, and myth, pigs have stood for pagans and prophets, chivalry, Christianity, royalty, angels, brutes, and the bourgeoisie. Eighteenth-century Germans reached the outer limits when they drew pictures of Jews gleefully suckling and wallowing with swine, and through bizarre mistranslation Martin Luther claimed that the Talmud sanctioned such *traifidic* intimacy.[7] A little further to the West, Spanish Inquisitors sniffed out Jews pretending to be Christian converts ("marranos," they called them) by challenging them to eat pork.[8] Of course, Jews themselves have also been known to slander in a swinological mode. In 1997 the Zionist zealot Tatyana Suskin was convicted of committing racist acts and inciting violence by, among other things, plastering Palestinian storefronts in Hebron with posters that depicted the prophet Muhammad as a Koran-trampling sow. Just about any modern American allusion has centuries of such pedigree. In the 1960s they were a source of insults for cops and in the 1990s for misogynists.[9] Kearney detects a common core to such stories:

> Ultimately the pig stands, more than any other beast, for mankind with its difficult antinomy of positive and negative characteristics. It seems no coincidence that in Jesus' parable told in Luke 15:17 the prodigal son "came to himself" precisely in a pigsty. As man comes to terms with himself in his representations of the pig, so his prejudice against the pig is both self-accusation and a desire to rise above his nature. . . . the object of man's peculiar cultural disdain for the pig is less the beast itself than man's own speckled soul.[10]

Since this is such heady stuff, I began to look for a serviceable shortcut, a pleasant beltway through the subject with off-ramps for key sights. An obvious route would run through the swine memorabilia that seems ubiquitous in America, once you begin to notice. It is tacky yet tangible evidence of a hog/human bond. Moreover, unlike most animal novelties, the pig ones are favored among people who share no obvious, single relation to the real thing or to each other. Surrounding yourself with the stuff is not a distinctly "gal" or "guy" thing to do. Los Angeles mass-marketers have at times visibly struggled to make their pig products asexual, prepubescent, or androgynous. For example, they hired a total of forty-eight gilts to take turns playing the lead role in the 1995 film *Babe*. Although Babe was supposedly a piglet without gender, only females were cast, precisely because the audience would not know enough about them to be able to tell. Male anatomy would be too visible, even to the untrained eye.[11] Americans who are black and white, who speak English or Spanish or Laotian, who may or may not have ever visited a sty, exchange homemade and store-bought swinalia with common gusto. You are as likely to find a sow-shaped flowerpot in the home of a farmer or veterinarian as by the window of an urban vegetarian or the neighborhood beautician.

My own office at the university now resembles a pig-tchotchke museum. Since hearing about the project friends have provided a steady flow of figurines, greeting cards, joke books, coffee mugs, posters, T-shirts, jewelry, and windup toys, all with snouts or curly tails and the like. Some of them even oink in harmony with a digitized recording that my computer plays back, courtesy of anonymous FTP. The university's parking lot for faculty in the humanities may be the only place that is thoroughly out of place for my truck with its "Proudly Producing Pork" bumper sticker. The fork-lift operator who gave it to me on the loading dock of the local feed store and who knows of my unusual job mix no doubt hoped for just such an effect. At last count, I had about a half-dozen sow pinup calendars at work and a couple more at home. When I began to pray that the supply of novelties might be depleted, I discovered a mail-order outfit that sold only swinalia and hence had a vested interest in new product lines and inexhaustible supplies. Whatever there is that draws Americans to hogs, whatever cultural work one does for the other, might be found in the tchotchkes.

I turned first to collectors. Everybody seems to know a particularly voracious one, and it was easy to acquire a list of addresses scattered across the United States. I started calling around.

"You know, there are a lot of people who collect pig stuff," I would say. "In fact, I've got a pretty big pile right on my desk here by the phone. I was wondering, though, how it works for other people, how it happens, and what keeps it going. Say, in your case, what is it like to collect pig things? How is it the same or different than, say, collecting owl stuff, Samurai swords, or zucchini recipes? Or collecting nothing at all? Is there something special about pigs or their fans, you think? Anyway, how did you get started?" An exemplary response:

> My uncle was a farmer down in Mississippi, and he had hogs at various and sundry stages. I can remember: One hot day, we came across a hog in a trough with just its ears and snout showing. My family and I thought they were very funny.
>
> I'm under the impression that hogs are more intelligent than other animals. They can be trained, and they're cleaner than people think. They're just nutty-looking animals that do amazing things. In fact, I was tempted to get one of those Vietnamese pot-bellied pigs, but I didn't.

It is time to push a little harder. I hope to find weighty meanings in this material but not by fomenting fantasy. Stick close to the things that expert folk—people who are extraordinarily familiar with the material—ordinarily do with it. So I reintroduce practicality: "Yes, but how about the actual collecting of stuff? How did that happen?"

> People—people at work and family—gave them to me. I never bought anything for myself.
>
> But there are photos and postcards. . . . (I really like those ceramic pigs.)
>
> I guess the funniest things are those hats, like Razorback-pig hats. (My mom lives in Arkansas. I *had* to buy one of those.)
>
> A lot of the stuff is in my office at work. Staff knew I was hot on pigs.
>
> I got cases of stuff: Hog Christmas cards from China, and . . . wonderful artwork, really.
>
> But I've more-or-less stopped collecting pig stuff now. You can fill an entire house with hog memorabilia, you know? I rarely look for it now.
>
> Just a phase. Mainly just people giving.
>
> I did, though, develop a way to make my computer terminal oink, and my voice mail oinks like a pig.
>
> But now I've turned to roosters, actually. They're beautiful, very pretty, but dumb and not really friendly.
>
> So it was just a phase. I still like pigs, but now more as bacon and ham.[12]

So much, I conclude, for master tropes and "speckled souls." There is little, if anything soulful here to be found, unless you absolutely insist on defying common sense. I am afraid that only high-wire semioticians (or others who do not get out very much) would glide a gossamer that hitches ten-dollar American statuary to the imperial treasures of Rome and Chichén Itzá. The connection is more enchanting than credible. A person could try harder to find one, but my forays into the subject certainly counsel against it. I still can barely imagine anyone with a straight face detecting biblical prophesy in porcine Precious Moments.

A half-dozen calls left a strong impression of pig collectors as quite sensible people, merely bemused targets for MBAs who specialize in impulse buying. A couple of times each year you want to find a small gift for a relative, a workmate, or a neighbor. You do not want to spend much money or time shopping around, but it should "uniquely" suit the recipient and incite a smile. You run out of ideas after the first year or two unless you can find a theme, and it is better to the extent that it is ever so slightly naughty. You are close enough friends to flout formality. Swinalia seem to fit just such modest requirements. A market and minor industry are born. Owls or roosters actually would do just as well.

But then, I thought, maybe the price range is the problem. Blowing a few dollars for a semi-obligatory gift or deciding to keep the one you receive hardly requires much thought. But what happens when the ante is upped, when the price is high enough to make most anyone pause? I wonder how the deliberations would go, say, if you were contemplating a Gucci boar brooch to match your designer gown. Or if for some reason among the American elite such a thing were considered a certificate of rank, the mark of what set you apart and above the prole? I imagine that you would develop an aesthetic or at least an excuse for the attraction.

It is not so far-fetched a possibility. Reputedly the oldest continuously operating and most prestigious social club in the United States is the "Porcellian," the Harvard University brotherhood of the pig. Most every year since 1791, the brothers have inducted fewer than a dozen new members for life, canonizing the bluest of blue bloods. Their members-only clubhouse (out of bounds even for brothers' wives, much less the ordinary grades of American life) is a five-story, walnut-paneled and pool-tabled, cigar-smoked and scotch-sloshed sty, still squat on Massachusetts Avenue, Harvard Square. For more than two centuries it has been a reserve for the nation's financial, political, scientific, and artistic elite: Supreme Court Justice Oliver Wendell Holmes, architect H. H. Richardson, Ambassador Henry Cabot Lodge, President Theodore Roosevelt, and a string of executive officers at the top of the world's largest banks, investment houses, and law firms. Supposedly Franklin Delano Roosevelt never quite recovered from the disappointment of having been passed over. The club song was written by Owen Wister, author of *The Virginian,* and he has been followed by the likes of George Plimpton and Nelson Aldrich, best known as the author of *Old Money.* The Porcellians would know. One of their sons, a Sedgwick, recalls:

> My father was generally oblivious to the animal world, but he did have an unusual affection for pigs. Around our house he had every possible kind except, thank heaven, the live variety. He had porcelain pigs, ceramic pigs, carved pigs, embroidered pigs, painted pigs. . . . They overran our living-room mantelpiece, swept over the tabletops, covered his bureau, popped up on his cuff links, watch chain and ties and even appeared on our drinking glasses and saltcellar. . . . Why all these pigs? Because my father was a Brother Porcellian—a member, that is, of Harvard's august Porcellian Club—and the pig is the club's emblem.[13]

Such high-class tchotchkes also helped mark an infamous nadir in the history of blue-blood solidarity. In 1937-1938, Porcellian brother Richard Whitney, reputedly "the last champion of the old Guard of American finance," was indicted and convicted for embezzling from the Stock Exchange over the course of several years. It was bad enough that he had been lining his pockets as the market crashed, wheeling and dealing for the brothers whom he confidently called "the best people." But he also had the supremely bad taste to appear before press photographers in a way that discredited the best people along with him. On the front page of newspapers, as at the bar of the New York County District Court: "Whitney . . . stood silent and motionless, his hands clasped behind his back, his head slightly bowed, his face quite expressionless, and his gold Porcellian pig hanging prominent from the watch across the vest of his dark-blue suit, while the indictment was read."[14] Everyone who was anyone was a Porcellian, and they were mortified. In fueling the scandal this image demonstrates that, at least for these sorts of folks, swinalia could be serious business. A man must live his life in a way that only brings honor to the tchotchke.

For better or worse, those sorts of buyers—the ones whom I imagined as more deliberative—are hard to find in my social circle. Even their secretaries resist taking messages from the likes of me, and calls are never returned. The Porcellians, in particular, are sworn to protect club secrets. But, as luck would have it, I stumbled onto a lead in one of those little ads near the back of an old *New Yorker.* An upscale jewelry "workshop" in Newport, Rhode Island, J. H. Breakell and Company, offered a one-and-one-half-inch "piglet pin." Next to a picture of the grinning brooch, James Breakell himself promised the "pick of the litter." Two other features signaled that this market would be more discriminating than impulse tchotchke-ites whom I better knew. One was the context for the ad, particularly the other products that competed for attention. On the very page following Breakell's pitch, for instance, was one for Inca Floats "expeditions" to Galápagos "by yacht" with "our licensed naturalist." The other was the price: $365 for the fourteen-karat gold version.[15] It is hard to imagine just picking one up on a whim or because a quirky workmate has a birthday coming. And since you could buy the pick of about seventeen live litters for the price of this little inert one, I saw good reason to assume that this was indeed a discriminating market. I decided to give Breakell a call.

As for the director of marketing: "I've been writing a book about pigs and American culture," I explained as usual. "And I have been wondering about the connection. What draws people to them, anyway? Do you sell a bunch of these pins? What are you appealing to when you sell them?"

> They are slightly popular. Not our most popular item, but popular.
>
> Since people order it from the catalog, we don't know much about them. Some collect pig things, like others collect hippos.
>
> A lot of the people who walk in the shop have a pig, one of those pot-bellied things.
>
> But there's no accounting for it. People are just into all kinds of things.[16]

So this path to the grand implications of swinalia proved disappointing.

Fortunately, an even higher-stakes, well-marked alternate route opened in U.S. district and appellate courts in 1995. Hormel Foods Corporation initiated a lawsuit against the Muppet masters, Jim Henson Productions Inc. The battle of briefs began when Hormel discovered that Henson's troupe was preparing to release a film parody of Robert Louis Stevenson's *Treasure Island*. In *Muppet Treasure Island* fuzzy seafaring characters encounter a tribe of wild boars who worship Queen Shak Ka La Ka La (played by Miss Piggy) and follow a particularly beastly high priest named "Spa'am." Although he eventually joins the happy Muppet crew, he leaves an unsavory first impression, even for a boar: "He has small eyes, protruding teeth, warts, a skull on his headdress, is generally untidy, and speaks in a deep voice with poor grammar and diction." In promoting the film Henson originally planned to plaster Spa'am's likeness on boxes of cereal and McDonald's Happy Meals, maybe even to sell some Spa'am dolls on the side.

But Hormel was not amused. Since 1937 the company had been using the trademark name SPAM to market more than 5 billion cans of "gelatinous luncheon meat." This pork product is a regular menu item in nearly a third of all American homes and familiar to many more. Through multimillion-dollar promotions, including a trademark character ("'SPAM-man,' essentially a giant can of SPAM with arms and legs"), Hormel corners about 75 percent of the canned meat market, and the company attorneys figured that Henson was about to infringe upon, dilute, taint, or otherwise imperil that success. Hormel was particularly galled to see its trademark linked with a "nasty pagan brute . . . evil in porcine form."

In its defense Henson asked Hormel to lighten up. SPAM has always been a target of teasing, anyway. Well before the cult of Queen Shak Ka La Ka La, jokesters have played to popular suspicion about the can's contents. Since people wonder if SPAM is really pure meat, why should Hormel resent a little extra association with pork? Furthermore, Anne Devereaux Jordan, an expert in children's literature, testified that Spa'am, the boar high priest, was actually a likable character, "childish rather than evil" or "unhygienic." Besides, the movie's humor requires viewers to distinguish between the subject and object of parody. If people actually confuse the two, there is nothing funny in their association.

These were among the arguments that persuaded federal judges and jurors to dismiss the suit and its appeal. Their decision: "SPAM is a luncheon meat; Spa'am is the character of a wild boar in a Muppet motion picture. One might think that more need not be said."[17]

At this point, I became convinced, too. The "best people" in America and the worst of us over the past couple of centuries probably would be no different if hyenas rather than pigs had leapt to proto-Porcellian minds back in 1791. The alternative title of "Hyenallian" would seem to afford no less promise for the sociality, influence trading, and condescension that followed. In 1938 Whitney could have provoked just as much scandal with a different gilded critter dangling from his watch chain. Parody and patent-infringement litigation seem to go together no matter what the species of the players.

You do not have to be a vulgar Marxist to see that these silly little objects, whatever the medium or the price, are about as purely in service of familiar sorts of human relations as anything could be. Besides, none of these things, whether tacky or elegant, is "really" a pig, live flesh and blood. In fact, most people, including Sedgwick and the collectors whom I have met, with good sense emphasize the difference. Swinalia might evoke thoughts about pigs, but at the moment of exchange no one knows or much cares what those thoughts are. As Americans often say of gifts, "It is the thought—simply having one, not what it is; the *thought*, not *the* thought—that counts." To learn more of substance, I would have to work with people who have devoted more attention to the genuine article. Certainly, relating to actual pigs would require more sustained consideration than does the trade in gag gifts, foyer egg-crating and fashion accessories. Hence I turned my attention to farmers, veterinarians, and sundry swinologists, assuming they would be different. In their case, I hoped, the thought would be closer to breathing, stinking life.

FIVE

HOG WARS AND PIG POLITICS

DURING THE LAST COUPLE OF DECADES of the twentieth century, pigs—the genuine article—actually did grab some headlines outside the Swine Belt. In the spring of 1995, for example, they set the tone for a visit to Iowa by the President of the United States. Iowa may be the capital of things swinely, but it is normally a flyover state for leaders of the Free World. Something powerful must have been afoot.

The obvious even if unstated purpose of the president's visit was far from porcine. Bill Clinton was shooting for enough profile points and then slam dunks in local caucuses to wow potential contributors to his 1996 reelection campaign. Iowans are accustomed to hosting such warm-up games every four years. But the excuse for this particular spectacle was agricultural. He was to lend luster to the opening ceremony of yet another Iowa State University conference on rural life. American policymakers often use such functions to pose as principled and academics as real-worldly. This one, with its furrowed backdrop, would be especially propitious for the populist pose. He could wax quotably about rugged individuals, the heartland, and other pastoral pieties. Orators have done so since the days of Thomas Jefferson and continued well after most Americans—among them, most Iowans—moved to town and took jobs behind a counter or a desk. But there was reason to worry that this president's photo opportunity might get testy, and the cause was resolutely porcine.

The president would be met, everyone knew, by protesters rallying to protect "family hog farmers" from "vertical integrators," the large, high-tech, multinational operations that took over poultry in the 1970s and that in the 1980s and 1990s set their sights on pigs. With statutes that are perennially reconsidered, the State of Iowa, like other states on the Plains, has long been hospitable to family farms (which diversify by raising animals as well as crops) and relatively inhospitable to factory farms (which diversify by trading grain futures, patents, and packing plants). Citizens were squared off, for and against change to accommodate "the big guys" of pork production. Hence otherwise calming clichés about yeomen or imagery drawn from *Little House* could turn incendiary. Iowa Senator Tom Harkin did his best to chill the crowd, introducing the president with a joke: "No one should be allowed to be president, if they don't understand hogs." Most everyone laughed, though likely for varied reasons.[1]

The tension that Harkin diffused was about the fate of actual animals and their caretakers, but even then they were cast as instruments of symbolic aerial warfare. Rather than evaluating the qualities of particular changes in pork production, people tended to line up on one side or the other of a single, exaggerated divide. From my vantage, their arguments, which began in the late 1980s and persisted through the 1990s, shed about as much light on farming as the World Wrestling Federation. Both mainly provided an occasion for a chorus of cheers or jeers as stereotypes were body slammed.[2]

Even when combatants met face to face, as they often did (not just in court) to work on their differences, depressingly predictable exchanges would ensue. In 1994, for example, with great fanfare the governor and some state representatives organized a public hearing in Creston, Iowa, where passions were running dangerously high. I was pleased that elected officials were trying to lead Creston back from the brink of blood feud to common interests, if not common sense. But proponents and opponents of corporate hog farming (a.k.a. "progressive producers"/"thieves in the night") seemed to agree only that they could not agree. They lined up on two sides of the room and spent the balance of the session trading invective.[3]

In substance, the divide in most of these disputes resembled that between the "cultural right" and "left" that I already found tiresome in my job at the university. I had hoped to find on the farm some respite from the "culture wars," as presidential aspirant Pat Buchanan dubbed them, of late-twentieth-century America. But allied forces tussled around the hog lot as furiously as they did around the ivory tower. The ammunition was standard bore.[4]

On one side you could hear the measured tones of manly "realism." People recall an inspiring past when—distinctly, supposedly—the best Americans (in this case, hog farmers) had the maturity to meet harsh challenges for the benefit of us all. The old days were great, though more in spirit than substance. (Who would want to go back to sod huts and bouts with yellow fever?) A reinvigorated, don't-look-back, enterprising spirit will continue to yield bigger and better things, as it always has, and help inspire confidence to face changes that, like it or not, the real world demands. Make way for the big guys.

On the other side the tone is more "populist" or "progressive." Underdogs or their self-appointed protectors see a less salutary "reality." The past requires pruning for style as well as substance. The first branches to lop are those whose fruits include pollution, arrogance, and injustice for most people and incontestable betterment only for a narrow elite. Hope might best come in restraining consumption and in better distinguishing the short-term interests of robber barons from the long-term interests of the public, the planet, and generations to come. Just say "Whoa!"[5]

For many of us in academics, these are familiar, even tedious battle lines. Back-pew arguments and letters to the editor of Swine Belt newspapers resemble the dialogue you might expect, say, between Jesse Helms and bell hooks or the Heritage Foundation and the Brookings Institute.[6] The realist line is most visible in feature stories or editorials facing farrowing-house flooring and dewormer ads in hog trade periodicals. The progressive line is more likely found amidst ads for New Age music or radical wear in the Soho or college-town press. For example, after a good deal of success exposing cruelty in veal-calf operations, the Humane Farming Association launched a "campaign against factory farming" from its office suite in San Francisco. The more mainstream press, of course, opted for "balance" of the on-one-hand/but-on-the-other variety, as if wisdom lay in fifty-fifty doses of resignation and reform, the journalistic equivalent of Solomonic justice.[7]

But even the mainstream press gave voice to Chicken Little. When, for example, *Time* magazine covered the hog wars in its "Business" section, the story began: "Colorado farmers Galen Travis and Jim Dober have seen the future, and it stinks. . . . From Colorado to the Carolinas, enterprising growers like [Ronald] Houser and agribusiness giants such as Cargill and Continental Grain are building such livestock factories to mass-produce hogs for packers like Hormel Foods and John Morrell. . . . The vast livestock factories are a long way from the here-a-pig, there-a-pig operations of traditional farms."

Note the forced choice between the Eden of yore and hell 'round the bend. Moreover, the invoked "tradition" makes sense only if your ag experience has been pretty much limited to summer-camp choruses of "Old MacDonald Had a Farm." There have been precious few here-and-there-a-pig operations in the United States (next to none capable of supporting a family) for at least a quarter of a century. The choice seems to loom so large in part because the past has been so heavily airbrushed. And the integrators who lead "the march of commerce" are dressed in jackboots: "Megafarms . . . turn out pigs as if they were piggy banks from football field-length buildings, where the animals are confined to small pens, fed, medicated and monitored with an exacting precision that fattens them to 265 lbs. in six months . . . [These] porkopolises [are] multiplying like rabbits."

Of course, *Time* does offer some information here that is worth crediting. Modern hog buildings are large, in fact, often even larger than football fields. Cargill and Continental Grain (although neither Hormel nor Morrell) have been among the key players, as they have been in just about every scrap of food grown, shipped, or processed since World War II. Yes, pigs reach their market weight in six months, but they have been doing so (give or take a couple of weeks) for decades. And animals are confined, fed, and medicated with increasing precision. But are we to gather that their care should be *less* exacting? Is the choice simply between the singular purity of what has been and the stink of what is coming? To hear *Time* tell it, at issue for citizens is nothing less than "a mechanized assault on their way of life."[8]

Such a background of hyperbole may help explain popular acceptance of hallucinatory scenes like the one that opened the film *Babe*. With film noir lighting and camera angles, viewers get the impression that Babe, the piglet, was rescued from a state-of-the-art operation that could pass for Treblinka under the *Schutzstaffel*. Sows, we are told, are routinely yanked from their suckling young (who are then nursed by robots), marched onto pen-side semis, and hauled to slaughter. Only through homespun miracles can Babe live out his/her days with a family that, we are to believe, can make it on home canning and pasturing a couple of dozen sheep (that, incidentally, never go to market).

"Big deal," you might counter. A business story in *Time* or a Hollywood movie (for children, no less) can only be expected to bloat its plot, given a distractible audience. But plots also are distended in otherwise staid periodicals. Look, for example, at the way the topic was covered in a 1996 issue of *U.S. News and World Report*, not in its "Business" section but in "Culture and Ideas" under the title "Hog Heaven—and Hell."

The story begins with a stock journalistic hook. An innocent (just like you, reader) vaguely recognizes a foul omen. In this case, that innocent is retired farmer Sidney Whaley who for months forbears "the nauseating odor and clouds of flies from 1200 pigs" on a big guy's farm upwind, Onslow County, North Carolina. Whaley patiently rocks behind closed doors and windows, waiting for a response to letters that he has sent, politely requesting relief from government regulators. He is a model citizen. And then the omen proves prophetic. On June 21, 1995: "After heavy rains, some twenty-five million gallons of feces and urine flushed from the buildings where the pigs were confined, burst out of the farm's eight-acre waste lagoon. The reddish-brown tide, more than twice the volume of oil spilled by the *Exxon Valdez*, poured knee deep for two hours across the highway between Whaley's red-brick bungalow and the First Church of God."[9] So even in the sugar-free, low-sodium prose that is *U.S. News* cuisine, Whaley becomes Job, and his suffering a signal from the Lord.

Actually, churches were receiving prophesy from the hog house well before Onslow County's Armageddon. Back in November 1994, for example, an ecumenical throng gathered at St. Augustine's Church in Des Moines to witness testimony under the title: "Community, Church and Large-Scale Hog Production." The list of speakers was a *Who's Who* of rural activism. Their names dotted front-page stories through the 1980s and 1990s. They were progressives—Methodist, Baptist, Catholic, and academic-agnostic.

Although I could not attend (I had to teach that day), I did buy four hours on videotape and got on their mailing list. And I could not avoid laughing along with the participants about the unlikely title of the conference and session subtitles such as "The Theology of Hog Confinement." But these were also people who had to be admired. They were soft-spoken and compassionate, obviously sincere, self-sacrificing, and committed to social justice. Many of them had been drawn into "the battle" by personal experience, growing up in a loving farm family that suffered greatly in prior farm crises. Key institutional participants included the National Catholic Rural Life Conference, which began with the great farm depression in 1923, and PrairieFire Rural Action, which formed in the foreclosure epidemic of 1985, a period of transformation for many of those in attendance.

Moreover, I greatly admired the savvy coalitions that they were able to build. Several conferees were influential in the self-designated Citizens' Task Force on Livestock Concentration. Their report, including detailed legislative and regulatory recommendations, was endorsed by an amazingly large and diverse set of interests, ranging from the Diocese of Sioux City to the American Federation of Labor and the Sierra Club. It was a model of good sense in countering the governor's version, which was more in-house and accommodative, less a grassroots, leadership affair.

We may not agree that the Lord has chosen to speak through hogs or that your average Midwestern farmer fits among the suffering meek of the world, akin to Mozambique refugees. I, for one, have a hard time restraining my cynicism when clerics put modern farmers at the head of a lineage stretching straight from Jehovah through Isaiah, Jesus, and Thomas Jefferson. They neglect to mention, for example, that the humble farm "community" to which Jefferson belonged was itself a model of vertical integration. And its meek were chattel slaves.

But, quibbling aside, it is hard to resist a clincher moral: "Hey, how many executives of ag multinationals (vs. family farmers) live next door to the huge hog barns and manure lagoons they are building? And if they won't, why do they feel entitled to stick them next to someone else?" It is, populists justly insist, a violation of the Golden Rule.

In general, I share their suspicion of high technology, monopoly capital, and rapid change, especially when the profits are so much more visible in suburban office parks than in the countryside or the city. I share their affection for neighborliness, greenery, and agricultural diversity that I, too, have lived with for most of my life. But my experience with "empowered local communities" has been less inspiring. After all, if locals really had their way, most of the people who plant cotton and rice in the United States would still be slaves, and my immigrant ancestors would likely still be living (or, even more likely, slaughtered) "back where they belong." Rural populists, in particular, have consistently included some of the most vicious bigots in American history. Every time commodity prices take a dip, the grass roots are ablaze in conspiratorial fantasies about Russian or Mexican intrigue and Jewish bankers. Thank goodness, cosmopolites have been willing and able to check some of the ugliest of agrarian impulses.[10] That which is traditional, family, local, and small cannot be so simply set against that which is new, corporate, distant, and large, like catechismic poles—Good and Evil.[11]

This recognition has fueled my determination to parse and evaluate more pointedly the charges and countercharges that arise whenever people get to talking about hogs these days. The commonplace practicalities of "pork production" inspire extraordinary environmental, social, technical, and spiritual alliances. The issues around which they ally can be very complex, each with its own cadre of specialists who have much to say, surely more than I could fully cover here. But they deserve sustained, critical—even presidential—attention. The following is intended to serve only as an opinionated primer on the most important issues, emphasizing their often conflicted, far-flung implications. The idea is to connect abstractions that pass for "culture" more precisely to circumstances on the ground, and vice versa.

SIX

FOOD SAFETY

ANY DISCUSSION OF LIVESTOCK must acknowledge a position best championed by people who, as a matter of principle, refuse to eat or otherwise "use" animals. Obviously, if that position were dominant, there would be no reason to argue about better or worse ways to raise them for slaughter. No way could be recognized as good, and herdsmen of all stripes would be discouraged. At the moment in the United States that position is far short of dominant. But it is ever available to buttress multifarious rationales for censuring traditions and resisting schemes that are hyped among the carnivorous. Since humans have been both fallible and predatory for eons, just about every calamity could be (and has been, at one time or another) considered a sorry side effect of people eating the wrong things.

Many consumers are now worried about the purity and safety of food in general and meat in particular. Nearly everyone acknowledges that Americans could improve their health and, quite likely, that of other peoples and the planet, if they were to curb their gluttony and increase the proportion of simple grains, fruits, and vegetables in their diet. Probably for the good of us all, this generation of grandparents may be the last to advocate daily doses of biscuits in sausage gravy with a side of titty bacon.

There is also widespread worry that meat is apt to be contaminated by "chemicals," unnatural ingredients such as the residue of medication that herdsmen and veterinarians administer. But this is one of the easier concerns for the industry to discount. Of course, those who distrust government regulations join vegans in remaining unconvinced, but others may be relieved to know that the safety of food in the United States is as high a priority of the federal government as ever and the envy of much of the world.[1] It is the responsibility of a number of agencies—chiefly the Food and Drug Administration (FDA under the Secretary of Health and Human Services), but also the Food Safety and Inspection Service (FSIS, which is part of the U.S. Department of Agriculture [USDA]) and the Environmental Protection Agency (EPA). Under the controversial Delaney Clause, regulations until very recently allowed "zero tolerance"—absolutely no measurable residue of drugs, growth promotants, or pesticides in edible portions of meat. In 1996, with the first revision of the Delaney Clause since its enactment in 1958, tolerance of risk was diluted from zero to something like one in a billion. USDA tests in 1991 showed that 99.6 percent of pork met regulatory standards, and they failed to find a single case of residue-related human illness in the United States.[2]

The Delaney Clause was controversial among producers and packers mainly because the ability to detect more minute quantities of residue has advanced at a faster rate than has knowledge of their effects. Even whole-grain flour is legally packaged with its likely harmless, small quota of rat hairs and other vile traces of its natural origin. Besides, those people hyping chicken breasts are allowed to market them with unlabeled ingredients, such as water added at the packing plant, that are barred in plain old pork. And who is monitoring those ex-hippies at "farmers' markets" or the rural folks trading eggs and raw milk from the back porch? Fairness and, as expected, "realism" are more at issue than safety.[3]

Moreover, contrary to popular lore and standard practice in the case of beef cattle, hormones have never been approved for ingestion or injection in market hogs. When pork producers talk about growth "promoters" or "promotants," they are referring to very small (if only because also very expensive) concentrations of "subtherapeutic antibiotics" that they add to rations. The main idea is to keep the lining of hogs' intestines from thickening (and hence, interfering with nutrient absorption) in response to common, low-level gut infections; the result: less illness, better nourishment, and a reduction in the feed bill sufficient at least to offset the cost of promotant. Of course, this is not to say that, in defending relatively unregulated use of promotants, the corporate types and realists are right or that, in demanding tighter regulations, the populists and vegans are wrong. The effect of such regimens on humans and animals is a tough call, and producers themselves have good reason for mixed feelings about them. Unfortunately, fewer producers than their critics have the luxury of considering long-term effects, even in the hog house itself. You cannot "wait and see" without cash flow.[4]

My main fear is that routine use of antibiotics will remain economically essential for hog operations long after it reeks havoc on the ecology linking pathogens and hosts. Yet, too, any change in public health routines, every protocol to combat infection, entails such risk. These are always tough calls, and the puzzle-ring of ecological consequence is apt to remain just that, a puzzle within puzzles. Vary the optimism of your assumptions and the breadth of your focus, and you can make almost any intervention seem like a pebble tossed in the ocean or a boulder in a puddle. In the meantime, producers have rushed into a voluntary program of "Pork Quality Assurance" to preempt threatened increases in mandatory regulation of farm pharmaceuticals.[5]

An even greater public concern conditioning all talk of pork is its contested "color." Is it "white" or "red" meat? Here the question is strictly human and dietary. Among many urbane consumers, when it comes to meat, only white is right. The nutritional problem with pork, which is the main stake in debating its color, is fat. Part of what people metabolize from any animal fat (whether in cheddar, chicken, or chops) is an intense dose of calories and "bad" cholesterol. At least in large amounts, it increases their risk of obesity, heart disease, and cancer. And despite a drone of public service announcements, food groups, and pyramid gimmickry, every day an average of about 100 million Americans eat pork. In part because citizens consume so much, pork can be considered a public-health problem. And since that consumption is not evenly distributed in America, it also represents a social problem. As elsewhere, Division Streets of class and gender are well paved. Heavy pork eaters are more apt to be male and have less income and schooling than the more female, wealthy, and educated people who have suspicions about the stuff . . . and who spend more per capita at the grocery store.[6]

One way that pork promoters have responded has been to scramble to the front of consumer resistance. The more readily people defile meat that is red, the harder admen insist that pork is really white. And they have been amazingly convincing. In 1987 about 42 percent of consumers said they preferred white meat; in 1995, 63 percent. But in that same period the percentage of people who considered pork "white" rocketed from 9 to 53 percent. By that measure, the advertising campaign, built around the slogan "Pork—The Other White Meat," was one of the most successful in American history. The phrase, sewn out of whole boardroom cloth, in just a few years became familiar to eight out of every ten Americans, a larger proportion than know the name of their senator or recognize the Bill of Rights. On the other hand, the slogan appears to have failed to whiten public images of the meat itself or to have exhausted the old meaning of "color." Most people still see pork, even if "white," as less nourishing and higher in calories, cholesterol, and fat than chicken. According to the industry's own surveys, "Consumers perceive pork as high in fat before they even look into a meat case."[7]

This is especially discouraging in the industry because producers have also put tremendous effort into making pork nutritionally less "red." They could not do much about the concentration of calories in a given cut, but in ten years

(1984-1993) they did manage to reduce by about a third the amount of fat they shipped to the grocer. Of course, lard is as poor nutritionally as it ever was, but the leanest cuts of pork now have less fat, calories, and cholesterol than the king of carnivores' lean cuisine, a skinless chicken breast.[8]

Unfortunately, it is very possible to place much of the blame for the corporate challenge to family farming precisely on the demands of health and waist watchers. Independent mom-and-pop farmers are too busy working the home place to monitor Yuppie trends. Marketing has always been the big guys' strength, and their profits trickle down poorly. As a senior economist for a Federal Reserve Bank explained, "To keep high-tech pork products on target for today's palates, the industry is abandoning its traditional way of moving pork from the producer's lot to the dinner table." Through packing-plant incentives, such as higher bonuses and stricter grading for lean yield, the big guys also changed the sorts of animals that would be bred and the conditions under which they would be raised. Lean, rapid growth was heavily rewarded to the neglect of nearly every other consideration that might benefit the animal, its caretaker, or the surrounding community. With a hint of remorse, a trade magazine explains: "For producers selling on a carcass merit system, the results are simple. There are three sire-line traits that impact profits: feed efficiency, growth rate and backfat [i.e., the leaner, the better]. While other traits may be important (soundness, meat quality, etc.), producers are not currently paid directly for them."[9]

This is but one of many cases where interests crosscut constituencies rather than lining them up along a singular divide. While everyone might benefit from a lean diet, various sorts of consumers and producers as well as their human and animal dependents do not share equally in bearing the burden of change. Efforts to move the market uptown, where "progressives" are more likely to live and jog, are also apt to be rough on family farmers; less so on the suits up at corporate headquarters.

Hence I could not help but find mixed messages in the ad campaign that followed "Pork—The Other White Meat." In 1995 the National Pork Producers Council (NPPC) explicitly targeted the wine-and-squab crowd. They launched a $3 million commercial blitz, including an astoundingly expensive, prestige spot on ABC's telecast of the Super Bowl. They featured precious cuts, fussy preparations, and Ritz-Carlton presentations. Their new slogan was "Taste What's Next." Michele Hanna, director of advertising for the NPPC, explained: "We're trying to bring an emotional tie in, a love of pork, instead of just saying it's okay to eat." Clearly, their work was cut out for them.[10]

SEVEN

SWINE RIGHTS, LIBERATION, AND WELFARE

MANY OF THE TROOPS OF HOG WARS love pigs as fellow earthlings, like humans, not meals in the making. Whether the chef is Epicurean or barbarian, whatever markets reward or consumers demand, any pork in the recipe comes from pigs with inalienable rights—simply as living beings—that need to be asserted. This is among the basic principles uniting modern American advocates of "animal rights." Their best-known representative is the organization called "People for the Ethical Treatment of Animals." "PETA People," as they are known with a bit of scorn around the farm, are related to hog barns more or less as CORE and SNCC were to racially segregated lunch counters during the civil rights movement.[1] In part because they have used 1960s-style protests to make their point, PETA is among the most controversial of groups challenging the industry. Probably its most famous stunt was smashing a pie in the face of the Pork Queen crowned at the 1991 World Pork Expo. Thereafter, rumors circulated that PETA might be secretly allied with the Animal Liberation Front, an underground outfit associated with vandalous, even murderous "direct action." When first organized in the 1970s, members of "front cells" in Europe and the United States mainly harassed laboratory-animal researchers, but by the early 1990s they were burning packing plants and firing potshots at farmers.[2]

Among the targets of PETA people are the rough handling of hogs and the trend toward lifelong confinement in dreary, crowded, noisy buildings with noxious air. Whatever your disposition, it is hard to imagine that the God of Genesis had gestation crates (or, for that matter, high-rise tenements) in mind.

Urges to nest, root, fight, play, wander, wallow, or just snooze in the sun are frustrated. With just a bit of exaggeration and anthropomorphism, the horror is overwhelming:

> If you aren't familiar with the confinement method of raising hogs, picture a warehouse—not a barn—housing animals who never see the outdoors. They live in individual pens inside buildings where the feed and water is completely mechanized and the need for human labor slight. . . . From an animal rights perspective, confinement is inhumane and unnatural. The animals often experience crippling because of the metal or concrete floors, and sows' legs eventually break down under the stress of being forced to overproduce piglets. . . . The pigs can be grown stacked on top of each other in crates. These unnatural conditions breed numerous diseases in the animals and necessitate dependency on antibiotics and sulfa drugs; how these drugs affect humans who consume the meat is unknown.[3]

Note how here, as elsewhere, each outrage is said to produce another, in a sort of allopathic social science. Every party to hog wars, in its fashion, tends to stand homologous issues in a line like dominoes destined to topple in a prophesied direction. The exact sequence connecting cause and consequence barely seems to matter. In this case, we are to believe, it is the "conditions" (rather than, say, regulators, carnivores, or Cargill) that "breed disease" that in turn "necessitates" drug addiction, cripples, and compromises food safety.

Such a causal chain is far from self-evident, at least to people who opt to invest in those buildings. Imagine, for example, if the scheme were so obviously catastrophic, trying to persuade a bank to lend money for it. As ag loan officers well know, sick and crippled animals usually do not survive, much less thrive, and even if they do, they fetch radically lower checks from the packer. And care for ailing animals is the single most expensive and time-consuming part of the business. Why would anyone think that getting them sick, crippled, and hooked on drugs saves labor and makes money? Only, I would think, if they assume that farmers are, as a rule, sadistic or stupid or both. The insult so strongly implied is awfully hard to accept, especially when it comes from some do-gooder who does not know the first thing about caring for animals as a matter of daily toil as well as self-righteous sentiment. And so the insult volleys go.

Of course, too, if the argument is strictly principled, the quality of the persons doing the arguing is beside the point. "Animal rights" are just a variety of "natural rights," objective entities that can stand up in court. They are plain as the nose on your face, and the challenge is to recognize that reality. Cranked up one notch, for example, they justify "animal liberators" who oppose the use of animals for *any* agricultural, commercial, medical, or educational purpose, unless the ratio of risks to benefits favors hogs as well as humans. To value humans over hogs in such ethical calculation is to succumb to "speciesism."[4]

It is tempting to chalk up much of this talk to silliness, decadence, or at least a combination of ignorance and wishful thinking that is easy to find in America. As illustrated, some of the animal-rights, just-say-whoa hyperbole employs a bourgeois, anti-urban, and anti-industrial bias. Someone else's heartless drive to save labor, mechanize, and make money (rather than, say, experience being tired, bruised, and poor or critics' own participation in a cash economy) is the culprit. Wouldn't it be nice, progressives seem to say, if those people would only stay on the farm with their cute animals, content to bolster primitivist fantasies about latter-day yeomen? While we ride the Fortune 500, you can be our pet premoderns . . . and make sure we do not hear of any pain and suffering in the process. Something must be turning horribly wrong out there, if we hear that pigs do not live like Babe.

But animal-rights talk has become too common among diverse sorts of people to be dismissed so glibly. By the mid-1990s, PETA prophet Jeremy Rifkin was close to a household name, and PETA spokesperson Robin Walker, with his mascot "Chris P. Carrot," were regulars on the talk-show circuit. On the *Today Show,* for example, they warned that unspeakably cruel "factory farms" were coming. As a deterrent, they advocated everyone opt for vegetarianism, including the family Doberman. Their slogan was "Eat your veggies—not your friends." PETA got a more substantial boost in the form of proceeds from a fourteen-song compact disk produced by Paul Mitchell Salon. The CD, titled *Tame Yourself,* included cuts from pop stars with conscience—the B-52's, The Pretenders, and Michael Stipe of REM. In 1994 a string of prime-time TV shows featured PETA lines. That was the season of *Beverly Hills 90210* when Brenda (Shannen Doherty) got swept up in a particularly radical animal-rights group. Their method, not their message, was what discredited them.[5]

Probably the largest live demonstration of resistance to high-tech hog farming came in the form of a rally near Unionville, Missouri, in April of 1995. It was both the kickoff for the National Campaign for Family Farms and the Environment (NCFFE) and an installment in Willie Nelson's continuing program, Farm Aid. About two thousand people (an impressive number for such an isolated place) gathered to hear country music and thirty-six speakers raise their voices in opposition to "megafarms," particularly Missouri's own Premium Standard Farms. At the time, PSF was the fourth largest hog operation in the United States and growing fast. Those assembled were a diverse lot, and there was much more at stake than animal rights. Cooperating organizations ranged from the Missouri Rural Crisis Center to the Rainbow Coalition. Yet they joined not only in opposition to PSF but also in agreement that hog wars were to be fought over "rights"—rural, animal, community, health, environment, church, labor, ethnic, civil—all of them, again, supposedly plain to everyone . . . except, for some ungodly reason, PSF.

Willie Nelson announced, "This is a classic example of why Farm Aid came into existence—to keep big corporate farms from running over the little people." PSF was characterized as "a modern day Genghis Khan," and their operation "a vestige of slavery and Jim Crow laws . . . here to rape and pillage our communities." After the rally, NCFFE began organizing boycotts and a six-day caravan across the Midwest to hand deliver a message of protest to the President of the United States, who just so happened to be opening a National Rural Summit in Ames, Iowa. That "Journey for Justice" is what got Senator Harkin worrying about candidate Clinton's photo opportunity. NCFFE aimed to shift some glad hands and limelight onto rights being wronged in swineland.[6]

As one might expect, producers hit back with heavy PR. For example, the Animal Industry Foundation pushed its eighteen-minute videotape, *Animal Agriculture Myths and Facts*. They got their "myths" directly from PETA's list of "facts." Similar promotions poured from ag schools and research centers. For example, the Missouri Association for Agriculture, Biomedical Research and Education announced new efforts to enlighten a public that had "become more urbanized and further removed from the food production chain" and hence vulnerable to animal liberation propaganda. "Education," to these folk, meant teaching "the importance of animals for food, medical research and products." Newspapers across the country got a steady supply of snazzy drawings with graphics and sidebars connecting pigs to lifesaving surgery, heart valves, skin grafts, miracle drugs, and the like.[7]

All that "rights" talk also encouraged substantial effort within the industry itself to raise livestock more tenderly. After all, "animal welfare" (the realist's answer to "rights" and "liberation") might be easily reconciled with profits. Humanely tended animals might be both happier (insofar as such things can be determined) and more productive. After surveying some of the difficulty in defining and enhancing hog "welfare," a standard, animal science reference book offhandedly concludes: "It is highly likely, *of course*, that many of the efforts that are successful in improving the well-being of the pig will also be reflected in increased efficiency of production and higher profit margins."[8] The same message is regularly featured in the trade magazines. In response to PETA agitation, for example, the editor of *Pork* magazine advised: "What you can do . . . to counter the animal activists' messages and protect your business. First, make sure your farm is operating flawlessly. That includes animal handling, herd health protocols and environmental management."[9]

Furthermore, many people in the industry sincerely commiserate with their livestock. For example, passion for open-air, "free-range" pigging is hot among producers in northern Europe and sympathetically covered in U.S. hog-trade magazines. Students in veterinary schools now tend to be critical of producer associations, precisely because associations, in a pinch, are more apt to promote the interests of members, who vote and write checks, than their animals. Vegetarianism is on the rise even in vet schools where meat inspectors are trained. So, once again, constituencies are divided in ways that handy fault lines (such as in vs. out of the industry, urban vs. rural, practical vs. sentimental, etc.) do not quite fit. Nevertheless, the rhetoric of hog wars is as often ad hominem as it is allopathic. Among the casualties in all of this is a chance to recognize common interest.[10]

Some of the worst quarrels come in figuring out what hogs make of all this. After all, it is *their* "rights" or "welfare" that is supposedly at issue. How do they feel, for example, about living in crowded buildings? Does it really bother them? The challenge of answering that question roughly resembles that of adjudicating "birth rights" in the debate over abortion. No one can be quite sure how to ask a fetus if it is "really a person" or to interpret potential answers. Instead, answers are deduced from natural rights syllogisms or inferred from observers' projections (e.g., "I may not be a fetus, but I sure would not want to be sucked out a hose.") Hence pro-lifers, like animal liberators, take upon themselves the ventriloquist's burden, speaking for creatures that cannot speak for themselves.

Hogs present an additional challenge in that they plainly are different enough from people that projections of the "if-I-were-one" sort are a radical stretch. Try as I might, for example, I cannot fully imagine being drawn to the smell of shit or engaging in other grossities that pigs normally elect when given a chance. Like most people who sing the song, I can more readily imagine "swinging on a star." But perhaps hogs under extreme conditions put out such clear signals that wholesale leaps of imagination are not required. The challenge might be a bit of translation rather than projection.

When, for instance, you walk into any swine barn, you are bound to see hogs chomping on the steel bars of their pens. Scrunch, scrunch, grunt, scrunch, over and over again. What does this "chewing behavior" mean? What does it tell us about, say, the animal's welfare or respect for its rights?

The Humane Farming Association, claiming support from "agricultural scientists" who study such things, says they know. Chewing is a "temporary measure of relief from the torment of crate confinement." It is "abnormal . . . simply the desperate expression of frustrated animals pushed to the point of madness . . . [that] resembles, in many respects, the development in humans of chronic psychiatric disorders."[11]

Since I am neither an animal scientist nor a psychiatrist, I may be handicapped here, but this is not the sort of interpretation that experience seems to counsel. I regularly see hogs (and "cribbing" horses) engage in such "endorphin-releasing stereotypes" in wide-open fields. Although the sows on the old-timey farm next-door have about a quarter-acre apiece to roam around, they ordinarily crowd together ("cheek-to-jowl," as people have been saying for decades before confinement was an issue) and chomp on the panels in one corner. Chewing on gates is a common enough pastime to suggest an analogy closer to biting your nails than banging your head against a wall.

I could understand why Bay Area vegetarians might leap to alarming conclusions, but I find it hard to believe that people who well know pigs would respond in kind. In fact, I was unable to find the "agricultural scientists" to whom the Humane Farming Association referred. The association's director would not respond to my telephone calls and letters requesting citations. In searching for those citations, what I found instead was strong support for my common sense. John McGlone, probably the single most respected pig specialist among animal behaviorist in the United States, monitored a controlled comparison at his research center at Texas Tech University and concluded: "Outdoor sows spend as much time chewing rocks and chewing in the air as indoor sows spend chewing bars. . . . We conclude from our studies that both behaviors are normal."[12]

Mind you, McGlone, his sponsors, readers, and allies identify with pig farming as a way to make money, a business. But it is hard to see how the profit motive in this case can be considered a source of distortion in general and insensitivity to animals in particular. If you are suspicious of any outfit oriented toward material gain, you will be hard-pressed to find an alternative. The margins in the hog business may be small and the economic pressures intense, but, judging from the trade literature, they generally argue for rather than against welfare concerns.[13]

For example, the largest part (about 62 percent) of the cost of raising a pig is feed consumption. When animals are stressed or ill, they eat less and gain less from what they do eat. Since the costs of housing a pig are minuscule by comparison (at least on a per-head basis over the life of a building or the payments on a pasture), there is no money to be made in overcrowding. When hogs have calming amusements ("enrichments," experimenters call them, like toys to play with), there will be a measurable decrease in the secretion of stress-related hormones and an increase in feed efficiency, daily gain, and immunity to disease. Any cause of stress—especially anything that gets between a hog and its nutrition, such as a feeder that is uncomfortable, too much of a crowd or a slippery floor on the way, too much heat or cold, even bright light or a draft—cuts down feed intake and the efficiency with which feed is "converted" to marketable flesh. It has been well documented that no amount of rations can compensate for overcrowding. So it is hardly the case that a farmer's search for profit and a hog's for a nice day are necessarily at odds. Up until the moment hogs leave the farm, well-being and profit are allied interests.[14]

This is more than my own hunch or the finding of a few Greened university researchers. It has long been an article of faith throughout the business, including the parts that depend on advertising revenues from the big guys. At least the cruel quotations that PETA people have found in trade magazines (and they, again, did not respond to my requests for citations) seem to me out of character. For example, an issue of *Pork '93* began its producer's guide to hog-house flooring with the truism: "Pig comfort is no longer a secondary consideration. An uncomfortable pig is a stressed pig. And a stressed pig is a non-productive pig." Flooring was evaluated by standards that, Dutch research shows, "piglets prefer." For similar reasons, all kinds of advertisers in the trades hype products that they promise make for "happier hogs." Of course, the merit of their claims of contentment is just as contestable as it is when attached to deodorant or detergent, but the industry's interest in the emotional life of their stock seems certain.[15]

Research on hog happiness is now itself a growth industry. It was among the imperatives, for example, that the USDA addressed in the form of several major grants, including $1.5 million to start a Center for Research on Well-being in Food Animals at the School of Agriculture of Purdue University. Distinguished researchers such as Julie Morrow-Tesch were employed to develop more reliable indicators of swine sentiment. It turns out chemically to be very difficult to discriminate "good" stress (which precedes mating) from "bad" stress (which makes an animal a less efficient converter of feed, a less dependable and more short-lived breeder, a less cooperative littermate, and a more likely victim of disease). Nevertheless, stress is readily visible in the cutting room, and packers will dock you severely for it. The culprit is PSE ("Pale Soft Exudative") pork, coral-colored mush that loses much of its weight in water and turns tough, dry, and tasteless when cooked. Consumers do not want it, and packers do not, either.[16]

There are several causes of PSE, but most of them can be glossed with words like "unhappiness." An animal that is challenged (e.g., because it is high-strung, hungry, scared, abused, or just slow in adjusting to a new environment) begins to change at the cellular level. Muscle cells prepare for action by releasing calcium, which aids in the conversion of stored to usable energy (e.g., to move or otherwise cope). In an anaerobic environment (where metabolism proceeds faster than the blood delivers oxygen), glycogen is converted to lactic acid, which begins to accumulate between the muscle cells. The greater the stress, the more rapid the rate of glycogen depletion and lactic acid buildup. You might recognize this stress-related chemistry in your own body—say, when muscles begin to shake or cramp during the final repetition in a workout or after a near miss of an accident. The same thing happens in stressed-out pigs. An animal slaughtered in such a state—which can be nearly permanent in some animals—will be in the midst of a lactic acid storm. In a warm carcass ("postmortem metabolism"), no matter what happened before death, the pH-level will begin to drop. The rate is crucial, because acidity denatures protein (the "sarcoplasmic and myofibrillar" or "myosin," in particular). The more rapid the drop below normal (a pH of 6.1 to 6.2), the more pale and exudative the meat will be. If it gets down to 5.8 or so in the first hour after slaughter, the storm's damage will be obvious. Meat will be too watery to cure and ineligible, by standard contract, for export. Since about 20 percent of the pork daily processed in the United States has been PSE and since stress is among the most obvious causes, greater hog happiness has clear commercial value.[17] After an extensive survey of research, meat scientists working with the Departments of Animal Science at Colorado State University and the University of Illinois reported a conventional wisdom to the NPPC that might encourage progressives as well as realists:

> It is believed that the most effective way to reduce incidence of poor-quality pork, other than through genetics, is to improve preslaughter management and handling. Reduced stress during loading, transport and handling, improved environmental conditions, and shorter travel times are some of the methods that can reduce the incidence of PSE pork. The use of improved preslaughter handling has proven to be an effective means of reducing PSE and unnecessary death loss.[18]

So, stressed-out pigs are a pretty sure sign of inferior meat in the making. Deliver to the packer a trailerload of such hyped-up animals and your check will be docked, if shipment is accepted at all. The next time the packer will be even less forgiving. The larger the operation and the more frequent the trips to the packer, the heavier the sanctions to promote at least minimal animal well-being. Small-scale, part-time producers may be the ones who less often confront the relationship between welfare and profit.[19]

This is not to say that animal-rights activists have no reason to work. The state of swinely art remains profoundly questionable from a hog's point of view. There may, for example, be a chasm between the conditions that make a hog profitable and the ones that serve its spirit, in understandable even if immeasurable ways. Those old-timey operators may be more likely to sense that a pen is too crowded even if things never get "overcrowded," so miserable as to show up in records of herd health, carcass quality, and profit.

Probably the most disturbing, frequent cases of cruelty happen at the packing plant, particularly for "downers." These are injured animals that may languish, suffering horribly for long periods, first on a truck and then on loading docks or in holding pens before slaughter. Even the NPPC has admitted that 90 percent of the suffering could be avoided. After a scandal prompted by "downers" at the Saint Paul Stockyards in 1991, most companies have rushed to implement better procedures, which are now mandated by the U.S. Congress and monitored by the USDA's Grain Inspection, Packers, and Stockyard Administration. But downers still represent a horrid, regular part of the business.[20]

Also weighing in on the liberators' side is the simple fact that food animals (and cultivated vegetables) from the moment of birth are headed for an untimely, unceremonious death. Every day thousands of swine surrender their offspring and an ocean of blood to become food for people, who could find other sources of sustenance. For animal-rights activists, that simple fact may be enough to render trivial the details of animal "welfare"—the fussing about optimal pen sizes, feed-conversion rates, grading systems, profits, and PSE. They might recognize, though too, that these are far from trivial for hogs and the people who have anything to do with them.

EIGHT

CRAPPING UP AND CLEANING UP THE ENVIRONMENT

YET ANOTHER BROAD STRAIN OF CONTENTION, when it comes to talking hogs these days, is their status as an environmental hazard. It is an issue that would seem equally to concern chops and sprouts eaters, no matter what their stance on food safety or animal rights. Imagine, for example, you are Raymond Stockdale. You and your partner, Kathy, have settled into a rural neighborhood with intense social ties. Maybe out of special respect for you (knowing, for example, that your cousin is president of Iowa Select Farms, one of the up-and-coming big guys), neighbors tend to keep their voices low when hog business enters the conversation, say, in line at the grocery store or filing out from Sunday worship.

One breezy afternoon you spot a surveyor down the lane. "Hey, how's it going? What are you up to?" you ask.

"I'm seeing if the manure lagoon is going to be far enough from your house," he replies.

This is a heck of a way to find out that, somehow your cousin failed to mention, Iowa Select is planning to build a giant farrow-to-finish operation basically next door. They are at the point of calculating how many feet will remain between your kitchen window and a lake that they will dig to fill with thousands of gallons of raw swine sewerage.[1]

An editorial in the *Chicago Tribune* summarizes the issue as generously as anyone could:

> Pigs are wonderful animals. They are smart, can make fine pets and—with regrets to People for the Ethical Treatment of Animals—are excellent when

> converted into chops, ribs, ham, and sausage. They also smell bad. No, correct that. Their waste matter, concentrated in hog lots, smells truly bad—one of the rankest, most persistent odors ever wafted by any creature, great or small.

The editorial was prompted by an International Round Table on Swine Odor Control that drew world-class scientists to Iowa for a few days in June 1994. Whatever their differences, scientists could agree that people like the Stockdales have good reason to worry. For example, Duke medical psychologist Susan Schiffman unveiled research that suggested a link between hog odors and a frightening accretion of ailments (multiple chemical sensitivity syndrome, MCSS) among people downwind. A microbiologist stated flatly that the pig is an "indefatigable and unsavory engine of pollution."[2]

Controlling the spread of such pollution has been among the most ready rationales for popular resistance movements. "Save the environment" was, for example, the main rallying cry in the grassroots campaign on behalf of House Bill 1151 in Oklahoma, which would have restricted the state's hog-house boom. Similar concerns moved activists across the country in the 1990s. Former Marine Colonel Rick Dove began organizing private pilots into a "Neuse River Air Force" to collect aerial evidence of hog-fouled waterways for citizen suits under the provisions of the Clean Water Act. At the first signs of agitation among the mainstream electorate, even Jim Hunt, the governor of North Carolina and old pal of the big guys, proposed an additional $2 million in the state budget to make enforcement of environmental codes around hog farms less dependent on citizen complaints. In the spring of 1997, after an unrelenting sequence of spills, stink, and public outrage, Hunt went so far as to ask the General Assembly to place a two-year moratorium on the construction and expansion of all large-scale livestock facilities. The Assembly agreed. Environmental protection was the clincher in legislating the ban.[3]

Assumed in these campaigns is not only that pigs pollute but also that people who raise them are flawed stewards of the air, water, and land that we share. Government regulations are required to impose responsibility that—however much they might be flattered, come stump speech season—farmers cannot be trusted to exercise on their own. Hence environmental regulations affecting farmers have grown to the point that the National Center for Agricultural Law Research and Information has to produce a separate hundred-page volume just to summarize them for each state. Nevertheless, farming is among the least regulated of major industries in the United States. Agricultural developments generally are excused from impact-assessment requirements. In the early 1980s most states passed "right-to-farm" statutes that protect livestock operations from nuisance suits under ordinary circumstances.[4]

Farmers can, for example, without inspections, permits, waivers, or whatever, daily collect and spread excrescence on fields in a way that would be crimes of the decade if perpetrated by a factory or municipality. It would be nice to think of this process as heavy-duty composting. After all, since the first humanoids settled down, they have been using manure to increase soil fertility. People rely on the bowels of livestock to refine and return to the earth some of the nutrients that their plants take from the air and the soil. Even if it stinks, spreading manure seems preferable to ever-greater amounts of the usual substitute, fertilizer made from fossil fuel.

But hog manure also contains things that can compromise soil and water quality and threaten humans and wildlife, not only near the farm but also far downwind and downstream. Perhaps the most frightening among them is a recently identified waterborne microbe, *Pfiesteria piscida,* that scientists dub "the cell from hell." Among its potential effects is sucking the life out of other cells that it encounters on up the food chain, including red blood cells in humans. In the course of its normal lifetime, it assumes elusive forms and exudes toxic fumes. It seems to thrive in runoff mixed with feces (human as well as hog), and it is on the rise in the rivers of North Carolina.[5]

Consider, too, less dramatic but important pollutants particular to hog shit—selenium and "trace minerals" like zinc, iron, and copper that are standard feed additives. Among the reasons they are added is that hogs depend on these nutrients for good health. At least in theory, they can get an adequate dose rooting around in a pasture, but often in fact they do not get enough, either because the soil where they root or where their feed was grown happens to be light on those minerals or because they are housed off the dirt (say, on concrete where they are also less likely to pick up parasites). But then mineral supplements that pass through hogs over the years can build up to dangerous concentration in fields where manure is spread. Hence the U.S. Food and Drug Administration has worked to limit the concentration of selenium allowed in livestock feed. The response has included predictable squabbles about what concentrations are "natural" or "really" bad for soil. Plus, it was soon obvious, if the concentration of mineral supplements were lowered, farmers would just have to serve larger volumes of feed, which result in larger volumes of waste and yet more pollution of other sorts.[6]

"The trouble is," John Gadd explains in one of his columns for the *National Hog Farmer,* "pigs are major polluters—worse than humans." You might think of it as the downside of their feed efficiency. Once they have so miraculously metabolized what is useful, what remains is miraculously foul. Since Americans flush so much water down the drain, their households produce a much larger volume of waste (an average of 240 liters per day per person vs.

10 per pig). But hog waste has a much higher BOD (biochemical oxygen demand). In other words, it is much tougher to decompose. Those few quarts of each hog's waste daily require nearly three times the amount of dissolved oxygen as those many gallons of the human stuff. Roughly speaking, a modest operation (a farm that raises pigs from a breeding stock of 100 sows) produces about the same amount of pollution as 5,000 people. These days state-of-the-art farms are more likely to count their breeding stock in the thousands. So from this point of view, a big guy's building complex down the lane has the environmental impact of 100,000 toilets. And they are not attached to a sewerage plant or even a septic tank, just a giant, open cesspool that they call a "lagoon."[7]

When the contents are sprayed on a field, problems can quickly spread. In addition to mineral traces, hog waste includes nitrogen and phosphorous in strong concentration. After enhancing soil fertility, excess nitrogen poisons air and water. Nitrates can leach below ground or erode into streams and ponds. Although phosphorous does not leach below ground, it moves with eroding soil into surface water, where it is very harmful.

Farmers and hogs stand to gain at least as much as the Sierra Club in minimizing such pollution. Among the greatest causes of excess, for example, is a ration that is too rich for the animal effectively to digest. More precisely adjusting nutrition to an animal's needs reduces the amount of waste it produces as well as the feed bill. In this respect "precision" has value that *Time*-style, traditionalist screeds overlook. Even animal welfare concerns can harmonize. For example, among the conditions that cause nitrogen concentrations in manure to rise (resulting in a higher BOD, lower feed conversion efficiency, and overall stink) are chilled pigs. Help them stay comfortably warm, and what comes out of their rear end will be a little less noxious.[8]

But it will remain noxious, no matter what you do, and farmers know that well, especially if they work indoors. When polled back in 1989, well before hog wars heated up, about 90 percent of farmers judged the air in confinement buildings "moderately hazardous" or worse. Scientists have found good reason for that judgment. Hog-house air includes a wide array of toxic or irritating materials along with asphyxiating gases that have synergistic effects. Hence, it should not be surprising that the people who work with hogs in confinement have all sorts of respiratory problems. After chores, you are likely to have a persistent cough combined with body aches, fever, and fatigue, a condition known as "organic dust toxic syndrome" (ODTS). It normally takes me about two days to recover from an eight-hour hitch cleaning under dusty farrowing crates. Thank goodness, it is seldom my job. According to veterinarian Kelly Donham, probably the leading expert on such occupational health hazards, about 58 percent of swine confinement workers experience chronic bronchitis. Poor air quality, of course, also compromises hog health and productivity.[9]

Such hazards are obviously no more confined to the building than is the stench. Most farmers know about disasters like that which took the life of Pete Harrington, who succumbed to gases while trying to fix a plugged pump in a manure pit out in the barnyard. His uncle, John Gergen, hurried into the pit to attempt a rescue and died beside his nephew ten minutes later. Farmers have learned, if only in media coverage of such catastrophes, that pit gas is an awful though natural by-product of bacteria. Its most deadly ingredient is hydrogen sulfide—a compound that is heavier than air and likely to escape recognition, because among its first effects is paralyzing your sense of smell. In its presence no filter or gas mask short of one attached to a scuba tank will protect you. In normal concentrations over a closed pit, it will kill you in about two or three breaths.[10]

Against such a background of scientific understanding, front-page news, and occupational lore—uniformly admonishing hog effluvia—protests tended to escalate in an olfactory mode. For example, people grabbed onto Professor Schiffman's report of MCSS downwind in North Carolina like a newly minted Gatling gun. The psychologist herself emphasized caution in generalizing from her findings, not least because crucial data were drawn from only forty-four people in a single state. It would barely impress a promotion-and-tenure committee back at Duke. But dissidents cited the report as if it could blow the big guys out of the water. Given the yawn that scientific round tables normally draw, there must have been raw nerves nearby. You would think this were the first sign that the foul smell of hogs really signals something foul in the air. Recall that nine out of ten farmers expressed a similar opinion years before.

As passions rose, propagandists made news out of the discovery that pigs smell nasty. For example, people who read about the swine boom in North Carolina began toying with the possibility of changing the state's slogan from "First in Flight" to "The Slaughter and Stink State." Pushed on the defensive, the National Pork Producers Council in 1994 declared odor control their "first priority" and set aside more than $300,000 for research to solve the problem. In the meantime, producers tried tossing megadoses of deodorizer and snake oil into their pits, hoping the damned stink would go away. Results, as you might guess, were unimpressive. As hard as it is to make a purse out of a sow's ear, it is even harder to make potpourri out of its poop. As the uproar gained momentum, the stink became increasingly difficult to justify as an unalterable fact of rural life.[11]

One of the first successful legal challenges to the swine business was won on environmental grounds. The defendant in this case was Norman Wolff, who operated an 800-head finishing unit outside the town of Newell in Buena

Vista County, Iowa. The challenge came from Dennis and Ruth Weinhold, who lived a half mile away. Suffering from the wind-borne stench, the Weinholds requested arbitration and then filed a nuisance suit in 1992. If the filing were just a few days later (when a change in the state code went into effect), Wolff would legally belong to an "agricultural area" in which such suits would be dismissed. The judge was forced to choose, then, between the Weinholds' claim of damage in the form of smell, declining health and property values and Wolff's claim that, but for a technicality, he would not bear responsibility for such things. The judge was swayed by the former, awarding the Weinholds $45,000. But Dennis and Ruth said that the environment, not the money, was the issue. The county should be *buena* in smell as well as vista. Thereafter the regulatory and judicial processes tended to privilege the environment over any another issue in deciding when pig people play fair or foul.[12]

For many pig people, especially those allied with the big guys, this was itself a breach of fair play. From their point of view, social, economic, and historical dimensions of hog wars are slighted when justice is referred to a mute, eternally frail Mother Earth. As one of the trade magazines put it:

> Like most of the hassles pork producers face today, at least part of the dilemma relates back to the fact that the nation is more urban and less rural. People are less tolerant of an inconvenience when their livelihood doesn't depend on it. Worse yet, some farmers are short-sightedly using the odor issue to prevent fellow producers from expanding. You might call it the "my manure don't stink" philosophy. Those who think odor is only a problem for large hog operations haven't driven through Iowa lately [where they are smaller].

An agricultural economist agreed: "Look at the residents of these small towns. They are more mixed today and bring different agendas with them. . . . Rural residents often believe that 'outside' money for livestock production creates units that smell works than those built by locals." In other words, the story behind the passion is less a decline in the environment (at least nothing new) than a decline in public sophistication about agriculture and in public tolerance of growth that is a normal part of "the free enterprise system."[13]

Industry-sponsored (although organizationally independent) pollsters have been able to find some evidence to support that interpretation. According to one, for example, the closer people live to big hog barns ("confined production facilities"), the less likely people are to object to them. The reason, the data seem to say, is that people who live near those facilities have a greater proclivity to balance negative (chiefly environmental) and positive (chiefly economic) values that they attribute to them. Even as the ballyhoo reached feverish pitch at that scientific round table, Professor Schiffman conceded, "Producers' hogs may not smell any worse today than they did a few years ago, but people appear to be more sensitive to these odors."[14]

Once variable sensitivity is acknowledged, Mother Earth seems a less objective arbiter of public dispute. Even when its quality is encrypted in legalese for the courts, culture-wars pedigree stands out. Take, for example, the much-publicized dispute in 1994 between the Thompson family and their neighbors in Hancock County, Iowa. The precipitating event was the usual one. The Thompsons entered into a contract feeding agreement with a big guy, Land O'Lakes. Fulfilling that contract would entail building five 900-head finishing units at a cost totaling about $630,000 in cash and, maybe more important, a yet-to-be-determined amount of local forbearance. The legal issues were also straightforward. State law exempts agricultural buildings from county zoning codes. Since just a year earlier Franklin County had lost its attempt to control big guys (in particular, A. J. DeCoster, previously of poultry infamy) through zoning, it was obvious that the Thompsons were legally in a strong position. At least in broad outline, the terms of the contract and their relation to land-use regulations were clear, regular, and predisposed to favor the Thompsons' prerogatives. Only the state's Department of Natural Resources (DNR) could be convinced to join the fray. About the only thing extraordinary in this case, then, was its privileging of a single, presumably objective question: Is a massive hog operation compatible with nature?

Those who answered "No!" did have environmental arguments to muster. Hazards on site would be frighteningly close to the Twin Lakes Wildlife Area, only about a half-mile away, and the Iowa River basin, just beyond that. Given the dimensions of the site, the new buildings would be 660 feet from the property line, awfully close to potential complainants when you imagine the stench. Fortunately for the Thompsons, their neighbor Richard Eliason, whose house sat only 2,000 feet from the site, was also a hog farmer. Since he annually marketed a couple of thousand head himself, his forbearance might be expected. But Eliason went public with his objections: "My hogs stink . . . you can smell them when you pull in the drive. But it's never going to smell like that one. . . . I might like to build something myself, but I wouldn't do it right under my neighbors' noses in a way that would damage his way of life."

Eliason's complaint is understandable. You do not call the Welcome Wagon when you spot a fleet of moving vans unloading swine just beyond your patio. But his objection is hardly on behalf of nature. The environment does not much care where property lines fall. Likewise, the appeal to neighborliness and maintaining "a way of life" starts down the conventional zoning or building code route, one that in the case of agriculture is legally closed. Moreover, Eliason was among those who would want to keep it closed, at least for the sake of protecting his own "right-to-farm" prerogatives. Only reluctantly would he ally with others who charged that the Thompsons were—literally, legally as well as figuratively—making a "factory" out of their farm and hence no longer protected through "agricultural" exemption. Lore and legal code unite here in implying that nature needs more protection from factories than farms. The only question is the category to which the Thompsons' finishing units would belong. In this way, an avowed concern for nature quickly devolves into a debate about the best ways to be neighborly, to protect private property, to conserve ways of life, and to apply a primitivist or at least anti-industrial bias.[15]

Thompsons' supporters drew from the same cultural well. They began by reminding people that this case was supposed to be about nature, not zoning, and that the Thompsons had worked extraordinarily hard to prove themselves reliable land stewards. For example, they volunteered to participate in the U.S. Environmental Protection Agency (EPA)-funded and Iowa DNR-administered program of test drilling around manure pits. They would do their part, beyond legal requirements, to monitor subsoil hazards. They also took this opportunity to stress their "family farm" status. The Iowans signing that Land O'Lakes contract were a couple of brothers, Doug and David Thompson, and Holly Thompson, David's bride. Land O'Lakes was itself originally the creation of independent farmers, a cooperative to defend little guys from big-guy competitors.[16] Sympathizers lent even more weight to their cause by reminding people that Doug and David were struggling to get back to the farm after years of drudgery working as machinists in an automotive factory in town. Thus, both sides harnessed the environment, in effect, to damn the poisons of urbanity and industry and praise their antidote, the family farm, as if it were nature's own repository of virtue.

When the court decided in favor of the Thompsons, the moral of the story remained contestable. Had nature with its mythic ally, the family farm, won or lost? The local press answered with a waffle: It all depends on who you are and how you look at it.

> The issue of developing large hog lots has been dogged by controversy. Supporters say the large hog lots are a *natural* evolution of the livestock industry, one that state officials should encourage . . . [lest] restrictions will send the industry to other states. Opponents of the large hog lots say the operations are little more than factories that ought to be regulated like other businesses. They say such lots foul air and water and spoil the pristine *nature* of rural life.

Both sides claim to have the interest of "nature" at heart—one the nature of farming as it pristinely evolves; the other of the countryside, in its eternal pristinity. One side could claim that the Thompsons' judgment was a victory for nature; the other that it was a defeat.[17]

Clearly, no one could ignore "the environment" in gearing up for hog wars, but everyone could tailor the trope to flatter a predisposition. Pork producers, for example, began cultivating their own Green image. Dale Miller, editor of the *National Hog Farmer,* cautioned against the environmental arrogance that crackpot realism implies. "The phrase, 'The smell of money,'" he said, "should be stricken from every producer's vocabulary." Instead, he promoted a lesson gained on European vacation: "After a whirlwind trip to northern Italy in early April, I returned with a new appreciation of three things: their incredible pasta, their wonderful wines and the Italian pork producers' obvious ability to control odor in their hog operations." This judgment was confirmed by Miller's own variety of environmental impact study, a pair of tests worth replicating. After a morning spent touring hog farms, he recommended: (1) "The restaurant test"—see how many heads turn to sniff when you go out to lunch; and (2) "the garbage bag test"—put what you wore on the farm tour into a garbage bag; then twist it shut. "The next morning, open the bag, stick your head in!"[18]

The trade magazines began pumping out advice for Green investment. Why not landscape around your buildings? How about turning some of that idle acreage into wildlife habitat? Wait for dry, windless days to spray manure. Maintain your lagoon. Sign on with your local chapter of Ducks Unlimited and Pheasants Forever. Plant trees. Even Murphy Family Farms, a giant company that is not well known for neighborliness, began planting hardwoods to supplement the forage crops it cultivated to absorb phosphorous, potassium, and nitrogen from hog waste.[19]

Part of this investment was, no doubt, motivated merely to mute public outrage and preempt regulation. After all, the absence of regulation can be considered a form of public subsidy to the Lands O'Lakes as well as the Eliasons of this world. Everyone else picks up some of the tab for noxious by-products of their profits. Regulations could shift back onto producers more of the cost of those by-products or of the hazards they increase.

But there are substantial incentives for such investment quite apart from government mandates. There is the simple fact that many farmers are themselves vulnerable to environmental hazards and accept responsibility for their neighbors more profoundly than urban activists might imagine. An incentive rests in body and soul. Yes, agriculture is a very tough business, and farmers may need government protection from institutions that force or encourage environmental "compromise" that in the long run is unsustainable. But it would be bizarre to assume that land and community stewardship are values that are alien to either side in hog wars.[20] Moreover, there are substantial financial penalties for waste in the business itself. The phosphorous that fouls waterways and the nitrogen (fed protein turned ammonia) that so stinks are signs of money down the drain. Cutting rations as close as possible to only what is effectively absorbed vs. excreted helps the feed bill as well as the environment. No conflict of interest is necessary there.[21]

The source of conflict is, I think, less a matter of difference in spirit than in the power of particular institutions and their stake in on-the-ground detail. Even as they stand accused of environmental arrogance, for example, advocates of more intensive agriculture often defend themselves as conservationists. By doing more with less—in particular, by producing more food from existing farmland, wherever it might be—they are reducing plunder of the planet. It is the traditionalists and xenophobes (e.g., the promoters of "national food self-sufficiency"), they say, who push farmers onto land that could have remained rain forest or affordable housing space for impoverished peoples. Sometimes, too, those "unnatural chemicals" and alliances with companies that produce them look pretty good from an environmental as well as economic standpoint. For example, those "in-feed growth promoters," in decreasing the population of bacteria in the gut, turn a hog's digestive tract into a more effective sewerage treatment facility of its own. Elanco Animal Health, the ag pharmaceutical company, claims that a 2,000-head operation—just by mixing a little Tylan into the rations—would annually use 31 tons less feed and 22,000 gallons less water. It also would produce 22,000 gallons less manure and urine with 1,200 pounds less nitrogen and 400 pounds less phosphorous. If true, this demonstrates the complex interaction whereby nature, economy, corporations, and chemistry do not easily sort themselves into two warring camps, at least in day-to-day principle.[22]

Where they butt heads is often in places like the courts, which do demand two sides. And, after heavy lobbying, statutes often predetermine which side has the easier row to hoe. For example, the *National Hog Farmer* greeted Iowa's 1995 Livestock Regulation Act with the headline, "Iowa Nuisance Bill Favors Producers." Granted, House File 519 was a crack in the vault of extraordinary legal protection that agriculture enjoys. Across the United States, largely in response to swine stink protest, state legislators struggled in the 1990s to concoct a special category for large livestock operations. They would be placed legally somewhere between "factory" and "farm," more protected than one and less than the other. Iowa joined other states in inventing a hybrid, the "AFO" (animal feeding operation). AFOs would be protected from nuisance suits without "clear and convincing evidence that the operation unreasonably and continuously interferes with a person's use and enjoyment of their property and injury or damage is caused by negligent operation." The producer was required to have an acceptable "manure management plan" (a complicated document, with various standards varying with size—but basically allowing existing norms to continue) and a small, one-time fee (plus a schedule of fines) to be paid to a "manure storage indemnity fund" managed by the Department of Natural Resources. Hence the bill neatly codified paperwork that would turn the status quo into a legally protected, "reasonable" future with state-supported insurance, should anything go wrong. Most important, the bill assured that the burden of proof in nuisance suits would be shifted onto the plaintiff, who must defray court costs and attorney fees, if the judge determines the case is frivolous. It is hard to see how Mother Earth stood to benefit substantially in any of this.[23]

While the various parties huff and puff and their lawyers jockey for position, the environmental damage accumulates. For the most part, it seems a glacial process, one that is poorly represented in the quakes that occupy the courts and popular press. Graphics in *U.S. News and World Report,* for example, gave the impression that urine evaporating today from hog lagoons will return as yellow rain next week. Even if nothing so calamitous occurs, there is cause for concern, and the higher the concentration of hogs on the surface of the earth, the more serious it is. Nitrate-polluted wells are already a common site in the countryside, and everyone knows where it is coming from, especially in areas where livestock density is high. On the western Great Plains, where the hog industry is newly booming, threats to the volume and quality of the Oglala Aquifer are clear enough to stir preventive action. Citing groundwater protection legislation, the Colorado-based Alliance Conserving Tomorrow was able to bar Ronald Houser's Midwest Farms from drilling a well to service its $80-million, 8,000-acre site. In North Carolina, where development has been particularly intense, preventive action is probably already too late. The hogs in that one state produce the equivalent in untreated waste of tens of millions of people. Those few counties where about 7 million hogs are concentrated abut a vital watershed, between Interstate 95 and the Outer Banks. "The rivers draining this hogalopolis . . . empty into Albermarle and Pamlico, sounds which provide half the coastal nursery waters for young fish from Maine to Florida."[24]

None of this seems easy to blame on card-carrying, anti-environmental villains. The damage from cases where blame has been easily assigned seems horrid but relatively minor in natural consequence. There was, for example, the case of an Iowa farmer who trucked manure from his 10,000 hogs over to South Dakota and dumped it in a sand-and-gravel pit from which it leached into waterways. And there was an absentee owner of a Kentucky hog farm who actually pumped waste into a sinkhole hydrologically connected to his neighbor's well. He was convicted of a felony, fined $33,000, and placed on a year of probation. But few farmers are such easy targets. Environmentalists awaited a bang that would call a mass movement to arms, but for a long time they could hear only whimpers and track discouraging trends.[25]

All of that changed with a giant shit tsunami in the summer of 1995. This disaster—the first lagoon spill of truly biblical proportion—was the one that Sidney Whaley watched pass between his house and the First Church of God in Onslow County, North Carolina.

The source was an eight-acre, twelve-foot-deep lagoon. One end received effluent from a 1,200-sow farrow-to-finish operation run by Oceanview Farms, a limited partnership. The other end was a wall of bulldozed and compacted earth. In between stood 25 million gallons of raw swine sewerage. When the dike broke on June 21, the dimensions of the spill were awesome. A wall of sliding sludge buried crops in neighboring fields and then wound to the headwaters of the New River. From there it flowed more than fifteen miles downstream, headed straight toward the Atlantic coast. Officials in the Division of Environmental Management (DEM) declared, "There's no way to contain and clean it." Considering damage to the watershed, Albert Little of the state Wildlife Resources Commission figured, "I think everything is dead in there. When you start getting eels and catfish, there's not much left." Headline writers across the country resorted to "Whoooweeee!" as if the damage were too great for ordinary adjectives to convey. Just as "Watergate" became the standard for political scandal and "the *Exxon Valdez*" for oil spills, the Onslow County tsunami became the model agricultural disaster. Nearly every report began by comparing Oceanview's spill to Exxon's and by hunting for cover-ups and payoffs that could attract a "-gate" suffix. This looked to be just the call to arms that activists awaited: "See what will happen if hog operations have their way?"[26]

Oceanview apologists did not have much material. There were some extenuating circumstances. In particular, rain the previous month was more than four times heavier than normal, totaling close to seventeen inches, the wettest June in a century. As promoters hoped, some farmers were forgiving. For example, Bobby and Celia Weston had fields directly in the spill's path. Their tobacco crop, already stressed by spring drought, was decimated. But still they said, "We're not mad and don't have any hard feelings toward Oceanview Farms. You see, we raise hogs ourselves under contract to Murphy Farms, and so we know how it is." They even offered, "The soybeans . . . were probably helped by the spill, since they can stand a lot more nitrogen." It is hard to believe that even such forgiving folks would count "more" by the tankerload. A larger number of neighbors—125 county residents—spontaneously formed a watchdog group to prepare for action against future porcine threats to the New River.[27]

Quite apart from nature's own challenge to the dike and the Westons' generous interpretation of its failure, Oceanview was obviously inept, if not downright sleazy. They failed to follow their own, state-approved management plan. They overfilled the lagoon; they failed—by more than half—to reserve enough land for disposing of its contents; and they failed to plant those fields properly. Therefore, they were fined heavily and enjoined from continuing operation by the DEM. Since eventually the fine was radically reduced (from $110,000 to $62,000 to be paid off in seven annual installments of only $8,831), Oceanview probably got off very easy. But damage to the industry's public image was as severe as it was to the environment.[28]

The main damage, of course, was due to the tsunami itself. Adjectives and aerial photographs pointed to unmediated horrors beyond their reach. Then there were a host of unpleasant ironies, beginning with the campestral name of the company, "Oceanview," and the slogan of the nearest town, "Richland—Land of Perfect Waters." In an instant their names became laughingstocks. The spill just so happened to be upstream of a national sacred site, Camp LeJeune Marine Corps Base. Moreover, it happened on the same day that a North Carolina House committee (chaired by a hog farmer) killed a bill, approved in the Senate, that would have increased environmental regulation of swine operations. Another spill, this one "only" 1 million gallons, occurred on a farm in neighboring Sampson County on the very same afternoon. The worst embarrassment was that the Oceanview lagoon was, prior to the spill, among the most conspicuous representatives of the state of the art. Only about a year old, it was the very first to be built and state-approved under new regulations that were supposed to prevent just such disasters.[29]

The obvious moral was that regulations and/or enforcement were inadequate. As Barbara Grabner of PrairieFire Rural Action quipped, "If this is state-of-the-art, I would hate to see slipshod." Even state legislators had to agree, at least if they came from the minority political party. Representative Howard Hunter, Jr., a Democrat from Northampton County, responded to arguments about unusual rainfall and "acts of God" with: "Well, I say maybe this is God telling us we ought to do something before things get worse."[30]

The North Carolina Department of Environment, Health and Natural Resources quickly dispatched sixty inspectors. They checked 3,643 lagoons, 3,100 of them on hog farms. They found nearly 100 too full and another 526 close enough to require immediate attention. There was now ample evidence that this threat was more than hypothetical. In North Carolina alone, just between the end of July and Labor Day, 1995, regulators were able to document lagoon spills totaling 35 million gallons, triple the usual measure of environmental disaster, the volume of oil that the *Exxon Valdez* spilled off the Alaska coast in 1989.[31]

The summer of 1995 was also spill season back in the old Swine Belt. In July 1.5 million gallons of manure leaked from a storage basin on a farm near Blairsbury, Iowa, down an abandoned drainage tile, and into the South Fork of the Iowa River (from which Iowa City, where my family and I work, gets its drinking water). Pollution traveled at least thirty miles downstream, and at least 8,000 fish were killed. The 700-sow source was a family farm. Under the name "SNB Farms" it was owned and operated by the Nelsons, brothers Doug, Mark, and Mike and cousin Jeff. So, in this case it was heartland kinsmen who had, Mark Nelson insisted, a "freak accident." But environmentalists quickly collapsed this case into the catalog of corporate, factory-farm evils. When another spill (also spread through neglected drainage tiles) immediately followed—this one in Fayette County, polluting the Little Volga River and killing at least 17,000 fish—talk of accidents was less likely to persuade. Yet another major spill in the same summer—this one when a pump broke on a farm in Alma, releasing manure from a 400,000-gallon storage tank, across a field and directly into Elk Creek, where it killed 16,000 fish—no one could believe there was anything freakish going on.[32]

Just when producers felt it could not get any worse, Premium Standard Farms of Missouri, among the most aggressively self-styled "good citizens" of the corporate swine world, had its own series of disasters. In just two and one-half weeks at the end of August and the beginning of September in 1995, PSF admitted six spills. Since there were no broken dikes, they were far short of *Valdez* or tsunami proportion. They were generally precipitated by an unserendipitous series of technical breakdowns, like blown pipes and broken bolts that escaped attention for an afternoon or so. But they were undeniably serious in consequence. The worst entailed 4,000 gallons of raw hog waste spilled into a river, killing fish as much as ten miles downstream.[33]

Then, within a few days, there was the R. H. "Moe" Mohesky affair. He was cited for violation of North Carolina DEM regulations. Actually, the spill was relatively minor. After a heavy rainfall, an overflow pipe in a twenty-year-old lagoon carried waste into a ditch that drained into the Conetoe Creek about two miles away and then into the Tar River. The out-of-compliance overflow pipe had been forgotten, buried in tall grass. The spill was obviously small and unintentional, but it was terrifically embarrassing because, at the time, Mohesky was president of the National Pork Producers Council (NPPC), the polity of the industry. Violations attached to his name invited a feeding frenzy like blood in a shark pool. About thirty protesters, members of the Campaign for Family Farms and the Environment, stormed the NPPC headquarters claiming proof at last that polluting, fat-cat megafarmers were in control of the organization.[34]

Spill Summer '95, then, determined beyond doubt that hog farms, whatever their state or region, pose environmental hazards. It was no longer possible to shrug, "These things happen." The frequency and scale of disasters were just too high. By the middle of the next year, policies began to change dramatically in both the private and public sectors.

Some of the large companies announced huge investments to reduce spills. In response to the Iowa River spill, for example, Select Farms started systematically checking around lagoons for hidden drainage tiles that could carry a spill great distances. They went so far as to hire contractors to dig test trenches around every old lagoon on their farms. Regulatory agencies beefed up their staff and procedures. For example, North Carolina introduced training and certification regimens, lowered permissible lagoon seepage tolerances, and increased setback requirements for hog buildings. Grudgingly bending "right-to-farm" statutes to benefit the environment, some state legislatures followed renegade counties in barring expansion of hog facilities altogether.[35]

What remained open to dispute in fashioning solutions was the importance that ought to be assigned to specific conditions: the size of an operation, its form of organization, or the good will of the operators. Should, for example, regulations be oriented more toward getting rid of the big guys or a few, presumably randomly distributed bad guys? Obviously, a bigger lagoon poses a bigger local threat than a small one, but might not the smaller ones collectively represent a greater hazard to the earth as a whole?

In a 1996 follow-up inspection of North Carolina lagoons (4,606 in all, 3,535 of them for swine), the DEM found "major violations" in 200 lagoons with 145 apparently willful rather than accidental. They concluded, "Very few violations of the existing standards were found at large company swine farms." Most occurred on smaller, independent operations. In Iowa, as well, where small independent operators are much more common, spills were either due to maintenance screw-ups on a family farm (as they were in the case of all three of the big spills in 1995) or brazen disregard for regulations by large ones. Again, the moral is contestable. Smaller, independent operators account for more numerous, minor spills; large operators for a few monstrous ones. On the one hand, the big guys do a lot of damage; on the other, they might be more easily corralled, if only because fewer inspectors would be required. Presumably, large producers have the resources better to deal with their own waste. Cursing them for concentrating hog manure is a little like cursing cities for producing sewerage. The yeomen's alternative is slowly poisoning the earth with scattered cesspools.[36]

What would happen if the disposal of farm waste were regulated like the human equivalent? At the moment it is not, but there is potential for change.

Traditionally, for example, storage, handling, and moving of manure has occupied a distinct legal category. Unlike condominiums or factories, most farms under the Clean Water Act have not been treated as "point sources of pollution."[37] Legally speaking, manure is "not conveyed to a navigable body of water by means of a man-made discrete device such as a pipe or ditch." Hence, every day that farmers pump out a lagoon and head out to a spread shit on a field, they do not need a permit from the EPA's National Pollution Discharge Elimination System. But in a recent case against a dairy farm in New York (*Concerned Area Residents for the Environment* [CARE] *et al. versus Southview Farm and Richard Popp*), the Second Circuit Court of Appeals ruled that manure applications (e.g., spreaders and swales) *are* point sources of pollution. In the court's opinion agricultural runoff is as subject to Clean Water Act regulation as other varieties of "storm water discharge." Opponents of the decision, such as the NPPC, argued that the application of manure was an effective, inexpensive, and less chemically dependent method of fertilizing that was encouraged by other government agencies, including the EPA and the U.S. Congress. But in ruling for CARE, the court signaled that the concentration of manure on large farms represented both a threat to the environment and an opportunity to do something about it.[38]

NINE

MEET THE BIG GUYS

WHATEVER THE DETAILS OR CATEGORIES OF CONCERN—food safety, animal rights, the environment—nearly all of them find their most horrid prospect in business-as-usual for "big guys." At least that is the rallying cry. They are the progressives' Darth Vader, a force of evil in the universe, inherently if not gleefully destructive of everything a caring human being, God, and Mother Earth hold dear.

Of course, few activists maintain this view when they are not pleading for resolve through a megaphone. My caricature of their rant is as limited as the ones they toss around when hog wars heat up and disown when they cool down. I guess I begin with wariness toward populists because their hyperbole strikes close to home. No matter what is actually coming through the megaphone, I usually hear people on the receiving end conclude that their suspicions about "Jewish money" have been confirmed. Plus, as the saying goes, some of my best friends work with big guys. Nearly all of them (and fewer of the protesters I know) actually grew up on Mom and Dad's home place. And many of them are damned glad to have a decent job that allows them to stick around or to get out. Moreover, my own university is itself an analogue—a vertically integrated student confinement system—but I would like to think it has redeeming value, much of it precisely because it is big.

What most sets me off is not the suggestion that evil is afoot but that it is arriving in a new form or that its carriers are alien. The United States itself was formed through hostile corporate takeover. For thousands of years, up to a few centuries ago, the hemisphere was home to dozens of native peoples who were themselves varyingly sovereign. It still is home to many, but through intrigue and bloodshed on a massive scale most lost title, first to European trading companies with crown-guaranteed monopolies, then to variants like the railroads, bonanza farms, and their descendants backed by force of the U.S. government.

> In reality, prairie farmers and other pioneer families owed their existence to massive federal land grants, government-funded military mobilizations that dispossessed hundreds of Native American societies and confiscated half of Mexico, and state-sponsored economic investment in the new lands. . . .
>
> The government got a bargain in the 1830s when it forced the Cherokees to "sell" their land for $9 million and then deducted $6 million from that for the cost of removing them along "The Trail of Tears," where almost a quarter of the 15,000 Native Americans died. Acquiring northern Mexico was even more expensive: The war of annexation cost $97 million; then, as victor, the United States was able to "buy" Texas, California, southern Arizona, and New Mexico from Mexico for only an additional $25 million.
>
> The land acquired by government military action or purchase, both funded from the public purse, was then sold—at a considerable loss—to private individuals.[1]

Bearing that history in mind, it is hard to think of family farmers as paragons of "independence" or their challenges as particularly new. No modern corporation could easily surpass pillaging precedents set long ago by the Dutch West India Company, slave traders, the United Fruit Company, or the U.S. Bureau of Indian Affairs.

The trade in foodstuffs has long been among the most vertically integrated and ownership-concentrated sectors of the economy. In the words of progressive rural sociologist William Heffernan: "The food system resembles an hourglass with many producers and millions of consumers, but the few firms that control the processing are in a position to control the food industry . . . 'from seed to shelf.'" Heffernan's colleague at the University of Missouri explains: "These giants are today's counterparts of feudal lords and nobles."

Economists might disagree with their definition of "control," but there can be little doubt that concentration is the rule. In 1993, for example, just four firms processed more than three out of every ten turkeys, four of ten pigs and chickens, and seven of ten sheep and cattle reared in the United States. The same few also milled more than 70 percent of the flour, corn, and soybeans. Probably the most familiar name ranking in every top four (even if occasionally falling to number two) is ConAgra. ConAgra sells feed, seed, fertilizer, and other chemicals to farmers; ConAgra then buys their meat and grain, which it can store in its own elevators; ConAgra then moves this produce with its own railroad cars and barges; and then ConAgra processes and markets it to consumers. Such firms obviously have been able to avoid shaving profit margins to maintain market muscle. According to the U.S. Department of Agriculture, food manufacturers between 1988 and 1993 averaged a 16.5 percent return on stockholder equity, a higher return than just about any other category of manufacturer in the United States. During the same period, American farmers were receiving less than a third that return. In commodity after commodity, when people notice that these companies are very profitable and farming remains perilous, cries about evil corporations increase. Thus, a pattern of concentration in the food system, profiteering, and outrage was already well established when, in the late 1980s and early 1990s, change in the hog industry accelerated. As soon as someone asked, "What's going on?" an answer was ready: "The big guys"—intensive, vertically integrated pork producers.[2]

Their uniqueness in many respects is arguable, but their location was entirely new. They boomed outside the old Swine Belt, where feed grain grows. Instead, these hogs were raised close to markets, their postmortem destination. The mold was struck in North Carolina, within a day's drive of over half the nation's population. The state was known for tobacco and broiler hens, but almost overnight the rolling hillsides were covered with swine barns that could dwarf Fort Bragg barracks. The rate of growth in one year, 1991-1992, surpassed the cumulative rate for the prior twenty. In the 1970s there were no significant hog markets in-state and next to no hogs. In 1994 there were 5.5 million. That was the year that North Carolina jumped to number two in the nation, dropping Swine-Belt stalwart Illinois to number three and moving fast on Iowa, which (with 14 million head) remained number one, but now with an anxious eye on the rearview mirror. From 1988 to 1996 North Carolina's share of the nation's hog herd jumped from 7 to 17 percent. Each new operation represented about $1 million invested, and those without integrators' backing quickly fell. In just five years, the number of hogs in North Carolina nearly doubled and the number of large producers (with over 1,000 head) tripled, but the number of producers was halved. Records fall almost daily, so it is hard to keep up, but as I write, North Carolina remains home to the largest packing plant on the planet. Smithfield's Carolina Foods in Tar Heel can kill, disassemble, and vacuum-pack pork from 32,000 hogs in a single day, every day of the year. Even if what was happening did not stink to holy hell, it would certainly attract attention.[3]

The closer you look, the more intense the change appears. The epicenter lies distinctly in just a couple of rural counties. The population of Duplin County quickly exceeded 25 pigs for every man, woman, or child. In 1995 this rural county hosted $18.5 million in new swinely construction, a figure dwarfed only by gross sales of $141 million in the same year. And it was obvious that this money was going into a small number of pockets. At the time, the largest 15 percent of North Carolina producers held well over 90 percent of the inventory (vs. about 50 percent back in Iowa).[4]

USDA statistics show that the shift had national scope. As early as 1992-1993, the trade magazines greeted without surprise reports that the structure of the industry was transforming. The largest producers (those annually marketing over 50,000 head) were increasing their production by as much as a quarter per year. Their growth was nearly perfectly matched by failure among small producers. In other words, total hog numbers remained close to flat; every gain for a single big producer was a loss for many smaller ones. Throughout the 1990s the number of hog farms continued to drop about 6 or 7 percent per year. Larger farms (those with more than 2,000 head), although only a tiny percentage of the national total (2 percent in 1995), were growing at lightning speed. In 1994-1995, despite legal and financial impediments, their number tripled (up 1,800) while the number of small farms (those with fewer than 500 hogs) plummeted. Market share shifted from one to the other by about 4 percent per year. You can imagine how this might strike Iowans, who still boasted the largest number of hog farms and hogs. They were losing fast. Four thousand hog farmers (more than one out of every ten) called it quits in 1994 alone, and the smaller the operation, the more likely it was to fold. Nearly everywhere you could see signs of panic in the countryside. When, for example, the state extension service, the state university, and county and regional pork producers organized a conference for southeast Iowa hog farmers, they titled it "Survive in '95." As I write, about 10 percent of the nation's 200,000 hog farmers raise about 90 percent of the hogs. The big guys may be far from the majority, but they rule.[5]

They barely stand still long enough to be counted. And then they change names; they merge, divest, and swap. A passing example or two (current as I write) illustrates the small number of firms involved and their tangled relations. Smithfield Foods, the owner of that record-setting Tar Heel packing plant, started as a father-and-son business back in 1936. As recently as 1974, it was like many regional packers on the verge of collapse. With $17 million in debt, the whole business was only worth about $1 million, and it was losing about that much each year. Then Joseph W. Luter III, who was grandson of the founder and who had been developing real estate in Virginia ski country and in the Caribbean, came on board. He started selling off everything that was not related to pork production and buying up everything that was. In the process he turned Granddad's packing company into one heck of a farmer. Smithfield is now the seventh largest finisher of fat hogs in the United States. It gets most of its feeder pigs (about 4 million per year) from numbers one, two, and five in hog production: Murphy Family Farms, Carroll's Foods Inc., and Prestage Farms, Inc. The president of Carroll's, F. J. Faison Jr., and the president of Prestage, William H. Prestage, as well as Murphy's chief executive officer, Wendell H. Murphy, sit on Smithfield's Board of Directors. In fact, Carroll's owns more than 10 percent of Smithfield's stock. Hence, when these big guys moved from North Carolina to Virginia, they started Smithfield-Carroll's Farms, which leased production facilities back to Carroll's, which in turn sold back to Smithfield hogs worth more than $60 million per year. Looking at the intensity of trade among these cronies and kin, a money manager complained, "It's like an inbred old-boys school, where the left hand washes the right."[6]

This sort of development, again, was hardly limited to the Upland South or even to the East Coast. To service better the California market, for example, the big guys set their sights on Milford, Utah. The developer was Circle Four Farms, which might be taken for a new player unless you ask, "Who are the Four?" Answer: the same Carolina crowd—Carroll's, Murphy (through its subsidiary, West Isle of Utah), Prestage, and Smithfield (through Smithfield-Carroll's). Circle Four got all of its genetics ("seed stock") from one company, National Pig Development, all of its feed from its own feed mill, its finishing from its contract feeders, and its packing and processing from its own plants. In five years, annual production rose from zero to 120,000 head. They are in the midst of adding 600 employees to help boost production to 2 million hogs per year.[7]

Among the rewards of this sort of integration—expansion in association with contract feeders—is easier access to capital resources. The outfit that owns the hogs and markets them does not have to tie up its own money building farrowing or finishing units. Instead, they just leverage the borrowing capacity of the contract feeders by guaranteeing a place in the market. "You, farmer, go to the bank, and we'll promise to help you meet your payments." It is a formula for success well blazed by fast-food franchisers and, before them, the church. Moreover, with or without the feeders, companies like Circle Four are going to impress lenders as more attractive risks—hence, deserving better deals—than anyone's cosigning Mom and Pop.

You have to bear in mind the difficulty of raising the capital to play with the big guys. Starting a fully independent, 600-sow unit requires about $1 million, up-front cash. Out of the question. Even a semi-independent contract feeder will need a couple of hundred thousand dollars to start, and farmers do not have that kind of money on hand, either. Lenders will not finance more than 70 percent, and even then, they figure percentage of appraised value, which is about 85 percent of the actual dollars required. You still need more than $80,000 in cash. Financial assumptions that would encourage you to think that you could make it—meet your loan payments, then recoup your investment, and then earn a living under the burden of such debt, not to mention the backbreaking labor—are not just optimistic; they are ridiculous.[8]

Enter the big guys in a growth curve, who need you to supply labor and pigs and to buy their feed grain. Thus, for example, Cenex/Land O'Lakes long ago initiated a "10-30-60" program for "select producers." The producer puts up 10 percent of the building costs (as opposed to 40 percent in the semi-on-your-own scenario); Cenex/Land O'Lakes takes a mortgage for 30 percent; and a financial institution puts up the remaining 60 percent. The feed company will help with the design and engineering, recommend suppliers and contractors, and represent genetics companies. Land O'Lakes even maintains model buildings, a showroom for off-the-shelf swine-house expansion. 10-30-60 is their way of maintaining demand for feed grain in the region and, in so doing, a well-greased path to more intense, more vertically integrated production in general.[9]

Other big guys have jumped right into capital markets with both feet. In 1994, for example, Farmland Industries and Yuma Milling and Mercantile Cooperative formed the Alliance Farms Cooperative and registered with the Securities Exchange Commission. Their Alliance Farm System was designed to move development capital for pork production through its headquarters in Yuma, Colorado, up to North Dakota, down to Oklahoma, and east to Illinois. Want to raise pigs? Buy a single share for $80,000, and Alliance Farm will provide twice that much in construction loans through CoBank of Wichita, plus a share of the big guys' action. Farmland Industries (which also markets nearly all of the hogs grown on Alliance Farms) manages all the genetics, logistics, production, and finance. It reports to the Alliance Farm Board of Directors, which is elected by the shareholders. To be a shareholder, you also must agree to purchase about a thousand 45-pound feeder pigs every nineteen weeks from the Alliance and to finish them in your own facilities. The price you pay per feeder pig is the sum of financing and operating costs per pig plus a production margin of $4.50, which is a "cash cushion" for Alliance. Surplus can be reinvested or returned at the end of the year. In other words, Alliance forges producers into a risk- and profit-sharing chain hitched to Farmland Industries.[10]

Concentration and standardization are also ruling principles in genetics. The days are just about gone when individual producers could earn big bonuses from buyers because, locals knew, they had an eye for picking out good gilts to breed back. "Reputation pigs," they were called. Neighbors could guess their bloodline from the mix of colors or the curve of their underbelly, the width of their shoulders, or the flop in their ears. Those idiosyncrasies were named for the traditions, people, or places credited with their origin, preservation, and improvement: Hamphsire, Yorkshire, Berkshire, Landrace, Duroc, Poland

China, Chester White. Instead, today a couple of genetics companies have developed crossbred boars and holy-momma sows that will pump out market pigs that are clean, lean growth machines but normally unfit for breeding back ("terminal sire"). Henry C. "Ladd" Hitch, a cattle rancher turned hog farmer from Guymon, Oklahoma, explains, "Hog breeds are pretty much a thing of the past. . . . They want their pigs to be like raindrops, and they're pretty close." Now a pig's variety bears the name of the company that patented its chromosomes. With these designer genes, you can market cuts of meat as if they were punched out of plastic, and you can mechanize more of an animal's disassembly. A carpenter who helped build a PSF packing plant in Milan, Missouri, told me how awesome their test run was: "You could snap a chalk line on the tips of their snouts hanging along the line."[11]

Pig Improvement Company (PIC) is the largest swine genetics company in the United States. It now owns National Pig Development Company (NPD), previously the largest in the United Kingdom and the sole supplier for Smithfield-Carroll's Farms in the United States and Mexico. Hence, when Circle Four turns to NPD they are also turning to PIC, which is in the process of going yet more aggressively global. For example, in 1994 PIC created PIC-Colombia in Medellin, in cooperation with Contegral (a livestock feed company) and Premex (a feed additive company). PIC-Korea has similar tentacles.[12]

There is much the same story throughout pork production inputs. As early as 1993, Murphy Farms' own feed mills could produce more than 25,000 tons of hog feed per week, enough for 180,000 sows and 3.5 million market hogs. Pfizer Inc., one of the world's biggest suppliers of animal medicine (e.g., Advocin, an antibacterial popular throughout Africa and Asia as well as the Americas) recently acquired SmithKline Beecham Animal Health for $1.45 billion, together posting annual sales of about $1.25 billion. A couple of years ago, when I was checking out the big guy action around Princeton, Missouri, folks at the local concrete factories told me they were too busy—dawn to dusk pouring new foundations for PSF—to schedule deliveries for anyone else.[13]

Ditto for outputs. More and more pork is processed in giant plants owned by a dwindling number of corporate colossi. Even the relatively old, well-known players were Fortune 500 subsidiaries: Oscar Mayer under Phillip Morris and Kraft General Foods; Morrell under Chiquita Brands; Purina Mills under British Petroleum. The new players are gargantuan in their own right. As of 1992 the top twenty companies processed about 90 percent of the nation's pork in just thirty-five plants. A mere eight plants handled a third of everything processed in the United States. In 1994 Iowa Beef Processors—for about a decade the largest cattle slaughtering firm in the nation—became also the largest hog slaughterer. IBP alone took over 16 percent of domestic production. Individual plants bloated accordingly. No one is likely to build a plant these days unless it can process 8,000 to 17,000 head—2 to 5 percent of the national total—on a single day.

As their CEOs would point out, growth not withstanding, meatpacking remains risky business. Even the biggest players had a rocky road while the industry restructured. For example, the three largest pork packers in 1978—Oscar Mayer, Wilson Foods, and Purina Mills—were out of the business less than twenty years later. But new, even bigger big guys took their place. In 1978 the thirteen largest companies slaughtered 63 percent of the pork; in 1994 their names were different, but their share of the business had grown to 73 percent.[14] As I write, plants that could rival Tar Heel are going up in Utah and Texas. In 1995, the 40,000 producers in Indiana, Kentucky, Michigan, and Ohio—via the Michigan Livestock Exchange with the help of Michigan governor John Engler—agreed to supply Thorn Apple Valley, a Detroit packer, with all of its hogs for the next decade, 4.5 million head per year. According to CEO Tom Reed, "This agreement represents the most significant strategic alliance between a producer-owned livestock marketing cooperative and a major food processing company to take place in the history of the U.S. swine industry."[15]

Such integrated growth is supposed to earn rewards in the marketplace, a lesson explicitly drawn from the U.S. poultry industry, now global in its reach. Among the things that companies like Tyson allegedly proved, back in the 1960s and 1970s when poultry restructured, is that market share (the percentage of every food dollar spent on a particular item) will grow only if the distribution system provides an utterly dependable, uniform supply. Wayne Purcell, an economist at Virginia Polytechnic Institute, explains: "You've got to have a presence in the market place to gain market share." In particular, you have to avoid the sort of fluctuations in production and price that occur when hogs are raised as a "mortgage lifting" or value-adding sideline on diversified farms.[16]

In other words, the restructuring of pork production was explicitly opposed to the sort of agriculture that typified the Swine Belt. Rather than increasing and decreasing their hog herds in response to grain prices (feeding many when corn is cheap; few when it is dear), farmers would agree to deliver a steady stream to the packer, the cost of inputs or profit in alternative produce be damned. In so doing, farmers would be the ones to commit to a higher-stakes gamble on feed-grain prices, a bet many would be reluctant to make. If your own corn crop were bountiful (and likely, then, the seasonal price low), you could not increase your hog herd to hold corn off the market (by feeding it) and increase its value. There would be no source of additional feeder pigs or destination for finished ones. On the other hand, if your corn crop were poor (and then, every bushel likely worth a bundle), you could not take advantage of the price rise by selling directly for cash what little grain you harvest. You would need it for feed. What is worse, if your crop were poor enough, rather than just farrowing fewer sows, you might be stuck paying great gobs of cash for someone else's corn to feed out contracted hogs at a loss. Normal daily life might be easier working with the big guys, but a single flood or drought, like the one you barely survived a few years ago, would be even more deadly once you wagered away your coping options.

The wager would also be hard for Swine Belt farmers to make because state statutes, passed in the 1970s and 1980s to protect them, limited their ability to obtain backers. Most of the distress that inspired that legislation was likely a result of increased production and falling prices around the world, particularly the so-called Third World that put pressure on agricultural real estate and credit markets in the United States. On the domestic front, it is not clear which happened first: a collapse in land values (about 65 percent in Iowa) or the recall of overleveraged loans. But the combination had a death-spiral effect. Lenders began popping the bubble loans on which farmers depend but for which securing collateral had disappeared. Although vertical integrators may have benefited from the crunch in the long run, they were related to the perpetrators very distantly if at all.

Nevertheless, "family farm" laws targeted food industry big guys, phantom conspirators at the time. The strongest example is probably Nebraska's Initiative 300, passed in 1982. It prohibited nonfamily corporations from farming or owning farmland. Since joint ventures were permitted only through general (vs. limited liability) partnership, farmers would be held personally responsible for every risk those ventures took. Hence, too, any individual member of a producer group would be liable for the debts of every other member, a risk no one of sound mind would accept. Slightly more porous versions of such family farm legislation are still common throughout the Swine Belt. As they inhibited phantoms more than a decade ago, they now inhibit genuine options that big guys offer as a lifeboat to family farmers whom they would otherwise drown. Even if you did not get your own lifeboat, at least you could get help buying one on layaway or a job rowing someone else's.[17]

It is understandable, then, that this legislation would become controversial in the 1990s, even among those whom it was designed to protect. For example, in 1974 South Dakota passed its Family Farm Act to prohibit corporate hog facilities. It sounds sensible enough, but when is a farm "corporate" as opposed to "independent"? What ensued, once people had reason to care, was a predictable squabble about the precise conditions that the words require. After all, the big guys still get the vast majority of their hogs through contracts with farms that they do not actually own.[18] At least since the nineteenth century, farmers have depended on relations with buyers and suppliers who are vertically integrated. How formal can those relations become and still allow certification of a farmer's "independence"? Would a farmer who uses Land O'Lakes to leverage borrowing capacity thereby become "corporate"? What if, rather than just securing farm finances, the grain merchant or meat packer actually buys the building for a farrow-to-finish farmer? Or supplies the feeder pigs? Or agrees in advance to purchase a fixed amount of production? Or to supply feed grain at a stable price? Or to guarantee the equivalent of a minimum wage? In 1995 Attorney General Mark Barnett opined that an operation is exempt simply if it engages in fewer than all three steps in production: breeding, farrowing, and finishing. This was good news for those who wanted to specialize and integrate via cooperatives but as a disaster for those who wanted to keep South Dakota an "independent" farmer preserve.[19]

Of course, preserves of any sort face unrelenting encroachments from the real world to which they are by design an exception. Confirmed realists will be undependable protectors. They will see the way the wind is blowing and clear paths for the big guys. The big guys will tack around financial or statutory impediments or hoist the blackmail flag: "If you want economic development here rather than across the border, you better let us in." They strike a deal in the capital, and locals find out only when it is too late. The idea is to preempt local resistance, a strategy that has been proven effective in causes that range from advancing civil rights to retarding gun control. Once the big guys gain legal protection at the national or state level, they have the full weight of "due process" on their side, and down-home protectors of independent preserves, even if democratically elected, are rendered vigilantes. The result might be a nightmare for corporate public relations, but it is a dream for staff attorneys.[20]

This was roughly the situation on the eve of that NCFFE /Farm Aid extravaganza that culminated in the "Journey for Justice" and those awkward moments for Senator Harkin and President Clinton. Premium Standard Farms was in the middle of a construction bonanza in Mercer and Putnam Counties, Missouri, just below the Iowa border. The legislature in Des Moines had earlier torpedoed PSF designs, just as ground was to be broken to the north. So this time, PSF made sure that it had iron-clad, advance approval in the capital to the south. The State of Missouri granted the company an exemption from restrictions on corporate agriculture. When folks back in Lincoln Township got wind of PSF plans to porcinize the neighborhood, they launched a petition drive. Local government responded with zoning codes that would limit PSF construction; PSF responded with an $8 million damage suit. Although no one can be sure, I bet PSF would have won or at least won rights to appeal long after local resources were exhausted. But in this case—once NCFFE, Jesse Jackson, Willie

Nelson, and the Sierra Club got into the act, with PSF stuck playing the larger-than-life villain—the suit was dropped. PSF could salvage some PR, save some litigation costs, and sweat the small stuff in more gentlemanly arbitration.[21]

As striking as each of these cases might be in scale and theatrics, I think they show how much the rise of the big guys of swinedom looks like American business as usual. The ideological and institutional resources for change as well as its resistance were all well in place before hog wars broke out. The big guys may be targeting a different industry and a new location, but the tactics are standard issue. We might object, but we should not be surprised by such a familiar economic and political process. In fact, much of the change is not even new in its economic particulars.

Back in 1959—more than thirty years before porcine realists and populists squared off—prominent Iowa hog farmer, Bernard Collins, wrote in the *Des Moines Register*: "Ever since the broiler business got fully integrated, a lot of bright boys, with a dollar sign looming very large on the horizon, have been busy trying to integrate our hog business out of the Midwest." Specialized, quasi-industrial swine operations (e.g., feeder-pig vs. farrow-to-finish) have steadily grown as a percentage of the total for most of the century. Some of the companies that only recently gained infamy were major players more than twenty years ago. For example, Murphy Family Farms has been contract feeding hogs, even in relatively hostile states like Iowa, since the 1960s; Tyson Foods, since the 1970s. To have attracted such passions anew, there must be more to these changes than normal capital movement and growth.[22]

TEN

THE TECHNOLOGICAL EDGE

THE BIG GUYS OFTEN JOIN WITH THEIR FOES in distinguishing themselves on more tangible, qualitative grounds. Their uniqueness is not just a matter of location, they say. It is not just their massive size or their integration with the food industry. It is not just their edge in capital markets or their penchant for widening regulatory loopholes. Rather, they stress the specific way they raise pigs. While critics lament the loss of a righteous difference between widget assembly and animal care, big guys boldly blend the two. They claim their production technology is an "improvement" on tradition. Hence, old-timers who fail are victims only of their own backwardness. After polling "pork industry leaders" the consulting firm Brock Associates reported a consensus "Vision 2000:" "You won't need a Ph.D. in swine genetics, nutrition, and marketing to run a profitable operation in the future, but . . . you'll need input from people who do." Hog farming has gone high tech: "As the swine industry changes so rapidly, the acronyms seem to never stop—All-In/All-Out (AIAO), Modified Medicated Early Weaning (MMEW), Segregated Early Weaning (SEW) and Artificial Insemination (AI), to name a few." These are all names for ways to take care of hogs that are supposed to produce, not just better profits, but better pigs.[1]

This is an important claim because it so flatters the cutting edge: modern, scientific, efficient, practical. An economist raves: "As a result of the productivity gains, hog farmers today can produce the same amount of pork as in 1980—the peak year for per capita pork production—using less labor, less feed, and an inventory of 20 percent fewer hogs."[2] What innovators promote, then, is technical prowess (and only coincidentally, they say, financial and legal maneuvers). Success that comes through such manual artistry—especially in a field as down-to-earth as farming—is supposed to yield a more honest living than wheeling and dealing. By emphasizing improvements in husbandry itself, the big guys and their allies echo an advertising slogan, "Our success came the old-fashioned way: we *earned* it!"

The technology deserves consideration, even if you never intend to raise your own pigs. One reason is to bring back into focus where pork comes from. All those tales of courtrooms and corporations seem alien, even to me, when going down a grocery aisle or grinding a load of feed. The ledgers and legalities are hard to reconcile with the fact that "production" remains a matter of shepherding living beings through pregnancy, birth, growth, and death. I am afraid that unexplained acronyms only further obscure off-farm visions of human/porcine relations, connections of life and death through food. Whether ultimately fair or not, there is a deal struck between livestock and farmers that is closer to being written in blood than any contract, writ, or receipt. Of course, all sorts of institutions are life-sustaining by some chain of cause and effect. But the link in farming is immediate, right there before your eyes and nose. When office or factory managers screw up, the company raises prices, lays people off, maybe files to reorganize under Chapter 11; when farmers screw up, lives are lost. Death comes fast, and it is hard to blame anyone besides yourself. Hence better production technologies also promise better ways to live with that heavy responsibility. Each innovation entails hard choices: Does it "work?" How? For whose benefit? Opting simply "for" or "against technology"—new, old, indifferent—is not possible because, like it or not, the fate of livestock—and through them, farm families and food for many others—hangs on the quality of inescapably specific technical decisions.

A second reason to dwell on the technology is to correct some popular misimpressions. Very little in hog farming has changed as a result of mechanical or electronic invention. There are no robots raising commercial pigs. One of the very few gizmos to have much effect is "real-time (B-mode) ultrasound," a sort of sonar that can be used to measure the "finish" (the amount, location, and distribution of fat relative to muscle) in a living animal. But so far only breeders, researchers, and an occasional buyer seem high on the device. People still scoop manure with a shovel and sort hogs by eye. Also contrary to Luddite fears, most innovations in hog handling entail both aiming for uniformity and responding to diversity among stock. Finally, it must be emphasized that the association between the integrators and "confinement" (raising hogs inside specialized buildings, where they may "never see light of day") is routinely overstated. Most sows have been farrowing indoors since the 1960s, and most feeder pigs have been finished indoors for nearly that long. In other words, the vast majority of hogs were confined (likely more crowded but also warmer, cleaner, and better protected from parasites and predators than they would be on pasture) well before companies like PSF or Prestage entered the picture. The big guys have been more likely to confine sows during breeding and gestation, but that is also neither new nor uncontroversial, even among the most high-tech operators. There are more open-front, fewer totally enclosed buildings in Swine Belt states like Iowa, but the percentage of animals affected is small. Of course, confinement might still be objectionable on principle, but the point is that there is little reason to think that big guys should accept extraordinary credit or blame for it. Overall there is not much difference between most old-timey and cutting-edge farms, judging from the use of confinement systems alone.[3]

A third reason to consider technology, then, is to figure out if there really is any front in hog wars that is new and distinctly porcine or even agricultural, anyway. If, as I have argued, the lords of swinedom zoom the same highway as Aramco, widgetry, and Wal-Mart, what is all of the squealing about? You have to get off the highway to see. As a matter of concrete circumstance, the late 1980s through the 1990s were undeniably years of profound change for hogs and their caretakers. Their daily lives were, I still think it is fair to say, substantially transformed. Also, as it turns out—important for purposes of this book—most of those practical changes were efforts to control contagious disease, such as TGE.

Most, but not all. For example, manure-targeted technologies (such as procedures for effluent storage and distribution or for rations to reduce pollution) have been innovating rapidly. But these innovations target what comes out of a hog house far more than the health of anything inside. Changes in manure handling probably are better considered a consequence of concentration in the industry, an attempt to redress sorry side effects, than a cause of it. Similarly, an array of improvements in nutrition—most notably "phase" and "split-sex feeding" (customized rations to accommodate the difference between barrows and gilts at particular ages)—entail significant changes in husbandry but do not appear to have much to do with disease. Yet more profitable but still quite disease-neutral are the innovations related to swine genetics. Integrators were quick to adopt bloodlines like those patented by PIC to produce rapid-growing and efficient feed-converting critters. They yield the lean meat that diet-conscious Americans and emerging markets (especially in Japan) demand. Even so, the story of swine genetics is hardly one of singular success. For example, the USDA's massive effort to genetically engineer a "super pig" (a.k.a. the "Beltsville pig," named after the research facility's Maryland hometown) was a notorious debacle. But clearly the hog business has changed right down to the DNA level.[4]

One way of accounting for the rapid change in genetics is an accompanying change in swine sex. Increasingly, rather than just turning a raring-to-go boar in with cycling sows, herdsmen are using artificial insemination (AI). This is one important instance where human handling of individual animals has actually increased, albeit at the expense of hog intimacy. It is a relatively simple, easy, and familiar procedure, even if new to commercial swine. "AIing" avoids the wear and tear, expense, and uncertainty that mount-and-hump breeding entails. Via FedExed bottle and plastic wand ("spirette" or "spiral-tip catheter"), a few prize boars can sire an astounding number of far-flung offspring. Buying semen is apt to be much less costly than purchasing and maintaining comparable boars of your own, assuming you could get them. Since quality boars generally produce terminal-sire offspring, specialized breeders—who also generally sell semen—are the only practical source. With AI, space on the farm that would have been reserved for boars can instead go to farrowing more sows and finishing more pigs. Moreover, in buying genetics by the vial, you are in a position to adapt better to natural and market forces. If genetic disorders appear or if packer bonus plans or consumer whims waver, you can change the whole herd almost overnight. Since you can be more certain when pregnancy began, you are also in a much better position to know "where," figuratively speaking, a group of sows stands (say, the best ration or temperature for them, given "how far along" they are). Their pigs are also much more likely to be uniformly ready to ship on schedule. And that is the sort of performance that earns a bonus from the packer.[5]

AIing does attract disease concerns, too. People worry that someone's McSemen might spread a previously unknown or newly mutated virus with hurricane force. Not surprisingly, some of these concerns can be traced back to ag pharmaceutical companies, who stand ready to medicate. Others admit the risk and then ask facetiously, as did North Carolina State University veterinarian Morgan Morrow, "How much sleep do you want to lose worrying . . . ? If you don't use AI because you're concerned about disease, what are you going to use—a live boar?" The point is that a catheter is much less likely to introduce disease into a birth canal than a penis. Every time you mingle animals that have previously been separated—as boars and sows must be, if they are to breed with regularity—the risk of contagion, sickness, and death increases. Sows that are spared being mounted tend to have fewer hip and leg injuries and to live longer. These days, too, it is popular to buy semen from a source that is guaranteed "SPF" (Selected Pathogen Free).

In short, artificial insemination is about as close as swine can get to safe sex. In an age of hysteria over STDs and emergent diseases like AIDS, both hogs and humans in America have experienced a redefinition of sex. It has become less a reproductive or recreational union and more a daredevil mingling of fluids that just might be pathogenic. Changes in pork production suggest conventional understandings of contagion and hygiene gone hog-wild.[6]

Nearly all of the major recent innovations share a focus on disease control. Rather than farrow-to-finish on a single site, most new operations these days dedicate separate sites to one and only one of the "three stages" of production: prefarrowing (breeding and gestation), farrowing (birth to weaning), and finishing. Rather than "continuous farrowing" (whereby new pigs are always being sorted in and out of pens that are kept ever as full as possible), littermates and coincident litters (born within seven days of each other) are now treated as an inviolate "group." The product of AIing, members of a group are close enough to clones that their nurturing requirements can be met with exacting precision. Insofar as possible, farmers will distribute groups to separate sites with enough space ("air wash") between them to contain accidental contagion, like watertight holds in a ship or firewalls in multiplex housing. Each building is emptied and sanitized before it is filled with one and only one group at a time. In this now-sacred "all-in/all-out" (AIAO) regimen, the flow of life should be strictly in one direction (avoiding "back flow") with a "biosecure" hatch slamming shut behind.

Biosecurity has become a visible obsession. An old-timey farmer might dash into the finishing unit to check hogs, say, between cattle chores and field work or before driving kids to school. If cautious, leave the dog outside and rinse boots. But state-of-the-art operators these days assign different hired hands exclusively to different units and forbid visitors. A herdsmen in Unit B3 will be fired if he helps out a pal in A2. I have met farm families who "modernized" by assigning different buildings to father and son who then, in the name of biosecurity, stopped eating noontime dinner together. The only way they could maintain biosecurity, as they understood it, and go back to sharing pot roast was if they expanded enough to justify hiring a full-time hand for each unit. Hands shower on the way in, don disposable coveralls, and shower again on the way out. Doing hog chores in this manner involves a whole rigmarole that is harder to synchronize with the irregular cash flow and work rhythm of a diversified family farm. The whole "shower-in/shower-out" compound is posted with giant "No Trespassing" signs and surrounded with barbed wire like a jail or a toxic dump. Such farmers are not trying to keep anything dangerous in; they are trying to keep pathogens out. Every visit is a breach and every visitor an unwitting "vector" of plague.[7]

There is relatively little argument about the animal science that puts these innovations on the cutting edge. Instead, scientists and producers split hairs: "Segregated," "Medicated," or "Modified Medicated Early Weaning?" The acronyms—SEW, MEW, and MMEW—punctuate the trade magazines. They entail slight differences in the timing and treatment of pigs during their first move from the farrowing house with mom to the nursery with peers. Even without getting into fine points, you can see how central disease control is to rationales for recent technological innovation.

Among the most wonderful of natural processes is the way that nursing newborns gain "passive immunity" from their mothers. Regularly ingesting colostrum, with very little stress, the baby gains resources ("factors") to fight infections that mother has endured. (This is, by the way, one powerful impediment to drawing analogies straight from the industrialization of poultry to pork. Since chicks thrive much more independently, the labor and technology requirements are very different.) This nursing process is especially essential for pigs because, unlike nearly every other species, they are born with next to no disease-fighting capacity of their own. For a little short of the first month, suckling pig and sow share a single immune system. Then, after a brief but dangerous interval when colostrum loses effectiveness, the pig will build and rely exclusively on its own ("active" vs. "passive") immunity. What is much less wonderful about this process for pigs is that momma sow not only shares immunofactors; she also sheds viruses like crazy. One might, then, look at this endgame of the "natural" suckling relation as, at best, running in place and, at worst, inviting infanticide. Momma normally helps baby fight off infection that she introduces. If momma happens to shed a heavy dose of pathogens in that interval between the litter's passive and active immunity, they get very sick. Attention shifts from feed efficiency to survival.

The various "EW" technologies, then, are attempts to maximize the disease-fighting benefits and reduce the disease-exposure risks that nursing entails. It is called "early weaning" because you separate offspring from momma, sometime (7 to 21 days of age) before pigs normally would favor a feed bucket over a tit. They do not yet have much in the way of active immunity, but they also have not yet been challenged by much of the disease ("gross contamination") that momma carries. In fact, if you move newborns into a sanitary facility, you might shortcut whole get-fight-shed infection cycles. Among the strongest claims for early weaning is that it is a way to "eliminate disease" without depopulating the herd. You segregate offspring before they become carriers.[8]

Even high-tech promoters admit that EW has its drawbacks. Moving "naive" pigs (those with little active immunity) requires special facilities, attention, and medication. Attendant "recurring expenses" total about $1.50 to $2.00 per pig. Potentially more serious are an array of persistent health concerns. Some sows apparently lose fertility. Without special handling, they may have more erratic, shorter, or less intense heat cycles and smaller litters. And some diseases (such as porcine reproductive and respiratory syndrome, *Streptococcus suis* and *Haemophilus parasuis*) seem utterly resistant to an early-weaning offense. They actually may get more deadly.[9]

But benefits make the trade-off look very good. Segregated, early-weaned pigs generally "do better." They tend to be demonstrably healthier, which is, of course, good at once for farmers, bank accounts, and the animals themselves. They may need medication earlier in life, but over the course of their entire lives they require a lot less, as little as a quarter as much as control groups. They are ready for market sooner (120 to 170 rather than a minimum of 180 days) with "more efficient lean gain" (an average of 120 pounds less feed for 7 percent more lean meat). One way to explain the advantage is that you have not "wasted feed" to help a pig build up resistance and fight disease that it could avoid.

Since sows can leave the farrowing house sooner after each birth, you might also "shorten the service interval"—the amount of time between farrows. Reducing the number of "nonproductive days"—when a sow is neither pregnant nor lactating—can effectively increase the size of the herd. You get more litters per year without investing in more facilities, thereby also decreasing the costs of shelter per sow and removing a potential bottleneck at the farrowing house. Exact figures tend to vary, but there is widespread agreement that early-weaned pigs are more profitable by something like 5 to 15 percent. An enthusiast (neglecting those fertility problems) calculated that a typical new operation (starting with 465 sows and a 96-crate farrowing house)—by just weaning six days earlier (after 18 rather than 24 days)—could increase efficiency and herd size sufficient to accrue an additional $93,000 per year.[10]

Obviously there is money to be made in going high-tech. The exact amount is hard to figure, since intervening variables are so dynamic. Even if trade-offs and market conditions were clear, uniform, and stable (which they are not), the cost of implementation would vary terrifically with the timing and mix of innovations as well as the financial resources of a particular operator. Roughly speaking, Illinois veterinarian Ralph Vinson estimates that going the whole, high-tech route could be expected to increase profits about $11 per pig:

$4.68 via improved genetics
$1.77 via segregated early weaning
$1.75 via all-in/all-out production
$1.20 via split-sex feeding
$ .90 via phase feeding
$ .58 via multiple-site production

No wonder, one might conclude, the big guys rule. Their expenses per head are lower. In fact, if you use USDA figures to plot the average cost of hog production as a function of herd size, the difference is huge. In 1990 herds of 140 required an average of more than $60 per hundred weight (/cwt.)—well over the market price at the time; herds of 10,000 required only $43/cwt. And costs for intermediate-size herds fell almost perfectly on a straight line connecting the two. Average production costs per head have been falling and herd size increasing ever since.[11]

Still, it is not entirely clear that the big guys' advantage is distinctly technical. A number of sources, including the National Swine Survey, conclude that, the larger an operation, the more likely it is to wean early and to have segregated-site production, an AI program, tighter biosecurity, and more precise feed management. However, it is not obvious that size and technological efficiency are as well correlated in actual individual cases as they are in abstracted aggregate. For example, in a Purdue University study of Iowa farms, the most efficient 20 percent were about average in size—neither unusually large nor small—and were, on average, more productive than the giants of North Carolina. In the 1990s, despite (possibly because of) tighter biosecurity, large farms were more likely than small ones to have disease problems that required the attention of a veterinarian. And differences in feed efficiency (pounds of feed divided by pounds of weight gained as a function of farm size) are in the nether lands of statistical significance.[12]

Hence Swine Belt activists, like Marty Strange writing for the *Des Moines Register* in 1994, could conclude that "research shows" technology is not a distinctive ally of the big guys. Family farmers take heart. But the very same researchers whom Strange cites also say:

> Farm records show that larger operations tend to have higher production efficiencies, lower per-unit costs [as much as 25 percent less] and receive more for hogs than small hog production operations. Large operations can gain scale economies in the use of feed facilities, manure handling equipment, managerial talent, trucking, grain handling and a host of other areas. . . . Operations that are at least 3,000 head of farrow-to-finish production can gain only about 75 percent of the economies of operations raising 10,000 head of hogs per year.

This message seems to contradict the spin that Strange gives it. Of course, it leaves open the possibility that a "small" (3,000-head) farm could gain a lion's share of the advantage of a large one, but only if it were more than twice the average size of actual Iowa farms at the time.[13]

At this point the lesson for me could be to go off and learn enough economics to derive my own mathematical model. One could, I suppose, credibly calculate a bottom line: All things being equal, how much of the big guys' success can be chalked up to technology alone? How much of that advantage might little guys also enjoy? But given the fact that in the real world "all things" are almost never equal, I am satisfied with rough estimates and mixed messages. I am easily persuaded that the big guys' success comes through strategic amalgam, part technical, part political and financial, part dumb luck. Their strength appears to be a specialized mode of production that rationalizes relations to capital, technology, input, and output markets. In some measure advantages are available to anyone, and in some measure exclusively theirs. Rather than making the "counterfactual assumptions" necessary to calculate those measures, I am drawn to investigate the moral calculus that makes the measures seem so important in the first place. We can assess reports from the hog war fronts to see how they might move, recruit, or at least comfort allies.

PART III

Stepping Out of It

Animals constituted the first circle of what surrounded man. . . . An animal's life, never to be confused with a man's, can be seen to run parallel to his. Only in death do the two parallel lines converge and after death, perhaps, cross over to become parallel again. . . . Animals came from over the horizon. They belonged there *and* here. *Likewise they were mortal and immortal. An animal's blood flowed like human blood, but its species was undying and each lion was Lion, each ox was Ox. This—maybe the first existential dualism—was reflected in the treatment of animals. They were subjected* and *worshipped, bred* and *sacrificed. Today the vestiges of this dualism remain among those who live intimately with, and depend upon, animals. A peasant becomes fond of his pig and is glad to salt away its pork. What is significant, and is so difficult for the urban stranger to understand, is that the two statements in that sentence are connected by an* and *and not by a* but.

—John Berger, *About Looking*

For I will consider my black sow Blackula
For she is the servant of the god of the feed bucket and serveth him.
For she worships the god in him and the secret of his pail in her way.
For this is done by screams of incantation at the appointed hour and lusty bites of daily communion.
For she stands with forelegs upon the top rail of the wooden fence in supplication.
For she grunts her thanks while she eats.
For she stands for the red boar with closed eyes at the appointed hour.

For having done she lies in mud to consider herself.

. . . .

For he keeps her well-fed and she breaks no fence.
For she grunts in pleasure from the mud when he scratches her ears.
For she is a tool of God to temper his mind.
For when she eats her corn she turns and shits in her trough.
For her master is provoked but hereby learns patience.
For she is an instrument for him to learn bankruptcy upon.
For he lost but four dollars each on the last litter of pigs.
For this is admirable in the world of the bank.
For every man is incomplete without one serious debt or loss.
For she provides this with her good faith.
For every farm is a skeleton without a mortgage.
For the Lord admonished black sows when He said lay up no stores of treasure on earth.
For she prohibits this daily.
For she is a true child of God and creature of the universe.

—David Lee, "Jubilate Agno"

ELEVEN

SO IT GOES

WHETHER HAPPY ABOUT THE BIG GUYS OR JUST RESIGNED TO THEM, realists explain, "So it goes." They prepare for populist offensives (in which they are apt to be rendered polluting, carnivorous brutes) and argue for calm. Maybe concentration and integration of production should themselves be considered natural. Industrial evolution cannot be reversed any more than the flow of the Mississippi. From this point of view, the best response is restrained and pragmatic. The public should shoulder responsibility to find a better place in a future that markets properly make. Wholesale resistance is itself immature or reactionary, a denial of opportunity like terrible-two tantrums or pissing in the wind. Maybe with due respect the know-how and institutional muscle that frightens can instead be flexed for public good.

Among the prerequisites for such a response is a bit of Prozac for Chicken Little. Pointing out, for example, how few of the big guys' moves are new or alien (and assuming that there has never been much of anything seriously wrong with industrial evolution), realists advise that there must be basically nothing to worry about now. With rare exception (such as the boom, 1975-1980) farms have been steadily declining in number and increasing in productivity for most of recorded history. So even if the sky is falling, it has been doing so for a long time. Why imagine that it is going to run out anytime soon? The lesson for old-time Swine Belt farmers is chill out.

Iowa, for example, remains even today by statute, custom, and demographic measure a family farm state as well as the undisputed champion of U.S. market-hog production (and, more than coincidentally, perennial number one or two for corn and soybeans, the main ingredients in livestock feed). Iowa's market share in 1992 (27 percent) was higher than it was in any year prior to

1982 (a peak) and followed three consecutive years of increase. Given the size of the inventory involved, a proportional increase in the early 1990s meant a whole lot more hogs in Iowa as well as North Carolina. But 65 percent of Iowa's 29,000 pork producers still market fewer than 500 head per year. So family farms are obviously more resilient than alarmists might guess.[1]

Still, change is undeniable. The number of Swine Belt producers, their market share, and the size of their breeding herds have rapidly declined. In order to operate at full capacity, Iowa packers now have to purchase 7 million hogs per year from out of state. Buck the integration trend, realists warn, and you force packers to relocate where supply is more dependable. Adapt, say by changing family farm and right-to-farm laws to accommodate more intensive contract feeding, and markets are protected. A bit of independence seems a small price to pay for survival. Besides, what sort of "independence" must cower behind statutory protection? Deregulation, as the saying went in Reagan/Bush America, will allow producers to seize the opportunity they need to be competitive in world of mobile capital and global markets.[2]

A realist tale—albeit in melodramatic mode—also can be used to explain change outside the Swine Belt. Governor Jim Hunt, who lorded over the porcification of North Carolina, was himself a farm boy who married a farm gal. He had obvious affection for rural ways. But his state faced a job shortage that was getting scary, especially in the countryside. With mounting threats to tobacco, the state seemed on the brink of agricultural demise. Leadership was desperately needed, but options were slim: Either cook up some pie-in-the-sky remedy or counsel despair. From this vantage the lords of swinedom seem less like invaders and more like saviors. Development on the western Plains can be given a similar spin. There the prospects were bleak for old farmers trying to survive or for young ones trying to start on the small parcels that were affordable, and extractive industries were in a tailspin. Enter the big guys on white stallions.[3]

Such episodes fit a predictable "shakeout" scenario. Every once in a while in a fitful rush, whole sectors of the economy innovate, then consolidate. Entrepreneurial zones are refigured, rationalized, routinized. Smaller, older, allegedly less efficient or more improvisational operators become the equivalent of ash and clinkers in the coal stove. No matter what the business, a periodic shakeout is only to be expected, a high tide that clears the beach. When, for example, cattlemen (uneasy sometime allies in the meat trade) discuss the troubles in swinedom, they find it a repeat performance of their own experience back in the 1960s:

> In that case, a group of entrepreneurs recognized that they could make big bucks in the traditionally small-bucks business of feeding cattle, if they did it more

> efficiently on a larger scale in a climatically-friendly area. Just 300,000 head of cattle were fed in Texas in 1958. By 1978, it would be nearly 5 million. During the growth period, feedyards were added in the belt of country reaching from Nebraska down through Kansas and Colorado to Texas. But when the price wreck of the '70s came about, the new feedyards simply changed hands and kept going. The older, smaller and less efficient ones in the Corn Belt were shut down. In a few years, the Corn Belt had lost its leadership of the cattle feeding industry. Much the same seems to be going on in the hog industry now. The Midwest, with its family farm bent, has been inclined to pass laws to keep the "corporate" pork producers out. So the producers have simply gone where they're welcome—North Carolina, dry, and rural states farther west like Missouri, Oklahoma, Texas, and Utah. Meanwhile, Iowa has seen several years of double-digit decreases in its breeding herd.

But even with déjà vu in mind, we still can wonder if there is something fishy about *this* shakeout in this industry at this time. Is the story of hogs wars best considered triumphant or pathetic?[4]

Of course, one among many possible answers is "neither." Change can be considered less a matter of making or bungling ethical choices than following natural law. Judging from demographic probabilities alone, for example, little guys are just about certain to suffer the bulk of the casualties. Since the vast majority of hog farmers have small herds, even purely random misfortune would hit them most often. Or the imperative might be considered primordial predestination. An invisible hand sets the pendulum to swinging, and ticks of the clock must follow. Once production methods are stable, competition assures that profit margins are, too. At that moment the reward for cost cutting is particularly great, and the earlier you get in on the action (before pricing fully reflects declining costs), the more you stand to gain. By many accounts, that is exactly what happened, first in the 1970s and then again in the 1990s. The winners of one round are losers the next, and both are mere messengers of the moment.

This is among the realist tales—a less melodramatic, more fatalistic or Zen variant—that economists often champion. In their annual surveys of the structure of the swine industry, for example, Professor Glenn Grimes and his associates at the University of Missouri find distinct, alternating periods of stability and change. The 1980s belonged to those who invested in intensifying technology (such as confinement and crossbreeds) new to the 1970s. Since for a few consecutive years their profit margins were good (20 to 30 percent) and their inventories small, investors could be easily convinced to finance expansion. (The bell rings for the next round.) The first casualties are those who depreciate their 1970s investment and hoist the white flag. Nothing newly ventured; nothing newly lost. As soon as profit margins drop in response to larger inventories, investors begin to migrate, and the bell rings yet again. Ron Plain explains: "The reason these guys wanted to be big is not so much that they made more money per pig. Everybody has been making good money. The big guys just wanted to make lots of money. . . . Being big works both ways. Last fall [1994], they were losing more money in a week than some guys invested in their entire operation."[5] From this point of view, the decline in small producers is just another turn in an autotelic cycle. In a sense, the worse things get for this round's losers, the more you can trust that the round is nearly over. The victors at this moment will soon suffer in the next.

John Lawrence, an economist at Iowa State University explains: "Most people see a window of 3 to 5 years to get into the business. If you get in while there are still some 'sorry' producers in there, you can extract some profits. After that time, everybody will use the same technology and only the efficient ones are left. They will expand out to their cost of production and drive the price down. Then all the fun will be out of it." And William Hoeg agrees. Through the realty company, Universal Livestock Properties, he puts together 2,000-sow expansion projects for Continental Grain: "The hog industry is a very fragmented, cottage industry in regards to production techniques. It is typically run as a family farm, not as a business. You still have people raising animals five miles from each other with seven different techniques. It is a fragmented industry where consolidation clearly is going to happen." Hence change is not something that people make for good or ill; it "happens." Shakeouts just come and go like phases of the moon.[6]

Nevertheless, the movers and shakers who benefit usually take credit. They boast "improved" technology to meet consumer demand for consistent, safe, lean, inexpensive pork. The losers are just overfat, sickly hogs and farmers suffering from "corn-harvesting disease." Rather than shrugging off market fluctuations, low-level infections, scours, and coughs, integrators are willing to redesign whole systems of production and distribution to control them once and for all. And what could be a better sign of success and the invisible hand's favor than the number of pigs that thrive birth to market? The USDA reports that over the past decade sows have been having more litters per year (2.3 vs. 1.9) and that more of the pigs in those litters were weaned successfully, especially on larger farms. Farms with fewer than 100 hogs saved about 7 pigs per litter; those with more than 2,000 saved about 8. Whatever the downsides of life on megafarms, they have been profitable in part because they have been a more promising place for a pig to be born.[7]

For high-stepping realists, size is less a precondition for success than a quasi-coincidence. Bulk, they say, is not an advantage—much less an unfair one—in itself. Technical prowess is their equal-opportunity benefactor. Even if little guys must remain at a competitive disadvantage, they still can make piecemeal improvements in a complementary spirit. Some of the technology, for example, requires little or no initial investment. Anyone, it is supposed, could steer the big guys' course or surf their wake. For example, University of Illinois economist Allan G. Mueller analyzed the differences in profitability among 705 hog operations in Illinois in 1991. He discovered that larger units were significantly more profitable ($389 vs. $296 per litter), but size (the number of litters farrowed) accounted for much less of the variation (only about 5 percent) than "other management factors," in particular feed efficiency (35 percent less feed required per hundred pounds of gain).[8]

Being bigger just makes it a little easier to afford to be better. Segregated site production and split-sex feeding are among the best examples. Barrows can be fed more cheaply (about $2 per ton) than gilts, but initial arrangements cost thousands of dollars (about $3,000 just for a second fifteen-ton bulk bin). That investment is hard to justify if you already have an operation up-and-running and small numbers. If your nursery still has some life in it, how can you abandon it to build a new one less convenient to the finishing unit? But if you are new and large, these investments will pay for themselves very rapidly, pennies per head.

An explanation for the success of integrators in places like North Carolina, then, might be simply that they started out new. They did not first have to adapt or depreciate old facilities. They could implement a full complement of technologies that had long been available, anyway. What investors exploited was just a gap between the most and the least efficient operators. Jumping that gap, they wagered, would be profitable enough to cover the cost of new construction and of shipping feed halfway cross-country. In the words of Wayne Snyder, director of livestock production at Farmland Industries, integrators seized an opportunity to "catch up with the demonstration farm."[9]

According to the usual fresh-start story, the move out of the Swine Belt should be considered not a social loss but a technological—especially a disease-fighting—achievement. In particular, hogs would reside in warmer, drier, more isolated, and hence healthier areas. The advantage of states like North Carolina, Oklahoma, Utah, and Texas is their "natural" biosecurity. Integrators bypass the north-central states in flight, not just from old-timey practices and regulations, but from the high density of airborne pathogens and the weather that favors infection. (Among the problems with this "natural migration" rationale is the fact that relocation was so uneven and its health consequence so arguable.

Growth in the southeast, for example, was confined almost entirely to North Carolina. Hog herds actually declined in the rest of the region, even though the climate, labor, and resource conditions were nearly identical. Furthermore, it is far from clear that hog public health has improved, much less that particular climates are distinctly inhospitable to major pathogens.)[10]

In addition to this somewhat lame rationalization of geographic change, the fresh-start story also provides an occasion for realists to discredit the standard populist threat, that pork will "go the way of poultry." Just about everyone agrees that such a change might, indeed, be more than farmers could bear. Since the mid-1970s egg and fryer production has been basically a wage-labor, highly concentrated, industrial enterprise. But the preconditions for that outcome were unique. Prior to the transformation, poultry was almost entirely a "backyard" industry. Companies like Tyson were not leaping a gap; they were making it. Almost overnight giant integrators developed production, slaughter, processing, and distribution sectors from scratch. But pork producers have long had those sectors, even if they were unevenly developed. This difference has an importance that even Tyson suffered for having neglected. The company failed in its own effort to extend the poultry formula. In 1995, after just two years in the hog business, Tyson traded its pork processing facilities to Cargill in exchange for broiler operations. Tyson concluded that profitable pork production had to remain a more mixed affair, at least for them. The company still would raise hogs, but it would rely on others for processing and distribution. In this story, the role of the big guys becomes that of leader rather than displacer. They can help the whole industry regularize financing and reduce market risks. They can help meet foreign competition and open new markets for the benefit of all.[11]

Naysayers, realists gloat, just forget that economic opportunity itself can grow. As painful as the shakeout may be for many people at the outset, it promises a new, larger, leaner, and more viable "structure" for the industry as a whole in the end. Such restructuring can benefit not only a few big guys with the vision and muscle to make it happen but also a contingent of little guys, if they will just quit whining and get with the program.

With a little loosening of family farm law, independent farmers can "network," cooperate contractually with each other, not just to secure better financing but also to achieve technical efficiencies otherwise monopolized by the integrators. They can agree to phase or split-sex feed among themselves, shipping groups that belong to the network from one biosecure farm to another. They thereby gain the benefits of separate-site production without having to buy additional sites. They can better manage their hog manure and veterinary services, buy at a discount and sell at a premium. The National Pork Producers Council itself has tried to encourage such "leveraged marketing" to give small producers access to more packers (who demand larger and more uniform loads), better prices, and reduced transportation costs. As early as 1994, nearly 20 percent of market hogs were "tied to some sort of network." Ag economists predict that through such strategies mid-sized farms are the ones best positioned to prosper in the next round.[12]

Integrators are especially proud of their success in developing trade with East Asia, Mexico, Canada, and Russia. During the mid-1980s less than 1 percent of production was sold outside the United States; now foreign sales total more than 3 percent. The price drop in 1994, as much as it hurt marginal operators, also enhanced the global position of American producers in general. In the first six months of 1995 U.S. pork exports increased almost 60 percent, with more than half of the increase headed to Japan. Exports to Russia alone, from 1994 to 1995, grew twenty-fourfold. The year 1996 was the first one since 1952 that the United States was a net pork exporter. In volume and value of foreign sales, the United States gained a rank of number two in the world. Such growth, integrators boast, helps all farmers expand and diversify demand for their produce and helps the United States with its trade deficit. At the very least, even if big guys hoard global gold, they leave little guys a smaller but more secure place in domestic markets as well as an opportunity to develop specialized "niches," such as those for laboratory animals or for "designer meats" in upscale shops.[13]

Nearly all of these stories are standard American resources in rationalizing economic change. I call them "realist" because of their common tendency to make observations about what has been into advice about what must or ought to be. Their morals are echoed in the slogans that punctuate motivational lectures in boardrooms across the country. In this case they address the relative decline of the Swine Belt and the rise of integrators in hog wars, but they are hardly so restricted:

> Coping is manly and mature; resistance is not. Life goes on.
> Shakeouts are inevitable. Opportunity knocks with every restructuring.
> Harness technology. Move with the action, or be left behind.
> Change hurts but benefits us all in the long haul.

And so on. I find most of these stories in the popular and trade press, but they are amply foreshadowed in the tapes, posters, and other motivational wares hyped in entrepreneurial, sales, and in-flight magazines. It remains curious, though, that these stories still accommodate despair as well as hope. Even given the alleged natural necessity of it all, are we supposed to be happy about the change or just resigned to it? Is "so it goes" said with a sigh or a smile?

About the only way to make pure virtue of this necessity is by taking some gloss off the alternative. Despite its popular mystique, the Midwestern family farm has long been far enough from utopia to make almost any change seem worth considering. According to the National Pork Producers Council itself (particularly its systematic study of the question in 1991, the date of reference in the following), hog farming under any circumstances is seldom pleasant or rewarding for those who are hired to do it. Farrowing managers and herdsmen generally work amidst danger, dust, and dirt for fifty or more hours per week, and they get only a couple of weekends off per month. Over one-third of the employees on U.S. hog farms get no weekends off, ever. Their average compensation is under $25,000 per year, and more than one in ten works seven days per week for less than $10,000. Moreover, gender inequity is the rule. Despite a general correlation between formal education and compensation, female employees usually have both more years of schooling and lower wages and benefits than males. About 80 percent of women in the business, but less than half of the men, earn less than $20,000 per year. Even more disturbing, these inequities are strongly related to the size of the operation, but in the opposite way that one might expect from family farm fables. The single best predictor of income, fringe benefits, occupational safety, and gender equity is herd size: the bigger the better. So, generally speaking, small Midwestern family farms, whatever their symbolic appeal or intangible reward, are also apt to be more dangerous, exploitative, and discriminatory to the people employed on them. Here is at least one reason to be happy that so it goes.[14]

Another soothing approach is to minimize the dimensions of the change. Recall that the upper Midwest remains a region of strong family farm production, and that integration and consolidation are old, familiar processes. Moreover, the lines between corporate and family farm, large and small, powerful and weak, new and old, seem to blur and cross. For example, many families even on small farms have incorporated, if only for tax, accounting, and liability purposes. "Corporate" does not necessarily mean large or impersonal, and "family" does not necessarily mean small. For example, more than three-quarters of the stock in the nation's largest farm are owned by the members of one family, the H. H. Bresky clan. Through Seaboard Corporation, they also control fifty-four foreign and domestic subsidiaries handling commodity shipping, processing, storage, and distribution, employing about 11,000 people and grossing over $1 billion per year. Hog wars hardly represent a significant threat to the Breskys. With this real world in mind, it is hard to see why families attract indiscriminate sympathy and corporations hostility.[15]

Realist tales may in the end be logically, politically, and morally unsatisfactory. You might perfectly understand shakeouts, pendulums, fresh starts, restructuring, biosecurity, and all and still find plenty that is wrong. But realist tales are helpful for taming passions. They encourage more circumstantial judgment. They balk before presumptions that certain categories of people or institutions are inherently virtuous and others destructive or unreasonable. Instead, realist tales make almost everything seem reasonable:

> What is the driving force behind this hog-expansion frenzy? . . . It's really quite simple—there's money to be made raising hogs. In 1993 the hog market averaged $45/cwt.; the cost of production in the best hog units . . . ran near $35/cwt. That's $25 per hog profit.
>
> Now imagine, with 50,000 sows averaging a reasonable 20 pigs/sow/year, you could market a million hogs; $25 million in profits tend to turn a few heads.
>
> Or say you have 5,000 sows. You do the math. . . .
>
> That's part of the reason large production units are springing up in Oklahoma and Texas. . . . Nestled among the geometric designs etched on the landscape by center-pivot irrigators, these wheat farmers will tell you they can make a lot more raising hogs on the dry corners missed by the center pivots than they ever could trying to raise wheat. . . .
>
> Think about it.
>
> The hog industry will go where people want it. Push them out, legislate them out, and the packers will follow them out. Count on it.[16]

Do the math. Count on it.

In the 1990s thousands of hog farmers pulled out their calculators, did the math, and called it quits. Others joined with their neighbors and a diverse coalition of activists to fight back. Yet others did their best to bend to "necessity." For example, Barney Rhodes, a North Carolina contract feeder for Murphy, expressed his satisfaction with technical support and price guarantees: "The company tells us everything we need to do for the pigs to perform properly. . . . There's absolutely no risk here for me." Jimmie and JoAnn Stroud, who also struggled to keep a home place going in North Carolina, voiced their realism more philosophically: "We tried to be independent with tobacco, corn and soybeans, but sometime you can just independent yourself into a hole. . . . I don't mind taking the direction of the company. I'd rather have part of something than all of nothing."[17]

TWELVE

SO WE'VE BEEN SCREWED

REALISM HAS ITS CLEAR-HEADED, GROWN-UP, NO-NONSENSE APPEAL. Yet, when applied to render hog wars mild or natural, it might instead sound gullible. Is the "something" rather than "nothing" that big guys preserve a fair share of the original? For whose benefit, skeptics wonder, has anything of value been saved? Even if the line between mega and family farm is short of crisp, surely gross differences exist. Tales ending with "so it goes" too often feature generic, composite protagonists—"the producer," "operator," "farmer," "feeder," or "herdsman"—as if the names were interchangeable and name changes inconsequential. The actual people who bear those names may have very little in common. Mom, Pop, and the hired man do not swap processing plants like Cargill and Tyson executives. No one would confuse them in a line-up. Some normally look and smell of sweat, diesel, and manure; others of dry-cleaning and freshly splashed cologne. Daily chores for some entail feeding and treating hogs; for others, plotting and closing deals. Calling them all "producers" merely obscures what is going on. "Restructuring" may be an apt term for a turn in their relations, but it hardly evokes the substance of anyone's experience.[1]

In a very few years an old class of "producers" (the one whose labor is more obviously productive) has lost, and another greatly gained. A larger chunk of the profit goes to people who never get their hands dirty or break a sweat outside the racquet club. They likely never heard of TGE, much less spent winter nights dealing with it. So, in general, the less you know about swine from hands-on experience, from feel rather than forecasts, the larger the share of the business that is yours. These are among the reasons that progressives bristle before hog war stories that treat integrator and herder—screwer and screwee—as if they were kindred "producers." Although concentration, integration, and shakeouts can be expected just about anywhere anytime, "naive" seems a more fitting term than "realistic" for those who belittle the consequence. With a populist eye on consequence, what has been going on is a massive rip-off.

To explain how it happened, you do not need to conjure invisible hands. The perpetrators pose for photos. They include executives in a cadre of companies—Carroll's, DeCoster, Prestage, Seaboard, and Tyson—who cross-trained in the chicken trade. There has been nothing secret or subtle about their mission. They successfully industrialized poultry, approached the limits of its integration and expansion, and announced that they would do the same for pork. In alliance with financiers, grain growers and mills, genetics, pharmaceutical and equipment suppliers, packing and marketing companies, they lined up the capital, drew up the papers, and started construction. Through overtly discriminatory procedures—undercutting or simply barring independents from the marketplace—integrators established a structure to which only their own ilk could adapt. Joseph Luter III, chief executive of Smithfield Foods, explains, "If we bleed, our competitors will hemorrhage." Such giant companies offer farmers a piece of the action only when they agree, in effect, to become piece-rate, wage laborers. Stories quickly spread of contract feeders learning the hard way to beware the big guys' penchant for fine print.[2]

Why, progressives ask, should integrators be credited with "leadership"—as in exports—when they so monopolize the benefit? With great fanfare in 1994, for example, the president of Smithfield International, Raoul Baxter, unveiled a $200-million, three-year deal in Japan. "Vertical integration," he explained, "gives Smithfield the first value-added, brand-identified pork program in Japan." But the supply was guaranteed to come entirely from big guys in North Carolina. Eighty percent would be produced on contract by just four of the largest pork companies in the United States—Murphy Farms, Carroll's Foods, Prestage Farms, and Brown's of Carolina—about a thousand miles from where most American pigs and their feed are grown. Every hog would be slaughtered and processed at Smithfield's own plant, Carolina Food Processors in Tar Heel (which would increase its daily kill from 16,000 to 24,000 head). Every chop and roast would be distributed exclusively by Sumitomo Corporation. Through this arrangement, Baxter promised, a single corporate clique could assure "control" of everything "from the hogs' genetics to where and how they're raised and processed." Only rhetorical gymnastics could make this control sound like community property, something that would similarly benefit dirt farmer and MBA. Mike Brumm, an animal scientist at the University of Nebraska, ironically observed, "Pig farming is now a white-collar job." What he meant, of course, is that the profit in farming is increasingly taken by people who do not actually farm.[3]

Progressives stressing this interpretation are apt also to dismiss the integrators' main counter: that they "earned" their share by perfecting production technology. As I have explained, data on this score are contradictory. Some studies support the conclusion that big guys grew because they went high tech; others that they went high tech because they were big; yet others that neither size nor technology adequately accounts for advantages gained. Statistics are available to support all three conclusions. As you might guess, populists tout those that show integrators to be no better than mediocre at the actual craft of rearing hogs. Supposedly that is one way to prove that they do not "deserve" their share; they must be liars and thieves.

Such statistics are easy enough to find. In particular, once you drop the really small, "hobby" operations out of comparisons, differences in production efficiency shrink or disappear. According to a well-publicized "Swine Enterprise Summary," over the past ten years the top third of producers made money every year, and the middle third all but two years (1985 and 1988), when they actually broke even. But the bottom third (which do tend to be small) lost money 70 percent of the time, mainly due to exorbitant production costs, about $25 per head higher than everyone else. In general, then (excluding that bottom third, which are barely commercial producers, anyway), midsize family farms and monster factory farms have similar enough records of feed conversion, sow longevity, and litter size to counsel looking beyond technology for sources of differential success.[4]

The progressive's most damning candidate is price fixing. Diversified farms have long relied on hogs as a dependable source of income. U.S. market prices have swung widely, but through the 1980s they were consistently at least 10 to 15 percent above the average cost of production. Even in the worst of times, when both grain and cattle were losers, hogs cleared at least a little. That is why they have been known as "mortgage lifters" and why farmers, even if they hate the animals, tend to keep some on hand. They are one of the very few farm commodities for which price parity or better has been the rule. In fact, in the spring and summer of 1987 (June to September, when prices normally peak) the market hovered around $60 per hundredweight (/cwt.) before settling down to about $50, as it did again in 1990. Since the typical break-even price is about $35/cwt., independent operators were making good money. But beginning in 1991, the good turned to adversity and then calamity.[5]

The dimensions of the drop in market price alone can explain why so many marginal hog farmers called it quits. By November of 1994, hog prices across the United States had fallen to only $28/cwt. That is $20/cwt. below the average for the prior decade. In just one quarter, hogs lost 25 percent of their value. The Iowa market price rose above production cost only one month in all of 1994. In that good month, net profits peaked at a mere $7 per head. Losses peaked at more than $33 per head. At the end of the year packers were paying less per pound than they had at any time since 1980.[6]

What made matters worse was a couple of other conditions that were especially tough on little guys in the Swine Belt. One was a simultaneous swell in production costs. Corn prices rose, allegedly in response to flood damage, 1992 to 1993. Since feed (mainly corn) is the chief expense in raising hogs, profit margins would be squeezed, even if other things remained equal. But other things got less equal.[7]

It was particularly galling that prices everywhere else in the pork industry failed to reflect the squeeze at the bottom. While hog farmers were losing money, pork prices at the grocery store remained strong. Obviously, somebody else was making more money. For at least a half-dozen years, the retailer's average share of the price of pork stood at about 47 percent and the packer's at 15 percent. But in 1994 the retailer and packer each took an additional 5 percent. The farmer's share dropped by about a quarter, from 37 to 28 percent, an all-time low. If farmers had merely received their normal share, they would have earned an additional $10/cwt., enough to get out of the red. But packers and retailers chose that moment to grab margins well above their long-term average. Since integrators boast control of those vertical layers, you can see why independent farmers might suspect that the squeeze was engineered to force them out. It certainly had that effect.[8]

Hence debates about whether the big guys earned or stole their fortune can hinge on explanations for the squeeze of 1994. Were prices fixed? Plainly, vertical integrators, unlike independents, could use profits in processing and distribution to offset losses at the loading chute. But were they just well positioned to survive a surprise challenge? Or did they arrange the challenge so that they would be the ones to survive? As with most matters in hog wars, it depends on whom you ask.

Economists, master realists, can spin amazingly ornate tales to make most anything seem ordinary, at least in hindsight. By whatever route, they all seem to end with an implicit "See? I told you so." Their tales of farm commodity pricing can be particularly (as if spitefully, in-your-face) baroque. There are, though, a few simple principles in the case of U.S. market hogs. One is the regularity of short (seasonal) and long-term (approximately four-year) cycles. Inputs and outputs waltz a familiar tune. The price goes up-down-up, step-turn-step. A second is the profound effect of supply. As a rule of thumb, each percent rise in national hog inventory depresses their market value by about $1/cwt. A third is the interdependence among meat markets. Consumers tend to spend only so much on meat and to choose the bargain variety. There is good evidence, for example, that chicken gained favor in the United States less because of changes in popular taste or nutritional standards than declining cost. Bryan Melton, an economist at Iowa State University, estimates, "Price dictates about 60 percent of meat buying decisions." With these three regularities in mind—cycles, supply, and interdependence—it is no harder to explain the debacle of 1994 than a cloudburst on a steamy afternoon.[9]

The science of cycles is well enough developed that, whatever the event, some economist will find alignments, rhythms, and extenuating circumstances to fit like a glove. Conspiracy theories, like those you are apt to hear from struggling farmers and populist protectors, seem far-fetched by comparison. Since, for example, times were good in 1987 and 1990 (maybe a little too good? a little too often?), things just had to turn bad ("correct") in 1994, especially with some prompting from floods. Cycles also help account for coincident change in vertical profit/loss sharing. In order to preserve pork's position in consumer markets—in particular, a stable price at the grocer's—processors and retailers ordinarily absorb much of the fluctuation in the hog market. They claim to have taken a beating when times were good for farmers; so they had to recoup when the opportunity arose.[10]

The supply picture also fits a so-it-goes explanation. One of the differences between large-scale, integrated contract feeders and diversified farms is their greater dependence on a single commodity. Facilities and employees must be kept fully occupied with swine, or the whole enterprise grinds to a halt. In other words, their contribution to supply is relatively "inelastic." By adding or culling a few sows, by reassigning the function of a shed or two, Mom and Pop can more readily adjust herd size to the market. Although total U.S. hog inventory was quite stable throughout the 1980s and 1990s, it increased by 7 percent between the springs of 1993 and 1994. To that extent, whether intentional or not, big guys must share blame for an unusually large and long-lasting bulge in supply. But by rule of thumb, that bulge would account for less than half of the actual drop in price. A $7 drop would have been tough (especially since costs were increasing) but survivable.[11]

Casualties increased because of market interdependence. During the mid-1990s meat of all sorts was abundant in the United States. The combined supply of beef, pork, and poultry ballooned, from 67 billion pounds in 1993 to 72.5 billion in 1996. A relatively small part (about 1 percent) of that growth came from pork (mainly from big guys gearing up for exports), but as usual herders of all sorts shared in experiencing price depression.[12]

Baroque and discordant as these stories may seem, I think, they are more compelling than price-fixing alternatives. It is worth remembering that even the most integrated companies require profit-making divisions. Certainly the various tiers of executives are unlikely to conspire against their own incentive plans. Moreover, the industry is not so concentrated as to be free of worthy rivals. When, for example, U.S. big guys plotted expansion on the Western Plains, they faced competition not only among themselves but also from overseas. Nippon Meat Packers, the biggest processor in Japan, had long been a major porcine presence on the Plains through its subsidiary, Texas Farm.[13]

In the end, the price collapse seems to have been too short-lived and harmful for big guys themselves to seem the fulfillment of their design. It severely challenged some integrators (such as Premium Standard) and compelled others (Tyson) to divest their integrating capacity. Realists in 1994 warned that in a restructured industry $40/cwt. was the best that anyone should expect. Their message: Quit, old-timers, while the quitting is good! Only big guys can push the volume to thrive on such small margins. But in 1996 the price peaked near $60/cwt. The very same realists were then recalibrating their cycles and predicting long-term averages of $51/cwt., close to record highs back in pre-restructured days.[14]

Yet no matter what the cause or lesson of each price fluctuation, obviously a vertically integrated corporation could handle it better than an independent or semi-independent producer. You do not need to identify conspirators or master cycle science to know that farmers are more vulnerable. This was among the lessons John Morrison, former president of the National Contract Poultry Growers Association (NCPG), drew from poultry experience. On behalf of the National Contract Growers Institute (a nonprofit foundation that NCPG began for feeders' self-defense), Morrison warned: "The contract system has turned many chicken producers into little more than low-paid employees of the large broiler companies . . . totally dependent on a single company for a stream of birds to sell and pay their bills and subject to being squeezed by the company whenever necessary. . . . Like with chickens, we know the honeymoon will be over soon with hogs."[15]

This fearsome story could easily be recast to make the ecology no less than the little guy the damsel in distress. Integrators freely admit that places like Oklahoma (where the number of swine doubled, 1992-1993) were largely attractive because pollution there would longer escape notice. Effluent would evaporate more rapidly; there was less surface water to carry it afar and fewer people to complain, if it did. Yet as with the farmer, the environment could absorb wholesale abuse only so long. During the honeymoon, progressives charge, the big guys gained by shifting costs onto both the little guy and the environment. And they did so with government encouragement. For example, the Oklahoma Department of Commerce, recognizing the state's overdependence on a foundering oil industry, actively recruited its pork integrators.[16]

The big guys obviously have friends in high places, including the nation's capital. For example, Don Tyson, magnate of poultry and pork integration, could swap more than Arkansas lore with the President of the United States. They were close enough friends for Tyson to convince Clinton to put exporting his company's produce near the top of the national agenda for discussion with Russian president Boris Yeltsin. In 1995 the Pork Industry Congressional Caucus included about a quarter of the full membership of the U.S. House of Representatives. The chair of the House Agricultural Committee, Pat Roberts of Kansas, won the NPPC's Bronze Symbol Service Award. Such friends have helped the pork industry win significant backing for its exports. According to NPPC estimates, the approval of the General Agreement on Tariffs and Trade (GATT) and the North American Free Trade Agreement (NAFTA) would yield an additional $1 billion in sales in four years. The European Union alone would be required to spend $172 million on U.S. whole-muscle pork. And guess which "producers" would get the lion's share of the business? The very first, government-approved supplier was an integrator, Premium Standard Farms. When big guys organized the American Pork Export Trading Company (APEX) to cooperate/conspire around nontariff trade barriers, the U.S. government exempted them from antitrust regulations. And the tax code provides advantages similarly tailored to large sizes. For example, according to the 1986 Tax Reform Act, "single-purpose livestock facilities" (such as monster, specialized finishing units) can be depreciated more advantageously than pastures or multipurpose structures where old-timers are more likely to finish their herds.[17]

Of course, like most Americans, producers of all sizes normally speak of "the government" as a pain in the ass—the source of pointless paperwork, excessive taxes, and overregulation. Human rights do-gooders impede favored-nation exports, and chummy diplomats accommodate trade insanity. By "realist" measure, these doings (unlike, say, the Pork Industry Caucus) are profanely "political." Politicians fail to build barricades against foreign commodity "dumping" and to blast the barricades that foreigners erect. From this vantage, integrators' influence could be a godsend. Through their friends in high places, we might all get some bureaucrats off our backs. What is so unfair about that? All that hog people ask is the same opportunity that competitors enjoy. For example, USDA regulations allow poultry packers, but not pork or beef packers, to remove "foreign matter" by washing rather than cutting with a knife. Since customers pay for the resultant water-weight gain, poultry enjoys a competitive edge of about $43 per head over beef and $16 over pork. Why should progressives shout "unfair" when such inequity is addressed?[18]

One reason they should is to slow the spiral trade in government-brokered "opportunity." In the name of "fairness," legislators and lobbyists auger ever more public revenue into corporate accounts. Although specific farm commodities may enjoy slightly greater, momentary advantage, they all have long enjoyed substantial federal assistance. A parade of "welfare" nutrition programs—Food Stamps, the Emergency Food Assistance Program, the Women, Infants and Children Program, and sundry other commodity distribution operations—may be touted before the pubic (especially in urban states) as humanitarian, but they are sold to Congress (especially to farm state representatives) as agricultural supports. When USDA-funded programs move farm surplus, they drive up the price. For example, the USDA's Agricultural Marketing Service (AMS) routinely buys millions of dollars' worth of pork products for school lunch programs. Over just three months in 1995, for example, $30 million in AMS pork purchases probably improved hog prices by 25 to 35 cents/cwt.[19]

Likewise, the government does more than get out of the way of production. In 1994, for example, U.S. taxpayers spent $6.5 million to maintain the National Swine Research Center at Iowa State University and $8.2 million for the national swine pseudorabies eradication program. Every year the federal government spends millions of dollars on swine-related research. When North Carolinians began to complain about the stink that megafarms were making, rather than tithing the source, the state took $85,000 from general funds and handed it to the state university's Animal and Poultry Waste Management Center to study the problem. Such government-industry-university joint ventures also provide on-the-job training for corporate leaders. For example, Howard Hill, director of Veterinary Services for Murphy Family Farms, gained his experience heading the microbiology section of the Veterinary Diagnostic Laboratory at Iowa State University. Although populists and realists debate the proper relation between private and public sectors, only a revolving door separates the two. Much of the technology that integrators embrace was developed with government-funded research conducted by people who do not stand for long on one side of the door or on the other.[20]

Beyond tax and treaty requirements, export policies are also far from laissez-faire. The U.S. federal government provides about $100 million per year to the Market Promotion Program, which funds the U.S. Meat Export Federation. Whatever its highfalutin rationale, the Export Enhancement Program (EEP) is also a conduit of government assistance for producers. It was not a foreign policy breakthrough in 1994, for instance, but the pork profit squeeze that led the Senate Agriculture Committee, chaired by South Dakota Senator Tom Daschle, and thirty-three members of the House of Representatives, led by South Dakota Representative Tim Johnson, to urge Agriculture Secretary Mike Espy to approve the sale of 20,000 metric tons of U.S. pork to the former Soviet

Union. Since the market was glutted and Eastern European exchange reserves short, sales were sluggish, but this EEP step was expected to raise the U.S. market price by $2/cwt.[21]

When fully assessed—at least over the long term in the abstract—government support may benefit large- and small-scale producers equally. But—in practice on the ground—it sure seems to favor the big guys. They are the ones who are better able to meet the delivery requirements of giant nutrition and export programs. They can better afford spanking-new technology and better use the developers who ply the revolving door. They have both the incentive and the capacity to see that fine print is added with them in mind. It could also be argued that USDA subsidies during the 1980s (particularly deficiency programs, which mainly benefited grain and fibers) in effect underwrote vertical integration. Until ceilings were lowered, those programs provided key players (such as ConAgra and Cargill) with the investment capital that integration required. Since those subsidies also helped assure that grain could be bought on the market at or below its production cost, diversified farms simultaneously lost their edge. The USDA helped narrow the difference between feeding grain that you bought and grain that you grew yourself.[22]

Lest all of these complaints also seem abstract, consider more blatant cases of government favoritism. State and local officials have been particularly prone to grant integrators preferential treatment. When, for example, Chiquita had trouble selling its outmoded John Morrell processing plant in Sioux Falls, the State of South Dakota donated one-third of the projected cost of modernization. The debt reduction package ($10 million over ten years) allowed Chiquita to unload the plant soon thereafter. It is hard to imagine little guys swinging such a deal.[23]

Seaboard Corporation has been particularly blessed with government favor courtesy of Guymon, Oklahoma. The city enacted a one-cent sales tax just to raise money ($8 million for fifteen-year revenue bonds) that would convince/bribe Seaboard to invest the $50 million necessary to convert a defunct Swift beef packing plant into a state-of-the-art pork packing plant. Seaboard agreed and then contracted finishers to supply the plant—among them, Hitch Enterprises, which was already the ninth largest cattle feeder in the United States. Paul Hitch explains how they erected twenty-eight 1,000-head units: "We built the buildings and leased them to Seaboard. We are more like landlords rather than involved in hog production." Once again, direct government subsidies grease integrators' path. The beneficiaries look more like extortionist investors than farmers.[24]

Seaboard also finagled subsidies in Kansas, which long had been inhospitable to integrators. Until very recently, that state, like others in the Swine

Belt, barred corporate livestock farming altogether. But in 1994 Kansas permitted individual counties to grant exceptions. At the same time the state also barred those exceptions from receiving/extorting tax-exempt "revenue" bonds. Seaboard, which used operations in southwest Kansas to supply the Guymon plant, soon qualified for exception and found a loophole in bond regulations. The state failed to bar tax-exempt "private activity" bonds. By that sleight of hand, in 1996 Seaboard secured $9.5 million in tax-exempt bonds to finance the construction of pork production facilities. Mike Jensen of the Kansas Pork Producers Council bluntly explained: "Large companies have always had access to sources of capital that other producers can't get." If so, why is commercial credit insufficient? Why do they also drain the public dole?[25]

In addition to the little guy, the environment, and the government, hogs themselves may be counted among the integrators' screwees. Swine surely suffer in some respects when phases of production are specialized and separated. The farther animals are transported, for example, the more frequently they are injured or die en route. The risk is minimal when shipping takes 10 to 25 minutes, but above a threshold of about 45 minutes, the risk increases greatly. Now that more and more hogs are reared outside the Swine Belt, where packers are more concentrated, about 80 percent of U.S. market hogs are shipped more than fifty miles for slaughter.[26]

Producers in general as well as hogs are still reeling from integrators' excess in engineering swine for the lean-meat craze. In the late 1980s breeders like the Pork Improvement Corporation (PIC) developed stock that grew significantly faster and finished leaner on less feed. They became immensely popular. Whatever their specific source, farmers called these giant-hammed, fast-growing animals "PIC-type" pigs. It slowly but surely became apparent that many suffered awful side effects. Judging from ledgers alone, "PIC-types" were "super performers," but too many were also incredibly high-strung, one might say, congenitally unhappy. They were very hard to handle and prone to suicidal panic. When startled—sometimes by as little as flash of light from an opening door or the rustle of your clothes as you approached—their whole body would begin to twitch in spasm. They might leap and crash into walls, go into shock, or just drop dead. The incidence of PSE (pale, soft, exudative meat) skyrocketed. A side effect of breeding breakthroughs, then, was a phenotype—known as "porcine stress syndrome" (PSS)—that cost hogs their well-being and producers a bundle of cash. The Livestock Conservation Institute estimated that, 1994 to 1995, PSS cost a typical packer $2,000 per day and U.S. processors collectively about $32 million per year. Even after substantial effort to avoid the syndrome, 12 percent of all market hogs were still PSS carriers. In addition to the suffering it brought those animals, in less than a decade it cost the industry about $44 billion in all.[27]

Given the cost and the public relations disaster for alleged perpetrators—for PIC in particular—big guys raced to identify the culprit and found it in the form of a "stress gene" (halothane or HAL-1843). Its main effect appears to be lowering the threshold and increasing the intensity of calcium release from muscle fiber. (Recall the role of calcium in glycogen conversion and PSE.) Hence PSS hogs are destined to be both great lean gainers and crazed at the drop of a hat. For a while, breeders tried to achieve one and avoid the other through terminal-sire, "monomutant" or "carrier" (HAL-1843-mm or Nn) pigs. Results were better than they were for "dimutant" or "homozygous-positive" (HAL-1843-dm or nn) pigs but still poor enough that the push was on to eliminate the stress gene entirely. Lauren Christian, director of the National Swine Research Center, concluded:

> The merit of the porcine stress syndrome (PSS) gene has been debated for a number of years. Nearly everyone agrees that in its homozygous mutant form (nn), the stress gene can have some devastating effects. If physical stress doesn't cause death, it's almost sure to cause pale, soft and exudative (PSE) meat under even the best of handling and processing conditions. . . . Hence, because the gene offers only a minimal increase in lean percentage while increasing significantly PSE frequency, it is clearly evident that the gene should be eliminated from the U.S. pig population.[28]

In 1995 PIC built a promotional campaign around its guarantee that all of its boars and 93 percent of its gilts were free of the halothane gene. Every PIC pig was sold with a bright yellow ear tag bearing the PSS-free pledge. Although stressed-out pigs and PSE remain a problem, the stress gene and PSS are now quite rare. Nevertheless, the whole episode was considerably more damaging to hogs than it was to PIC.[29]

Progressives also charge that big guys threaten the well-being of the communities that surround them. Imagine, for example, their effect on tourism. That is exactly what internationally syndicated humorist Dave Barry did in response to newspaper coverage of one of those lagoon breaks in Iowa during the summer of 1995:

> The story said that state officials were especially alarmed because the manure spilled into a section of the Iowa River considered to be "one of the most prized canoe areas of the state." I can see where it could be a real crimp in a person's canoeing vacation: You're paddling peacefully down the Iowa River when you hear this faint rumbling noise, which gets louder and louder until it sounds like a freight train, and you turn around, and there, thundering right at you—this

> would be just like the tidalwave scene in *The Poseidon Adventure,* only more aromatic—is the dreaded, biblically prophesied Wall of Swine Doota, and at that instant you realize that even if you do survive, you will never be welcome in an elevator again.[30]

In actual circumstances, as when negotiating/extorting favors from local government, integrators routinely exaggerate the "multiplying" benefits they return. A study in Missouri showed that integrators provide about one-third fewer jobs than comparable independent producers. A study in Minnesota found hog farm size and local spending inversely related: The bigger the operation, the smaller the percentage of its income that reaches businesses within a twenty-minute drive. These are among the reasons why states like Kansas discouraged communities from mortgaging their future to integrators and why you might pity those that did, even if under duress.

Rural towns in most of the United States have long been struggling. When the local meat packer teeters, Main Street merchants—who lived off the packer's payments to farmers and the union-scale wages of its employees—see only doom ahead. In hope that integrators will restore that base, "community leaders" (generally those very same Main Street merchants) will grant anything they can—revenue bonds, tax and zoning code exemptions, cut-rate water or sewerage services—to attract big-guy saviors. When the integrators break ground, the local economy booms. Construction trades cast a honeymoon glow on the whole affair. But when construction is complete, the jobs and the money start flowing back to their distant source. The hoped-for "multiplier effects" migrate from town square jewelers, hardware stores, and used-car lots to Lexus dealers and prep schools far from the stink and the crumbling court house. Local communities can be added to the list of integrator screwees.[31]

As might be expected, the premier parables of little guy, environment, government, animal, and community abuse come out of North Carolina, where integrator growth has been most intense. Populists, muckrakers, and investigative reporters like Pat Stith and Joby Warrick of *The News and Observer* in Raleigh have assembled impressive evidence of Tarheel intrigue.

Among the initial requirements for the state's porcification was suitably gargantuan packing capacity. In 1990 and 1991 Murphy, Carroll's, and Prestage urged Smithfield to respond by building its Carolina Foods plant on the banks of the Cape Fear River. By all appearances, it would be very hard for Smithfield to secure permits. The plant would daily dump 3 million gallons of treated waste into the river. According to the Department of Environmental Management, the river was already "at or near capacity" for receiving industrial discharge, and Smithfield's environmental record was hardly stellar. At the time, the company held a national record for water pollution fines it accrued in Virginia. Nevertheless, Governor Jim Martin's environmental managers decided that the risk was so obviously negligible that they waved the requirement of a detailed environmental impact study. By consent decree, Carolina Foods also won protection from fines and lawsuits for water violations. In its first couple of years, the plant flourished, providing 2,000 jobs—albeit low-paying ones (starting at $5.75/hour in 1995)—plus a stream of fecal coliform bacteria and other contaminants, hovering around the legal limit.[32]

Few would deny that North Carolina politicians had a personal stake in porcification. When Craven County commissioner Gary Bleau was asked if there might be conflicts of interest on pork issues, he answered, "Hell, everybody up here has a conflict." At the time Bleau himself was both chairman of the commission that rules on land-use regulations and a hog farmer working for Murphy. Among the most suspicious characters might be Republican U.S. Senator Lauch Faircloth. In jeremiads as president of the Alliance for a Responsible Swine Industry, Don Webb made Faircloth a favored target. It is easy to see why he would. When Faircloth became chairman of the Senate subcommittee on Clean Water, Wetlands, Private Property and Nuclear Safety, he was among the partners who owned Coharie Farms, the thirtieth largest pork producer in the United States, and an investor—with stock worth more $100,000—in the two largest packing plants in his state. All of those farms and plants operated in areas strongly affected by wetland regulations (or the lack thereof) within his committee's jurisdiction. It is hardly surprising that he volunteered his name to those petitioning the USDA to subsidize pork exports to Russia. But it is absolutely baffling that the Senate Ethics Committee could see no conflict of interest. Should such Tarheel officials forget where pork interests lie, political action committees (PACs) stand ready to remind them. Between 1992 and 1994, as the honeymoon for integrators waned, pork PACs more than doubled their contribution to campaign coffers, dwarfing the amount donated by labor unions and environmental groups.[33]

No single Tarheel stands more ready to be demonized than Wendell Murphy, founder and chairman of the board of Murphy Family Farms of Rose Hill, North Carolina, the biggest pork producer in the United States. In 1995 alone, his 600 contract feeders in the Midwest and South brought in $200 million. His prospects were built during a decade of service in the state senate. As vice-chair of the Agricultural Committee, he helped enact bills with environmental, tax, zoning, and building codes that benefit interests like his own. Other pals in public service filed close behind. Fellow Democratic state senator Charles W. Albertson, chair of variously named committees on environment and agriculture through the 1990s, was an old friend and dependable supporter.

Likewise Murphy's counterpart on the House side, Republican representative John M. Nichols actually began raising pigs for Murphy during his tenure, as did U.S. Senator Faircloth and County Commissioner Bleau. Murphy and Governor Jim Hunt were friends since they were classmates back at N. C. State University. Decades after graduation, they could still be spotted catching an occasional basketball game together at Reynolds Coliseum. Regardless of party, branch or level of government, Murphy's foxes guard the hog house.[34]

In the 1992 gubernatorial race, Murphy covered his bets by having his name or that of members of his family or executives in his company prominent on the list of contributors to both candidates, the loser, Republican Jim Gardner and the winner, fellow Democrat and Wolfpack fan Jim Hunt. Hunt even flew from Raleigh to rallies aboard Murphy's company plane. One day, during a cash crunch just before the primary, Hunt's campaign received a stack of $2,000 checks from Wendell Murphy and sundry kin—his mother and his wife, siblings, children, and in-laws—totaling $20,000. They sent another $20,000-stack three weeks before the general election. In so doing, the Murphys met the letter of the state's legal maximum ($4,000 per contributor per election) but exceeded the spirit of it by tenfold. Murphy is probably the biggest single source of hog-tied campaign funds, but he is hardly alone. The PAC of the North Carolina Pork Producers Association is a reliable contributor, as are other barons of North Carolina swinedom: Joyce Carroll Matthews, chairman of Carroll's Foods, the Maxwell family of Goldsboro Milling, and William H. Prestage, president of Prestage Farms.[35]

Here you have, then, all the necessary ingredients for outrage rather than resignation in the face of hog wars. Given a progressive spin, hog wars become high melodrama. Corporate titans trample the humble and hapless—family farmers and their communities, hogs, the environment, and the public trust. The story achieves its most dramatic staging in North Carolina. The bad boys are characters like Albertson, Bleau, Faircloth, and Hunt who bend to Beelzebubs like Murphy.

THIRTEEN

SO WHO'S RIGHT?

They've been a godsend.

—Mayor of Milford, Utah, Mary Wiseman,
describing Circle Four Farms

It's like the devil came to Milford.

—Milford farmer Joey Leko,
on that same operation

My family is together because of that farm. . . . that's a fair trade for a little smell.

—Patty Cherry, Milford waitress
and mother of Circle Four employees

WHETHER THEY PROMOTE, resist, or just tolerate the latest round of agricultural change, people are right, I think, to see its high stakes. At issue is the way Americans affect the land and its inhabitants, the quality of their daily lives, of powerful institutions, and of environments they share. The crossfire of hog wars illuminates a dreary terrain: peaks of privilege atop American society, blight in a human-hijacked biosphere, consumer decadence and producer arrogance, corruption in government, technocracy and chicanery in boardrooms, delusion and despair at large. The main point of contention is whether the big guys are better considered part of the problem or the solution.

Integrators do not promise perfection, but they do regularly pose as model citizens. Allies pledge that concentrated pork power will better tailor production to consumers' satisfaction, improve animal welfare, reduce pollution, promote innovation, and circulate profits among far-flung partners. Of course, owners and executives stand to make a bundle, but only if they also succeed in raising healthy, reasonably contented hogs and in delivering inexpensive, high-quality produce. While cutting waste and helping feed the world, they also can provide employment for rural residents, vitality for their communities, and more balanced trade for the nation as a whole.

These claims are highly contestable, at least when you get down to specifics. Through a pair of characters—the integrator-excusing "realist" vs. the anti-integrator "populist" or "progressive"—I have tried to show how each of these specifics might appear from opposing hog war sides. In fact, these characters (like the "war" itself) are figurative, cobbled from found material. No one, I suppose, is a certified, card-carrying realist or populist. I do not know how many people would embrace either label, where they live, or what else they have in common, though I can guess from the press. I doubt any of them would represent themselves precisely as I have in the preceding pages, but I hope they could find themselves—or at least you, reader, could find yourself—consistently rooting for one character as opposed to the other. The two are less "typical" of anything than "archetypal," representative in the manner of a Socratic caricature or an ideal type. These figures should diverge from reality, then, only in that they are composites. With a bit of allowance for stridence, they are thereby better advocates of each predisposition—more assertive, clear, informed and articulate—than any particular person we would ordinarily meet.

Among the reasons to so inflate garden-variety reality is to find the leaks. Under pressure, for example, the presumption that animal or environmental "rights" are transparent, plain as the nose on your face, or that producers and nature have inherently competing interests just does not hold water. Likewise, the invisible hand turns out to be a visibly biased arbiter of producer-consumer-government relations. With slight amplification, it is easy to hear the bellow of sacred cows and the rhythm of standard song in otherwise droning, technical detail. Each realist assertion seems to move to a driving, capitalist, macho backbeat; each progressive one to a New Age tinkle, bourgeois, primitivist, nostalgic.

Of course, it is possible that these tones were introduced in the bloating process. A mere two categories can forge incongruous coalitions. For example, people who want purer ham sandwiches may have very little in common with PrairieFire or PETA people, even if they are similarly suspicious of industrial agriculture. In labeling them all "progressive," listening for the tinkle, we might easily forget the figurative constitution of the category in the first place. The distinction between "populist" and "realist" still seems to me sharper than the one between "factory" and "family" farm. At least it is sharp enough to pop some dogma-filled balloons. But it may be too dull for dissecting everyday experience.

The forced choice—which side are you on?—may also introduce spurious symmetries. Whenever one side alleges virtue, we expect to hear about vice from the other. But the real world does not so swing on a gate. People may unwittingly disagree about the subject or simple fact of their disagreement. For example, producers assign central significance to the control of swine disease. It is among their main justifications for getting big in the first place. They invest hundreds of thousands of dollars and risk their own or employees' health to work sites that are designed as physical barriers to infection. But few populists address swine disease at all, except very obliquely, mainly through concern for their own health as pork eaters. Animal rights and food safety literature is full of accusations that factory farmers callously invite infection that they then overmedicate. No "save-the-virus" or "free-range pathogen" campaigns have been launched, even though a populist/realist pairing would lead us to expect it. In many cases, as in this one, the search for "sides" presumes too much common ground. Often hog war troops are so out of touch that they lock on targets that are better considered hallucinations than friend and foe marks.

Recognizing these limitations pushes me toward two concessions. First, the question "So who is right?" probably is best dodged. It is unlikely to yield an answer that would convince anyone who was not already favorably predisposed. Besides, there have already been too many victims of friendly fire. As Lee Wirth, president of Cargill's North American Pork, explained: "The game used to be who could outfox who. We both lost that game. We need to move beyond that now." I would only add that the "game" was misguided from the outset. There were always more than two sides, and losses have been far from equally shared among them.[1]

Second, the players could use more particular, human faces. After nearly twenty years working with family farmers, I think I came to know theirs pretty well. I could confidently characterize the ways they talk about hog wars with relatives at home or with neighbors at church, after a school event, or while fixing fence. But my impression of big guys has been more affected by press releases. Through friends I got to know a few folks on the payroll of integrators, but for fairness sake, not enough. I wanted a more detailed, on-the-ground sense of their workaday lives, one more like what I knew of neighbors. In fact, those few whom I did meet seemed cut from similar cloth. Their life course—typically from farm kid to ag college student and then "manager" of sorts for an integrator—was a variant on local, rural and small-town tradition. They were as fond of those traditions and as ambivalent about the big guys for whom they worked as classmates and kin were back at the home place. In fact, many of them still helped Mom and Dad on weekends or caught corn, come harvest. But

I did not know enough about any one group of these people to be confident that my impression was an institutional reality.

So I had to pick out a corporation to get to know better. Of course, in its particularity (which is the payoff) it could not "stand for" big guys in general. Warring caricatures do that better. I wanted to focus on an integrator that would be unique in an instructive way, one that might evoke a more grounded sense of the potential for other such firms. Given, for example, great suspicion about how well integrators can relate to their natural and social surround, what is the best we really can expect? In actual, favorable circumstances, how have an integrator's distinctive qualities (its scale, structure, or technology) played out, not just as a matter of statistical or syllogistic probability but as a matter of fact?

With those questions in mind plus a bit of serendipity, I early on chose Premium Standard Farms (PSF). The reason was analogous to the populists' in trailing Murphy or DeCoster. They are by consensus the bad boys of the integrator lot and a safe-bet source of horror stories. From the outset, PSF vigorously styled itself as their foil. If there were a chance for redemption, a possibility that integrators were truly model citizens, PSF was the most likely candidate.

As Vice President Don Skaburg explained, even the name was selected to mark the high road they had taken. "Premium" was to indicate their market ambition—branded pork products of such top quality as to draw export demand. "Standard" was to indicate unrivaled precision and consistency. "Everything is standardized at PSF," Skaburg boasted: "Genetic programs, feeding programs, animal health programs, management procedures and facility designs. . . . When we started, we were a Missouri hog farm. Now we are a global food producer . . . conception to consumption." Speaking for competitor Murphy Family Farms, Randy Steoker agreed: "They [PSF] approached the business with an attitude of spare no cost to do it in the best possible way the first time out." Even opponents readily recognized that PSF was the "flagship" of the integrator armada.[2]

There were, though, troubles on its maiden voyage. The company was formed in 1988, and Iowa was to be its home port. The proximity to feed crops, packers, and technical support was an obvious draw. But in 1989, after substantial PSF investment, the Iowa Department of Natural Resources denied the company a license to center operations too near Ledges State Park. PSF angrily pulled up stakes and headed to the town of Princeton in Mercer County, Missouri, just across the Iowa border.

The State of Missouri granted PSF regulatory exemptions so that it could intensively porcinize northeastern counties. PSF discreetly purchased parcel after parcel, eventually accumulating the equivalent of an entire county in total acreage. It controlled enough terrain to position hog buildings on hilltops with a God-given air wash in between. In addition to a more favorable government reception, this element of biosecurity was among the main attractions of the Missouri site. PSF still would be close enough for convenient trade with Iowa's grain growers and packers (and close enough to thumb its nose at the state government) but far enough from its concentration of hogs to make contagion a little less likely.

Yet another major attraction was a crying economic need in the five traditionally agricultural counties that PSF most directly affected: Grundy, Harrison, Mercer, Putnam, and Sullivan. For at least a decade or two, their economy had been stagnant, and during the 1980s it fell into a tailspin. On the eve of PSF's arrival, after years of struggle, most of the families who farmed there in the 1970s had called it quits. Average income for farmers who remained had fallen to just $10,000 per year. In Mercer County the number of hogs dropped from 37,000 in 1971 to just 6,000 in 1989. In 1983 the last of the local banks closed, and in the next four years 63 percent of Mercer's hog farms failed. For long-time residents prospects were worse than bleak. School-age children matter-of-factly said that they "knew" they would soon have to leave if they hoped for a better future. Home was for losers. Of course, what this meant for PSF was a relatively cheap chance to make winners and a profit. It was a buyer's market for property and for workers with relevant skills. It would not take much to revolutionize the place, and PSF's ambition was gargantuan.[3]

Its scope went far beyond one corner of one state. In 1994, for example, PSF brought its pork invasion to Texas cattle country. It purchased and expanded operations that previously belonged to National Hog Farms near Dalhart. With $200 million and 800 employees it could pump pigs from 16,000 sows. Just north of the town of Hereford—named for the cattle breed and home to the nation's biggest cattle feeding company, XIT Ranch—PSF put a $45-million pork packing plant. Once again the advantage was the support of local government, isolation, surplus labor, and convenient transit to Midwest granaries. By 1995 boosters anticipated sufficient breeding stock—some 80,000 head—to turn North Texas from cow to sow country. At the same time, arrangements on the other side of the globe included contracts with Marubeni America Corporation to market Premium Standard brand pork to Japanese consumers.[4]

But operations and growth remained firmly centered back where they began in Mercer County. On its own PSF increased pork production of the whole state by a third. From 1989 to 1995 its inventory rose from zero to 80,000 head on 37,000 acres. It built its own mammoth milling, trucking, maintenance, and fabrication facilities. By obvious, gross measures it represented an economic reprieve, especially for those five counties. In 1994 Dennis DiPietre and Carl Watson, economists with the University of Missouri's Extension Commercial

Agriculture Program, calculated that in its first half-dozen years, PSF pumped about a $1 billion into Missouri. That investment, DiPietre and Watson figured, would "create" 2,639 permanent jobs, adding $199 million to personal income in an area where 20 percent of the population had been living below the poverty line. The construction phase alone yielded $119 million in personal income, about three-quarters of it in the immediate vicinity. At the time of the study, the company had an annual payroll of over $20 million for 1,100 jobs, 90 percent of them within a sixty-mile radius. The lowest-paying jobs earned about $13,000 per year, higher than 30 percent of the household incomes in the region before PSF arrived. The mean salary was about $20,000. The average wage in Mercer County doubled.[5]

Of course, these gross measures may overstate prosperity and the credit due PSF. The salaries of transplanted executives—bloated to cover "hardship" when compared to latté land—certainly skew the averages. Likewise, construction hiring would diminish once the basic infrastructure—hog houses, office and service buildings, tract homes plus a few well-Jacuzzied mansions—was complete. No doubt, some economist could tweak cycles to make the turn around seem illusive or self-generating. But it was easy to find laborers at the Princeton tavern who counted PSF among their lucky stars in helping them come home after a year or two of job-hunting exile in places like Kansas City.

In nearly every respect, PSF was working very hard to earn a good-citizen reputation. Management was proud of its preference for local investment. For example, it issued press releases on the 65 to 70 farm kids employed every summer to mow, rake, and bale forage crops and to manicure company grounds. PSF helped pave Princeton sidewalks, fill roadside planters, and install snazzy streetlights. When compared to the 1980s or even the 1970s, nearly all public spaces were spruced up, more alive, as if the company had hit the town with giant defibrillator paddles. When the local boom created a housing shortage, PSF built its own complex with air conditioning, a clubhouse, and a swimming pool, all of which were open to residents, even if they were not employees.[6]

PSF also aimed to be a conspicuously good steward of the land. The company hired three full-time environmental managers and announced an aggressive policy of cooperation with the Missouri Department of Natural Resources. To supply drinking water for its 80,000 sows, PSF built an independent source, a network of thirty freshwater lakes. Effluent would, of course, be a problem, even more so because of the procedure that PSF used to improve sanitation and air quality in its hog buildings. Every two hours, wave machines flush the floors in production units with over 1,000 gallons of water cycled through "single-stage anaerobic treatment" (i.e., pumped in and out of an open, twenty-two-foot-deep lagoon). Mainly because of this flushing routine, about 95 percent of what went into a lagoon was actually water. Solids were allowed to settle and bacteria to do their work. Periodically, the top two or three feet of the lagoon would be decanted and sprayed onto forage crops. To minimize runoff, PSF announced that it would limit each application of manure to a quarter of an inch per acre and do no applications at all from mid-November to mid-March (when ground is apt to be frozen) or on fields that were saturated or that sloped more than ten degrees. Even the treatment of dead pigs and hogs was to be environmentally first-class. Rather than piling up and putrefying, as they often did on Mom-and-Pop farms, casualties would be trucked to a refrigerated holding area and kept chilled until they could fill a truck headed for the company's own rendering plant in Milan, Missouri. Protein and bone meal byproducts would then be recycled in PSF feed mills.[7]

While Missouri family farms still have a Tobacco Road look about them and Murphy buildings one of accelerated depreciation schedules, PSF's ooze Aspen ambiance. The company headquarters, with heavy timbers in natural finish, set amid undulating gardens, resembles the sort of compound that draws investment bankers on retreat. It houses management, research, communications, and training facilities well-appointed in urbane, corporate style. Everything from mixing rations to hosing out trucks was to be treated as a profession, honed in gleaming demonstration labs. The massive hog buildings themselves were scattered off in the distance, set far enough off paved roads and surrounded by enough greenery that you might drive through the area without seeing one. The architecture of the processing plant could pass for middlebrow if it were a university library. Of course, though, what went on inside was considerably messier, a state-of-the-art death and dismemberment operation. Workers could turn 8000 animals into packaged pork every two-shift day. But, PSF crowed, the production pace was determinedly slower than that of competitors in similar facilities. At 650 head per hour, the plant could accommodate 200 quality checks, nearly four times the number required by the USDA.[8]

These are among the qualities that drew me to PSF. Unlike most of the big guys, it sure seemed to have food safety, animal welfare, the environment, and the community as well as profit in mind. Even if the "good citizen" was more a public relations posture than a heartfelt conviction, it certainly seemed preferable to the blatant Beelzebub alternative. Screw-ups, like those spills in the summer of 1995, still would have to be expected. They were happening everywhere. And, once burned in Iowa, PSF would surely play hardball with government officials, like everybody else. Admittedly, it was always possible that the impoverishment and depopulation of rural America, in process for at least a century, would somehow abate on its own. Government programs, like those that turned Indian lands into family farms in the first place, could be rejuvenated as they were in the 1930s. Public resources that favor Boeing and ConAgra might be redirected to prop up the little guys of Middle America again. But I would not hold my breath. And I had doubts about the singular virtue of little guys, anyway.

So PSF seemed an ideal choice for case (one might say, best-case) study. Of course, too, PSF also seemed an attractive focus to me because it was damned convenient, an easy drive south from my home. Since Iowa was still a family farm preserve right at the company gates, you could overhear analysis of the difference in ordinary conversation. For PSF and its neighbors as for me, hogs were more than an academic curiosity.

But the timing of these realizations—1993 to 1994—was awful. I had determined to give the company the fairest possible hearing just as Willie Nelson and Jesse Jackson, environmentalists, community organizers, rural activists, trade unionists, liberation theologians, sustainable agriculture and animal-rights advocates, and vegetarians were forming coalitions with nearly the opposite purpose. They already had made up their minds. For the benefit of God, humanity, and the planet, they aimed to expose PSF as the agent of pure evil. The 150 citizens of Lincoln Township and PSF had just begun trading snipes and lawsuits. Since these neighbors were communicating through attorneys, I suspected it would be rough for me to kick back with people over cups of coffee.[9]

Although company officials might trust my farm side, my university side could be an obstacle. As always, I had no desire to pose as anyone other than the open-minded eccentric I understood myself to be. I wanted the consent of PSF employees to help me, only if that consent were truly "informed." In fairness to them and in fear of my own liability, I wanted all of the cards on the table. But I had serious doubts about my ability to overcome potential suspicion that I was a spy, secretly in league with the populists.

The problem was a combination of that contentious moment and the reputation among many agricultural folks of my full-time employer, the University of Iowa in Iowa City—rather than, say, Iowa State University, the land-grant institution in Ames. In this part of the world that is a difference of major proportions. My college town is the one known for hippies, protesters, and liberal Democrats. "Nuclear-free Zone" signs are still posted at the town line. There is a socialist potter on the city council and quite possibly the oldest lesbian softball league in the nation. When you go somewhere to eat, people may well give you the benefit of the doubt by assuming you do not eat meat. Ames, on the other hand, is more Republican, straight and carnivorous. Iowa State is home to some of the world's leading programs in agricultural science. In fact, PSF has regularly contracted with ISU staff for veterinary services. My pedigree would be a problem.

It was fortunate, then, that I had done my homework. I had already looked up enough friends of friends throughout the industry that I knew some important "pig people" not only in Ames but also at other institutions, schools of agricultural science and veterinary medicine around the United States. Among their friends and students were people employed with PSF in a wide variety of capacities. Some of my contacts had grown up with PSF executives, herdsmen, farm managers, mechanics, carpenters, buyers, and truckers. I even knew childhood playmates of people in the company's public relations office. By the spring of 1994, then, I felt that I was as prepared as possible to give those contacts a try. I hoped they would help me talk with people and look around, to learn whatever I could about pigs, disease, and culture according to Premium Standard Farms.

FOURTEEN

MORE SHIT HAPPENS

I REMEMBER THE LAST FEW MONTHS OF 1993 and the first few of 1994 as a blur of activity. One of the guys on the farm was hospitalized, and my responsibilities increased accordingly. Every day, rather than once a week, I had hog chores to do. It was not a big deal, just a couple of hours of dirty duty before noon. Feed, clean a little, and check to be sure everything is okay. Moving and sorting stock, castrating, hauling manure, grinding feed, and most repairs could wait till Friday, my normal full day for such things. Meeting early-afternoon classes, Monday through Thursday, would still be manageable. But preparation time would be tight. It was hard to get used to planning around daybreak commutes to the farm, rush trips from farrowing house to feed store, midday showers, then academics in the afternoon, family time, household chores, and heaps of manure-scented clothes to clean at night. One sick sow or a broken waterer, normal fare in winter, was enough to throw the whole week out of whack.

I was also finding the shift from shit-covered floors to ivied walls and from hired hand to professor disorienting when repeated daily. I distinctly remember one morning when a burst pipe in the farrowing house prompted panic. I wondered how emergency plumbing repair (did I have to run to town for a fitting?) would affect my ability to finish rereading a book that students had to discuss that afternoon. Since the book was about rural life in America, I was in the odd position of trying to limit experience with the subject so that I could teach it better. That irony and the press of the moment were too extreme to be amusing. On the in-town side, I also remember struggling to understand what a student meant in aiming to "combine" two ideas. I was stuck with this image of an International Harvester picking and shelling them, like ears of corn.

I did not get it. Only when the student repeated herself, placing a strong accent on the second syllable (com·bine´ versus com´·bine), did I fully shed mental coveralls and get back into the classroom groove.

Sometime in the midst of all of this (early March 1994) in a fast-food restaurant, I bumped into a colleague whom I had not seen in a few months, Mike Chibnik from the Department of Anthropology. I began telling him about the challenge of PSF and my preparations for full-scale fieldwork there once classes were over. "I suppose you know," he said, "that Paul and company are doing research on pigs, controversy in North Carolina or something."

Actually, no, I did not know. And the thought of yet another complication threw me into just about total tizzy. I immediately knew that we had better straighten out our relations or we would be tripping over each other very soon.

"Paul and company," Mike explained, included three names. One was Kelley Donham, a veterinarian and professor of preventive medicine at the university's Institute of Agricultural Medicine and Occupational Health. I had read some of his research and heard farmers and veterinarians speak highly of him. We had talked a couple of times on the phone about ag-work environments and health technicalities, but we had no more personal contact. For the sake of diversity, I had been concentrating on scientists farther from home. Good experiences with them and mutual friends encouraged me to believe that Kelley and I could easily get along. Second, there was Kendall Thu, a new research scientist in Kelley's institute who just earned a Ph.D. from the university's anthropology department. We had never met, but guessing from the bent of his likely teachers and the title of his job, his disposition could well be more hard-data, macho-scientist than mine. In case of friction, I thought, be prepared to apply a little extra goodwill grease. The third name also belonged to a university anthropologist, Paul Durrenberger. It brought instant terror to my heart.

Although we have shared the same campus for most of the past twenty years, Paul and I have had very little to do with each other. In attempts at collegiality, I have found him unpleasant, as I suspect he has me. His scholarly swagger, among other things, incites a competitiveness that I wish were not there. Differences with people who are "subjects" of study usually seem to me intriguing, even attractive. I welcome them as a chance to learn. That is why I seek them out in the first place. But I am much less patient with people "like me." If we have similar privileges and responsibilities, I wonder: In what way have they applied them? Was it better or worse than mine? The greater the similarity in our background and the difference in our way, the higher the stakes. With Paul and me, they reach the house limit fast. We are both intense, white-guy baby boomers, overeducated during the 1960s. We both style ourselves politically progressive. We are both professors trained in cultural anthropology (though he more exclusively than I), even in similar subspecialties back in graduate school. But from that similar beginning we have moved in very different directions. It probably never bothered him as much as me.

This sort of estrangement is hardly uncommon among academics, and people who read research should make allowance for it. Human differences show up in the work. They are more than mere matters of personality or expungable bias.[1] For example, compared to Paul, I am confident that I worry much more about being "nice." Even though we both fashion ourselves in opposition to the status quo (especially when it wears a suit), I do not think I relish the position with his enthusiasm. And since it bothers me that, even when restrained, I strike some people as coarse or judgmental, for their imaginary benefit I am outraged by anyone more abrasive than I. This difference in disposition is evident in our methods as well as demeanor. In particular, it is much tougher for me to move from cultural observation to criticism, from finding my feet with people to stepping on toes that seem out of place. I first want to be sure I have given everyone the benefit of the doubt. Paul has no such hesitation.[2]

I also fear conflict with him because he is more proficient at professorial one-upmanship. He pursues academic prominence and embraces it as just reward. I am more reluctant to seek it and feel absurdly wounded when it goes to someone else. It does not help, either, that we have a different hold on our credentials. Colleagues in American Studies give me some credit for being in tune with anthropology, but I hold an interdisciplinary Ph.D. Whatever fields I favor in practice, I remain resolutely ecumenical toward the denominations of academe. So Paul and I both know that, if there should be a difference of opinion, diplomas are waiting in the ammo box: "Well, if Horwitz were *really* an anthropologist, he would know better." At the very least, I know that from the moment I arrived on campus, many times I sought his expert opinion, even sitting in on his classes, but he has not once asked for mine. Since his exchanges on committees and in print have their share of intrigue and bile without me, it is probably just as well that we work apart. In such ways the building blocks of knowledge come in unmatched sets and stack in parallel play.

But given common interest in pig politics, a quarantine would be hard to sustain. Sooner or later our paths would cross. He would interview people I knew, or vice versa, and I wanted to be sure that no one would confuse us. In particular, I worried that he (in being mistaken for "we") would jeopardize the goodwill that I had developed among industry leaders. I needed their referrals for work with PSF (and, I suppose, their assurance that I had been "nice"). If he were as new to the subject and the subtleties of American history as I suspected, he might easily ally with one of the "sides" that hog war combatants avow.[3] He might be ripe for populist picking, the very side that could close PSF doors. Of course, all three of these people were free (in fact, well paid) to form their own opinions. I had neither the ability nor the desire to control them (though, you might think they would welcome my advice).[4] Rather, I wanted to make sure that we agreed to respect our differences, in particular to make it clear to informants that we were working independently.

Even if Paul and I could summon sufficient civility to agree on that objective, accomplishing it would be tough. For most people, if you have seen one professor, you have seen them all. By definition, they assume, "academic" differences are without substance. In this case as in many others, that is a poor assumption. Research is less immaculate in conception than Jesus, and some folks have serious doubts about him. Furthermore, since so few University of Iowa, "Sin City" types (vs. ISU profs) would study swine, a reasonable inference—one potentially fatal for my work with PSF—would be that we were working in tandem, maybe even in secret, good-cop/bad-cop conspiracy. Of course, at this point my concerns were based only on generalities and tizzied speculation.

As soon as I finished lunch with Mike, I began calling around. I first checked with old contacts to see if there already was damage to control. I called a few prominent people in swine-related policy, research, teaching, and consulting operations of USDA and land-grant institutions as well as the offices of the national and a couple of state producer associations. Most of them already knew or knew of Kelley, but they knew nothing of the other two. No problem. I then called Kendall, Kelley, and Paul to see if we could meet. How might we help each other out or at least stay out of each other's way? We agreed to chat over a couple of beers at Joe's Place, a bar nicely perched halfway between Paul's office building and mine. I agreed to send them a long and a short version of the proposal for this book, and they sent me their press clippings.

One glance at their press, and I knew that my worries were warranted. They had penned an op-ed piece for the *Des Moines Register,* the state newspaper of record, under the title "Large-scale Hog Farming vs. Quality of Life." So, two sides.[5] In case there were any doubt about which side they were on, another headline settled it: "Large Pig Farms a Threat to Jobs, Researchers Say." Journalists credited Paul and Kendall (no mention of Kelley) with these findings and portrayed them as sober, disinterested scientists. But they sure suited some interests better than others, and actual research did not seem much of a constraint. After their "look" in North Carolina, for example, they were quoted as knowing that a claim of a Continental Grain executive (that a $50 million investment would create jobs) was "laughable," even though he was speaking of northwest Missouri.[6] James Romesburg, one of the organizers of Neighbors Against Large Swine Operations (NALSO), took credit for helping arrange for Kendall and Paul to visit North Carolina. Then he boasted, "I had no idea they were going to back us 100 percent. Exactly what they said is exactly what we've been trying to tell the whole community."[7]

The clippings made it clear that this was basically a Durrenberger/Thu, professor/former-student operation. Their visit to North Carolina just a few months ago was very short, and it was their very first foray into the subject. From the start, their exposure was supported by organized integrator resistance. Despite some assertions to the contrary, I believe, they bought the two-sides story and were drafted or volunteered for one-side service. In doing so they fashioned themselves at once as high priests of neutral science and as bold allies of good little people locked in life-or-death struggle against greedy, dishonest, home-wrecking titans.[8]

I am surely overstating their published position, but parts of it were just as surely incendiary. No doubt, some friends (unfortunately, mainly the ones who also go on about "New York bankers") would love their line, as would more constructive, beleaguered progressives. They need all of the help they can get, even if the quality is questionable. It always is, anyway. But most of the farmers and industry leaders whom I know, if they read this stuff and thought it came from the likes of me, would stop returning phone calls. Of course, they understood that I was not obliged to sing anyone's praise, but they could trust that I would not seek headlines for myself or feed popular passions with scientistic fuel.[9] At least if I thought they were liars they could trust that I would both base that conclusion on something I knew about them (vs. a caricature or statistical category to which they belong) and give them an opportunity to respond before it went to reporters. I was seriously worried that these researchers were already backed into a self-righteous, populist corner and that Paul would come out swinging. As I headed to Joe's Place, I was hoping that the press got it wrong or that, in any case, beers would lighten the mood. Maybe I could preoccupy myself just getting to know them, as if they were also research "subjects." I aimed for my more forgiving, ethnographic mode, but the risk to my pride and to my interest in PSF kept me on guard.

As we assembled in a backroom booth and sipped a beer or two, the conversation flowed easily. Kelley Donham was as amiable in person as he was on the phone. It was immediately clear that he would not be very hands-on involved in whatever pig projects evolved around him. I figured that Kendall would be the one to help earn his salary by spinning off grant proposals (say, on this pig business) and that Kelley would contribute his name and administrative support.[10] Such arrangements are routine among academic entrepreneurs. Since completing path-breaking work on the poultry industry a decade ago, Kelley had earned a reputation for careful research on pressing health issues in agriculture. His name carried weight in the world of funded science. Despite a quasi-populist op-ed piece in a veterinary magazine (that struck me as naive, although far from uniquely so), he could easily agree that hog wars had more complex implications and that opinions were polarized prematurely. He mainly wanted to learn if there were ways to get on with the work itself.[11]

I also found Kendall pleasant, modest, and generous. Even though he was Paul's student, he obviously had independent judgment, and he seemed genuinely interested in sharing experience and confronting bias. Like me, he had grown up with affection for rural ways in the United States. His prior professional experience, however, was almost exclusively European. There, in the familiar ethnographic style that is also mine, he had started working from the ground up, hanging out with dairy farmers in Norway to understand agricultural issues. Unlike me, however, he decided that on-the-ground details were a potential "distraction" from understanding "larger processes and forces." So he jumped directly to "the top," to policy elites whose influence, he concluded, was easier to track, decisive, and poorly understood at the local level. Of course, I disagreed, arguing that work allegedly on behalf of the grass roots can hardly bypass them. It would take some powerful persuading to convince me that you can opine expertly on dairy production while discounting the experience of tending cattle. I am more reluctant to consider social-science abstractions "larger" than everyday life. He generously acknowledged that other researchers, perhaps like me, would have taken a different tack, but he was pleased to have learned as much as he had in his own way and to extend his experience to other places and branches of agriculture. He would be happy to learn as much as he could about hog farming, especially because they had barely started on the subject and were still figuring out what to do next.[12]

For about a half-hour, even after Paul showed up, the conversation slid along an easy, getting-to-know-you groove. Kendall and I exchanged views and joked around, while Kelley looked on with quiet interest and Paul with mute scowl. He slouched in a corner staring at his glass. Although his silence and irritated expression gave me pause (was he plotting something?), I was glad to babble on about swine disease and such, doing my part to clear the air. We quickly agreed to advise informants that there was more than one group of field-workers from the university and that we might well be different. I still had my doubts whether such advice would suffice, but I was greatly relieved that we at least got that far. When I pressed Paul for his explicit approval, he grunted in assent. So far, so good.

That item of business—mainly mine—complete, Paul suddenly became very animated. He turned our attention to ways that we might cooperate in the future. Might I, for example, want to join a panel of presentations that he was organizing for an anthropology meeting?

It was a generous offer. I was grateful, but what, I wondered, would it be about? Would I have to be associated with a "side"? After all, even after working quite a bit longer than they, I was not ready to promote any party line. At the very least I wanted to explore the complications and to assess carefully first on my own how much PSF might (just might!) be avoiding the North Carolina horrors. The work I had been doing entailed methodically working "up" the ladder of agricultural expertise. Get to know the farm work itself, and then pursue connections in agricultural equipment, medicine, science, and policy, with due regard for the diverse people involved and for complex modes of interaction. Since this has been a fairly conventional strategy, a defining one in ethnography for most of the twentieth century, I did not expect it to get me into trouble.

But the ensuing discussion did not go well. Paul quickly let me know that he did not think much of me, or at least what I had learned and the pig people who taught me.[13] As far as he was concerned, we were all likely deluded. Elites are in the business of hoodwinking the public and, since they surely succeed, it is the job of Serious Social Science to set everyone straight. "Hard data" rather than limp, "mere psychology" will set us free. Hence, a quick breeze through the countryside on the way to tabulations would be more than enough.

In fact, at that point that was about the full extent of their data. Prior to the summer of 1993, they had not given swine a thought. A lawyer, organizing and representing opponents of big hog companies and seeking prestige endorsements, contacted Kelley's institute. She invited researchers to spend three or four days in North Carolina and judge for themselves. Kendall was in charge of the response, and he contacted his mentor, Paul. They would be the two to go. Four citizen's groups, also organized to fight the integrators, arranged an itinerary. Of course, these sponsors would allow them to move freely and to form their own opinions. Also during those few busy days, Paul and Kendall supposedly "randomly interviewed" people in spontaneous side ventures and tried—albeit without success—to meet some integrators. But activist hosts expected returns that would bolster their side, and they were not disappointed.[14]

I do not mean to imply that anyone did anything dishonorable. God knows, my own data are soiled, and I share their progressive bias. It does not take a Ph.D. to smell some serious shit happening in North Carolina. It has been well covered in the popular press and trade journals for years.[15]

But dressing this up as holy Research seems an abuse of authority, like Dr. Science or Irwin Corey without a sense of humor. For the record, their three-day excursion became "the summer of 1993." Their off-the-shelf itinerary became "a fact-finding trip" and "an ethnographic survey," thereafter cited as *vox populi*. With their activist guides Paul and Kendall "drove over six hundred miles" to meet with "over one hundred farmers, business men, retirees, homemakers, factory workers, pastors, construction workers, mayors, and many other rural folks." That sounds like a lot until you imagine how well you would get to know so many people in a few days, especially if they were all that diverse and if much of the time were spent getting from one place to another. More than half the hundred came from a single meeting where the agenda, courtesy of local populists, was discussing large-scale hog operations. I gather that Paul and Kendall were surprised to find that people who attend such a meeting have complaints. From this base and a little subsequent numbers crunching, the two were coming to the breathtaking conclusion that giant companies were powerful and that local communities were not benefiting as much as promoters promised. They took this finding on the road to meetings, op-ed pieces, and radio talk shows.[16]

On the plus side, I thought, at least in some respects we both value meeting the folks we study. Maybe this could be a basis for cooperation, something that would appeal to academics attending a conference. After all, I had spent nearly twenty years working with people who had spent a lifetime practicing something like what Paul and Kendall aimed to survey in a few days. But when I offered to share some tips on what I learned—not from hypothesis tests, briefings, or proto-rallies, but from the workaday world of swinedom—Paul muttered, "Bullshit."

Now that the conference session was out of the question, he offered another idea: What about working together on a grant proposal? Turning to Kelley, Paul asked, "What sort of research design would work on the odor problem?"

I found it an intriguing question because, purely by happenstance, I know a little about odor research. As it turns out, Amos Turk, a pioneering smell chemist at City College of New York, was a longtime neighbor of my mother-in-law.[17] Apparently, her discriminating nose qualified her as the research volunteer of choice in his work. She told me all about it. Amos would call her down to his makeshift laboratory in his home on Tarrywhile Lake: "Pass your nose over these sample vials, Shirely. Sniff and tell." So, the state-of-the-art laboratory smell-o-meter (well, twenty years ago) was, in fact, my mother-in-law's schnozzolla.

Since Paul was eager to get a reply from a real scientist, I resisted the temptation to share this anecdote. I doubted he would see the humor in it. But Kelley did not have much to say yet, either. He offered bromides about control groups and funding.

No doubt with childishly spiteful sarcasm, I began to talk about Dr. Turk and Shirley's nose, but midtale Paul pushed the discussion back to the grant money.

Kelley knew that, under hog war pressure, tens of thousands of dollars would soon be available to people doing research on livestock odor. Litigants and regulators wanted objective measures. The first round of grants competition had something like a $40,000 prize.

I said, "You mean, you guys want $40,000 to figure out if shit smells? And what *I* do is bullshit?!"

"Yes," Paul said. "You need the data. That's what they listen to, and if you think otherwise you are just naive."[18]

I started to explain why I thought *that* was naive. "Who do you think is going to set those 'objective standards' in the end, anyway? Who has the money to buy the science and to pressure the policymakers and regulators, right? Why go down that tired road? Why not, for example, as an expert, tell them—whoever 'they' are—that odor is subjective? Things stink when people say they stink. That is all there is to it. Why foster the fable that numbers and procedures will make human differences go away?"

"Bullshit," he said.

That was pretty much the end of it.

For the next few minutes before awkward good-byes, Kelley and Paul discussed sample design and such, pretty standard "Intro to Sociology" stuff. I sat there, quiet, hurt, and angry with myself for letting something so predictable bug me. After those two left, Kendall and I chatted awhile longer. He could tell I was upset. Even though he was eager to get home to his wife and their newborn, he took time to offer comfort and to emphasize his interest in sharing what we learn in the future. It was very kind of him.

I was pleased, too, that Kendall followed through. He sent me some of their findings even before they went into print. As I sat down to read, I reminded myself to be fair. After all, like Paul I often disagree with research subjects, even occasionally find them deluded. I, too, am outraged by swine news from North Carolina. I, too, depend on hard data for understanding. If forced, God forbid, to choose between generalizations based an anecdotes or press releases and those based on systematic surveys, I, too, prefer the surveys. I am indebted to people with the boredom tolerance to conduct them.

But even with that provision, I was shocked by howling holes in what I read. Apart from inflation in claims about their trip—that three-day, populist junket turned summer of disinterested research—there were a host of technical problems. Kendall and Paul purport to test the claim of integrator boosters that they enrich the countryside. It is a crucial question because, if big guys do not spread their wealth, why should the public welcome them? And Paul and Kendall suggest that the public should not, because the claim is wrong. To prove the point, they show a statistical correlation between intensified production and indicators of, not prosperity, but poverty in rural Iowa counties. In other words, the higher the ratio of hogs to humans, the larger the share of the surrounding citizenry that are poor. Such a pattern is too strong to be reasonably supposed the consequence of pure chance.[19]

I had some misgivings about the way they arrived at the pattern in the first place. Little technical things, like the way they defined variables, gave me the impression that they were cooking the books. But then again, maybe I was jaundiced from reading too many contradictory studies of this sort. And there was that chip on my shoulder. To be fair, I should be no more than a little suspicious that some of their measures (such as those they used to define the difference between large- and small-scale operations) were questionable. Reasonable people disagree about these things.[20]

But even if they are right about the existence of a pattern—that big guys and poor neighbors go together—why should they suppose that one caused the other?[21] As every elementary statistics text warns, the leap from correlation to causation is a big one, infamously prone to abuse. There are all sorts of reasons why any two things might coexist. But instead of considering all of the reasons that poverty and intense production might share terrain, they just start inflating verbs. First, they wonder if the two "are related"; then they find they "are positively correlated"; and finally: "By virtually every measure, large-scale hog operations consistently create or intensify negative socioeconomic relationships resulting in heightened food stamp usage and more poverty. . . . Farmers' skepticism of the local benefits of expanding swine production has empirical support."[22] For me, anyway, that leap from relation to creation is short on the landing.

A plausible alternative comes to mind. Perhaps poverty set the stage for intensification in the first place. In other words, maybe it is not that the megafarms "created" the poverty but that the poverty attracted megafarm investment. The big guys readily admit that they prefer to locate their facilities where there is a surplus of rural labor and relatively low land values—in other words, poverty. In fact, unless Kendall and Paul analyze data from more than one point in time, they cannot discount the possibility that, as bad as things are around large-scale operations, they were nearly as bad or even worse before.

Or there is another alternative. Maybe, for example, large farms are the ones that keep poor people around, while small ones force them to leave. Sons and daughters who do not inherit Dad's farm or who cannot find wage work nearby migrate to the city, where they become "urban" rather than "rural" problems. Megafarms might offer just enough encouragement to keep poor people in place or to attract workers who are short on options. Under such conditions, poverty does not grow or diminish so much as it maintains or changes address. Only in primitivist fantasy is suffering eradicated when it moves to town, but Paul and Kendall allow just such an illusion. By their measure, if a poor person leaves a rural county, poverty is "reduced"; if that person has a crummy job (say, working for a contract feeder) or enough public assistance to stay, poverty is "created." To distinguish between the migration and the creation of need, Kendall and Paul would have to track individuals rather than counties and beware urban-rural relations.

Admittedly, it would be difficult to assemble data that would credit one interpretation over another. A historical or longitudinal study, tracking actual humans rather than counties across time and terrain, might be prohibitively expensive and still short of convincing. Better something than nothing. But I am less tolerant of their glossing over alternatives with a rhetoric of creeping probables. What connected their correlation with "empirical support" for skepticism was at least as smelly as anything I was scooping. And at least I did not claim to be doing holy science and accuse opponents of laughable lies. I readily admit to people in the industry that I am suspicious of the Cargills of this world, and they appreciate my candor. Such feelings are common enough, even among megafarm executives, that they are comfortable dealing with them honestly. But Kendall and Paul were coming on like crusaders, and I suspected that would piss off people at PSF.

So, I concluded, I had better not let my list of PSF contacts wait till summer. The less time for things to go wrong, the better. The first call went to the communications director, Charlie Arnot. I figured that I would introduce myself and see what sorts of arrangements would be best. Since I could not follow through for a couple of months, there would be no pressure at all. We could take it easy. Mutual friends assured me that we would get along just fine, anyway. When messages left with a secretary went unanswered for a couple of days, I did not give it much thought. He is a busy guy.

In the meantime, I went ahead and called some other folks whom I knew at least indirectly, starting with a couple of farm managers whom friends recommended. But when I reached them, every one told me the same thing: "You'd better go through public relations. Everyone will tell you that now. We are not supposed to talk to anyone on our own." I knew right then that something was seriously wrong.

When I finally did manage to get Arnot on the phone, he confirmed my every fear: "No, no. Sorry, no. Why should I talk to people from [the University of] Iowa, anyway? When you are going to misrepresent what we do? After we welcome you guys in here? And you print out-and-out lies in the newspaper? No, I won't speak to you people or allow anyone to speak with you until I can be assured that your work will be sound."

I backpedaled a little and then pressed. I had every intention of producing sound work. I did not know whom he met before (though I could guess). In any case, I work solo—no grants, no sponsors, no itinerary, no strings. I, too, would make up my own mind, but only after giving PSF a fair shake. I was interested in the company mainly because I thought it might be better than its Tarheel competitors. So insofar as bias was an issue, it favored PSF. I would not publish anything without a formal release, anyway. I could show him proposals and such, whatever he needed to make up his own mind. I gave him the names of people whom I know he trusted and who would vouch for me.

But he held his ground, and you could hear the anger in his voice: "We already had two guys down here from your place, Durrenberger and Thu, and . . . I won't do that again. I can understand that people might disagree with us. But they were irresponsible. Utterly irresponsible! We welcomed them here, showed them around. And then they turned around and went very public, misrepresenting facts that we gave them. I can understand science, but what they were doing was just bad science. I tell you, ridiculous, bad science. Totally irresponsible! So I have to be very reluctant to let anything like that happen again."

I eventually convinced him to look at proposals and samples of my writing about the industry, but in the end it did not work. The director of communications for PSF determined that there would not be any.[23]

When I hung up on that last call, I was beside myself with anger. A couple of fly-by geeks had fouled the nest that I had been tending for years. Struggling to channel my rage, I decided to call Kelley. He should know that his associates had just cut off a source of data that he might want for that $40,000 prize. They had blown the only decent shot I had to give integrators a fair hearing. Besides, why should university people, especially from a tax-funded institution, treat the public, heck, anyone so shabbily?

Kelley just let me rant on like that for several minutes. When I realized that I was going overboard (and that he might be holding the receiver away from his ear), I stopped, and he responded. "Yes," he said, "it's too bad. Frankly, I cannot be too surprised. In fact, a bunch of producers went to the president's office and complained. You know: 'Is the university trying to kill the hog business?' But, as I told the vice president for research, yes, maybe things could have been handled better. But I bet something like this would have happened anyway, sooner or later. We had the same experience with poultry. First, the companies are willing to cooperate. But when the press does not go their way, they get paranoid and stonewall. That's just the way it is."[24]

I had to admit that Kelley was probably right. Mixing academics with hog shit will likely produce offensive material, regardless. Paul and Kendall probably just hastened the inevitable. Besides, there was nothing they could do about it now.

But what could I do now? I did not know of an equally "top-end" integrator, much less one within a day's drive and with a staff whose friends would offer introductions. I had planned it all so perfectly, building up to the ideal case that might return some humanity to a debate turned jihad. It was, I still thought, such a good idea.

As I was poking through those last embers, in came yet another bucket of cold water. Rumors had long circulated that PSF was in financial trouble, that its good-citizen pose was just too expensive to maintain. Since that notion implies rampant insanity among the company's many investors, I never took it very seriously. Apparently we all should have. On July 2, 1996, Premium Standard filed for protection under Chapter 11 of the Federal Bankruptcy Code. The precipitating event was a cash-flow shortage. On March 15, 1995, a $7 million bill for interest on loans came due, and PSF could not meet it (even though up to that point owners/managers, Dennis Harms and Theodore Gordon, Jr., had received annual salaries of more than a quarter of a million apiece plus bonuses). In the file, it became obvious that this flagship had been listing to port badly for a long time. Despite marketing half a billion pounds of pork per year, there were $570 million in liabilities and only $540 million in assets. So the company was $30 million in the red. Their couple of million Missouri hogs had lost about $50 per head in the prior year. According to the proposed reorganization, PSF would, at least temporarily, become "Newco." Harms and Gordon would no longer own or manage the company. Instead, creditors (who had supplied 99 percent of the capital, anyway) would assume 97 percent of the management. The job of rescuing the initial investment would fall to previously silent partners: Putnam Investment Management, The Prudential High-Yield Fund, Loews Corporation, Cypress/GEM Capital Management, and Hanawa Company, Ltd., hardly model yeomen.[25]

So my flagship exemplar—the best possible face that I could find for the integrators—was hard aground and taking on water so fast that pilots were jumping ship. Activists were joining mutineer investors, albeit for different reasons, to keelhaul officers, who had little chance for satisfaction beyond stonewalling the likes of me. Yet again, progressives were licking their chops: "See, we were right! Integrators cannot deliver." And realists were hunkering down: "See, you cannot think pie-in-the-sky! Citizen, shmitizen, PSF has proven what a hardball game it is." What a mess.[26]

Fact is, my heart was never entirely in hog war coverage, anyway. It was essential for the stories I wanted to tell, but it was terribly prone to cliché. Coverage from the cultural left and right seemed equally unenlightening, yet another round of Scrooge or Magwich vs. Tim and Pip. No matter what I had to say, people just wanted to know which side I was on. While I was aiming for cultural nuance, they were asking for loyalty oaths. The idea of gearing up for yet another go-round was just too much to bear. So my bitterness toward Paul and Kendall became mired in funk. I was angry to find myself barred from turf that was unattractive, anyway. Then, of course, I regretted the absurdity of such anger.

It is embarrassing to admit how long that mood stuck. More than a month passed before I began to note parallels between such academic frustration and variants around the farrowing house. I began to see how my PSF case study was like a sow sick with TGE. I had allowed myself to become preoccupied with finding someone to blame or a phantom cure. Farm experience should have taught me to live better with the likely futility of both. As Roger and Phil remind me, you might as well recognize: Sometimes shit just happens.

Late one night I awoke from another nightmare about Paul and Kendall, realizing that a story of the ruin of my PSF study could be just as useful as the study itself. In a way, they had done me a terrific favor. Setting it up would allow me to provide most of the nuance I wanted. It also would provide an opportunity to clarify ways that my own understanding of farming, disease, and such differs from that of other "experts"—why, for example, I am so reluctant to advocate a hog war side, even one with scientific endorsements.

Although I am as convinced as anyone that I know things that others ought to consider, I am less confident that those things especially deserve "implementation," at least in advance of a chance to kick them around in light of diverse, day-to-day experience. In this respect, Paul and Kendall are ready foils. Most of their pronouncements ooze a presumption that science is intrinsically superior to lay experience and common sense. Given enough time and money, they say, the bright light of pure scientific fact will burn off the sooty clouds of folk belief. Then we will really know what to do. "The downtrodden" particularly deserve such devoted, disciplined disinterest.[27]

Of course, this item of devotion is itself a pretty common article of bourgeois faith. When people get to screaming at each other, cool, pragmatic, and principled heads, like those of the academy and kindred bureaus, have work to do. Among the worthy traces of such noblesse oblige in America is the tendency of academics to amplify voices that are easily overpowered (albeit by the same process that reproduces academic noblesse). Furthermore, no one could doubt that many sorts of questions are answered effectively through methodical measurement or experiment. What are the health risks of ingesting one thing or another? How much does a particular technology cost? What kinds of regulations have best accomplished particular goals? These are fairly limited, straightforward questions that deserve answers in kind. They make great anecdotes about the past, ones that could well be consulted in the future. But it is, I think properly, the creative process of democratic sociality rather than the dicta of scientific priests that turns these anecdotes to prophecy. Matters of folk belief cannot be "factored out." They are necessary to render these anecdotes relevant and instructive in assessing the quality of our lives. In this respect, scientists are neither better nor worse than the rest of the folk. Sometimes shit happens; sometimes it doesn't.[28]

Anecdotes from the history of social science itself counsel such wariness. The number of times that anthropologists or sociologists have intervened effectively on behalf of the oppressed is small, comparable with the number that have gone horribly wrong. The struggle of Amish Americans for segregated schools or of African Americans for integrated ones stand on one side and U.S. policy toward native peoples on the other. It is small comfort that such expert influence has been decisive only in combination with other pressures. During World War II, for example, the Office of War Information (OWI, later the United States Information Agency) employed academic units like the Foreign Morale Analysis Division to analyze enemy culture and anticipate the consequence of Allied tactics. At issue was not just the quality of life in a few hog towns but the survival of hundreds of thousands of people, and the government, for once, actively solicited scholarly advice. OWI head Alexander Leighton assembled the largest number of elite social scientists ever to serve the U.S. government. But historians have been unable to find any evidence that policymakers ever read, much less followed, the division's recommendations. Leighton himself concluded: "The administrator uses social science the way a drunk uses a lamppost, for support rather than for illumination."[29]

Sound methodology seems less important than a convenient conclusion in making for "influential" social science. And on that score, Paul and Kendall do have something going for them. Most Americans form their opinions from a plainer potpourri of information—from conversation with family or friends, Sunday sermons, Hollywood movies, and TV shows more than T-tests—but they are similarly prepared to blame social ills on rich and powerful outsiders. It just baffles me that Paul and Kendall, like many populists, think that they are bucking or rescuing dominant ideology in distress rather than riding it. Of course, it may be the only ready ride around, but I hope to have helped readers consider options beyond go and whoa.

It is fitting, then, that in thinking about how academics best relate to agricultural change, we use contrasting historical precedents. The model for Paul and Kendall is a book, *As You Sow* by the anthropologist Walter Goldschmidt. It evaluates the well-being of California communities under the increasing influence of corporate agriculture in the 1940s. Thanks to the big guys, things ain't what they used to be. The good and old got beat up and went. With charts and tables and all, the book is, I think, pretty dry reading, but its stand-up-for-the-little-guy moral and the subsequent struggle for and against suppression of its publication make for appealing drama. The big guys went after the messenger in a big way. But Goldschmidt, the earnest, tough-minded social scientist, was up to the challenge. For more than a decade he battled corporate America and its government cronies. He would not abandon the meek suffering back in the hinterland of Lotusland. With determination, virtue, and a couple of liberal U.S. Senators on his side, Goldschmidt prevailed, at least in fighting off his censors. The book rightly remains a classic a half century later.[30]

The historical precedent that I prefer also entails the assessment of agricultural change and social science brouhaha in the 1940s. As you might guess from the popular version of its title—*Pamphlet No. 5*—the publication I have in mind is also far short of a thrilling read. Though mercifully short and inexpensive, it never had wide circulation or much consequence for agriculture itself. But the mere idea of the pamphlet stirred passions and changed lives, especially in academia. The whole episode became known as "the Great Oleo Controversy."[31]

In the early 1940s various government agencies tried to free up whole milk for the war effort by encouraging American consumers to substitute margarine for butter. If only because oleo was a newfangled food, consumers were a bit resistant and, of course, dairy farmers were appalled. Their livelihood was at risk. Since at the time Iowa was a dairy state, the governor asked the president of Iowa State University (then ISC) to round up a blue-ribbon panel to assess the potential impacts. Since at the time ISC also had the most distinguished faculty in economics in the United States, the president knew exactly whom to ask. Everyone awaited what they assumed would be a story fit for Chicken Little, a call to arms.

What they got instead, though, in the form of *Wartime Farm and Food Policy, Pamphlet No. 5: Putting Dairying On a War Footing,* was a calm confession of faith in the invisible hand with which free markets mercifully (presumably) rule. Basically, these world-class economists figured, everything would be fine in the end. Not to worry.[32]

But dairy herders, the Iowa Farm Bureau Federation, and the governor went ballistic. They demanded the withdrawal of the pamphlet or the head of ISC president Friley or both. What they got instead was a public relations nightmare centered on the issue of academic freedom. Writers for *Time* magazine, *Reader's Digest, Harper's,* and *The New Republic* portrayed the academic authors, Oswald Brownlee and all, as beleaguered heroes and their dairy-defending persecutors as petty rubes. Of course, no one much discussed the economic interpretation itself.[33]

Eventually president Friley was able to broker a retraction of sorts, but at considerable cost. Although most of the key players denied that the oleo fight was decisive, six faculty from the authoring department left just six months after the brouhaha began. Within two years more than half of the department's twenty-five members were gone, including Brownlee as well as the chair of the department and every member of the review committee for the pamphlet series.

Among the reasons this story is worth remembering, though, is a fortuitous twist it takes. As it turns out, in their disaffection these economists did very well for themselves and for their field. Theodore Schultz, who was the head of the ISC department in 1943 and the most vocal defender of the pamphlet, moved to the University of Chicago, where he also became chair of economics and won a Nobel Prize. Following him from Ames to the Windy City were at least a half-dozen former students and colleagues (including Brownlee) who helped make that university the new champion of academic economics in the United States. As it turns out, too, Iowa farmers did pretty well. Dairy farming declined, but soybeans became a prize, dependable crop, which it remains to this day, in part because margarine took hold. About half of the oil in oleo comes from soybeans.[34]

Of course, the Great Oleo Controversy does not prove that hog farming should blow with the big-guy winds. As I have explained, there is still good reason for resistance of many sorts. Public policy should include a skeptical, watchful, and protective eye. But I hope we can also agree that tracking economic winds is short of an exact science. There is room for both resistance and a good deal of happening shit between patterns of the past and the future.

Moreover, the recent past has been far enough from utopia that its passage can be considered a mixed blessing. People who know and love farming are among the first to emphasize that it is also dirty, dangerous, risky business. Even want-ad recruiters admit that traditional, hands-on hog work demands people who are "young and fit. . . . Much bending and lifting involved. Tender loving care and record keeping . . . in freezing conditions."[35] As the average age of American farmers advances, old-timey styles lose what little glamour they had.

We cannot be surprised that many of them, albeit under pressure, are willing to consider previously profane alternatives. Maybe it is time to network with neighbors and contract with the big guys, get big enough on your own to hire a kid or an immigrant to handle backbreaking chores, or just get out of the business. In fact, these were among the very alternatives that my neighbors, including the Stutsmans for whom I worked, were weighing around the time that PSF filed for bankruptcy protection. Since my role had evolved to that of a bit of a swine specialist, these alternatives struck home. If the Stutsmans decided against networking or expansion, I could lose my job. Since my wife and I still had other jobs, our family could handle the loss of income. It never amounted to much more than mad money, anyway. But I was deeply worried. Farm work meant the world to me. But that world was changing. Megafarm executives like

Murphy and DeCoster had it by the tail, and I could get the boot. Nevertheless, even as these changes cut so close, I could not consider them pure cases of either progress or pathos.[36]

It is, I still think, American Yuppies more than yeomen who cling to the belief that food traditionally—normally, properly—comes from perennially happy, independent bumpkins. More people want their food to come from peasants than want to live like one. They seem shocked to discover that agriculture for urban humanity entails massive institutions, staggering amounts of money and alterations in the environment. Everything grown for consumption is tied to the soil; everything you ingest is "dirty." Everything just might get sick, and everything dies. Every meal and every job is connected through food to the city as well as the countryside, to death as well as life, to money, boards of trade, government, petrochemicals, pollution, packers, and pathogens. One way to temper the outrage of hog war activists is to recognize that those connections can be changed, but they cannot be willed away.

Of course, it is always possible that the populists are basically right, that the big guys threaten to impoverish us all. The world might be a better place if we just say "whoa." At least I might be a little more likely to keep my job on the farm. But I am far from confident that current arrangements are any better suited to the long haul with its intricate connectedness than the one that integrators hype. Maybe I am too fearful of biblical warnings about idolatry to invoke a Messianic Age for anything. Whether the age is allegedly next year or yesteryear, the idea seems equally vain. But I think I learned my wariness less from reading any scripture than from trips to the farrowing house.

Hog farms may remain among the best places to learn how shit happens and how to slog through it. They offer plenty of lessons for the folks who grow up there as well as academics, integrators, and activists, if they pay due attention. Apparently even the high-tech titans of Premium Standard Farms were lax in that regard. Press coverage of their bankruptcy proceedings indicate a near-fatal combination of vanity and happening shit. Their huge size alone made them vulnerable to fluctuations in feed and hog prices. Just a ten-cent rise in a bushel of corn would cost PSF $1.4 million per year; a $1 drop per hundredweight of hogs, $4.7 million. But the cost of corn rose 1994 to 1995, not ten cents but more than $2 while the price for hogs fell by $15. PSF was also vulnerable to its rate and method of growth. Executives were trying to tough it out with highly leveraged capital, "junk bonds." They were having trouble finding cash or short-term credit to cover extraordinarily high interest payments, and they could not keep qualified personnel. But they also had problems of the same sort that plague Mom and Pop. Despite huge investment in state-of-the-art expertise and technology, PSF had not achieved the size litters, rate of growth, or feed efficiency to justify the financial risks it took. Despite vigorous implementation of all those swinely acronyms and tireless biosecurity, disease remained a problem. In explaining how the company failed, "PSF also noted serious outbreaks of TGE in 1994" and a recurrence complicated by PRRS in 1995. Yet again viruses are last-resort reminders of the humility that husbandry requires.[37]

PART IV

Stepping Back: Hog Farm Memories, Fair and Foul

Cultivators of the earth are the most valuable citizens. They are the most vigorous, the most independent, the most virtuous, and they are tied to their country, and wedded to its liberty and interests, by the most lasting bonds.

—Thomas Jefferson writing to John Jay
from Paris, 1785

I think it would be quite untrue to suggest that barnyard creatures are dependent on each other. The barn is a community of rugged individualists, everybody mildly suspicious of everybody else, including me. . . . I do hope, though, that you are not planning to turn Charlotte's Web *into a moral tale. It is not that at all. . . . It is, I think, an* appreciative *story. . . . It celebrates life, the seasons, the goodness of the barn, the beauty of the world, the glory of everything. But it is essentially amoral, because animals are essentially amoral, and I respect them, and I think this respect is implicit in the tale.*

—E. B. White anticipating the film
translation of *Charlotte's Web*

Man found that the rationalization of the hunt disappeared when he started penning in animals as a dependable meat source. Domestication had attendant pitfalls. It bred familiarity, even love. And . . . it occasionally bred vegetarianism.

Vern The Pig Man is not a vegetarian. He loves his pigs—literally to death. "How can you do it?" I asked him one afternoon.

"How can you get so attached to these pigs and then slaughter them?"

Vern's answer was to pick up his horn and play "You Always Hurt the One You Love." . . . He was being facetious, but it had a kernel of truth in it. The corollary of love is hate, and, given the right circumstances, pigs can generate a lot of the latter.

Moving pigs from one location to another is a case in point.

—Karl Schwenke, *In a Pig's Eye*

FIFTEEN

FARMING AND CYBER-SWINOLOGY

MOST MORNINGS I COUNT A JOB IN THE IVORY TOWER among my blessings. That admission may seem odd because I so enjoy farm work and people far from hallowed halls. I am much more suspicious of professors, of figurative than literal bullshit. If forced to choose, I would rather be bone-tired from work outdoors than nerve-wracked from high-serious blather, and that sentiment is well-known. For example, I interrupted an academic year spent in Hong Kong (where it was "The Year of the Pig") for a brief return to Iowa. A neighbor called to ask if I wanted to help sort from a crowded feedlot about a hundred cattle to ship to market. He hardly needed my help; at best, he "could use" it. And he knew that "university people" would be madly trying to reach me. My stay was short, jet-lagged, and busy. But he also knew that time with balking and charging steers, curses, and kicked-up manure would help me feel at home. Students as well as friends from the university who were eager to see me understood that, even as I might regret taking time away from them, I would jump at the chance, and I did.

Some of the lure is surely old-fashioned male bonding, "getting down and dirty with the guys." But not all of them are guys, and limits to the lure are just as predictable.[1] Farming is perilous both physically and financially. It is more likely to maim or kill you than coal mining, and the return on investment is apt to fall short of the savings interest rate at your local bank, which charges five to ten times that much when you need a farm loan.[2] The guys at the bank and a bevy of lawyers will get their money without leaving the office, much less getting dirty, even if you lose everything.

Profit and loss margins vacillate wildly, often in response to conditions that are distant, uncontrollable, or inscrutable: a drought in the Ukraine, a change in Japanese tariffs, a cyclical adjustment on the Board of Trade, a mutation in a pathogen, or the height that corn happens to reach when borers swarm through. As often as not, the voice on a radio in the café or aboard the tractor belongs to an agricultural economist who explains why, yet again, the market has taken a predictable plunge, even though he failed to predict it last week. The one thing we all can predict is that the market for farm expenditures, for borrowed money or new equipment, will fluctuate in anything but sympathetic rhythm with farm commodities.

Agriculture would certainly seem less attractive if my economic stake in it were normal. Among the blessings of an academic job is confidence that someone else (likely, I regret, my neighbors) will absorb the bulk of the blow from a change of two cents in the price per bushel or of one sentence in conservation regulations.[3] In a university that is ag-state funded, such changes also might incite a worrisome response—say, cautionary memos about thermostat settings or photocopying—but I know that none bears comparison with the prospect of losing credit, not only for next year's crop or for kids' tuition but also for groceries next week. And as much as I like work in open air and changing seasons, I know how hard it can be to face the Iowa version. Day after day in January, the thermometer might not get above five below; in August, under ninety. Thank goodness, university colleagues have encouraged my maintaining a "farm side" that is only a side. But the ones who have been supportive have generally seemed as out of touch with the substance of it as the cosmopolites who scorned it. For both their sakes, I hope I can communicate a bit of what farm life is "really like."

A lot of my impressions were formed well before I contemplated this book. Many years spent making a home, raising a child, and working in the countryside led me to believe that, even if I were hardly "native" (a contestable standard, at any rate), I was relatively knowledgeable. My family, for example, has long been included in farm community events: birthday, wedding, and anniversary celebrations, tours of local test plots, trips to check equipment auctions or new breeding stock, quilting bees for neighborhood newborns, babysitting at the church, dealing with runaway cattle, the PTA, 4-H, and T-ball. We have exchanged intimate confidences, seeing each other through bouts of ferocious weather and medical crises as well as hours of simple pleasure, sharing meals, watching TV, trading rumors and shopping ideas, running errands for each other, and the like. Of course, we have shared boredom, disappointment, and irritation with each other, as well. In these ways, farm families are more than part-time workmates or research subjects; they are loved ones. A head count at just about any of my birthday parties would substantiate the point: Farm folk typically outnumber academics by about two to one. When I whined yet again to my wife that after fifteen years working two jobs—one part-time for minimum wage with farm neighbors and the other full-time for high pay with high-power professors—it was getting to be too much, she advised me to cut back at the university. Her advice has proven sound. And in drafting a will, when we had to name an Iowan whom we could trust to handle our deaths, to take care of our son and everything we own, we named a farmer.

Of course, I doubt that these people feel "the same" (another contestable standard) about us. Although many of them also have read and traveled widely and do plenty of business, even work part-time in town, their ties to the farm, extended kin, the church, and the surroundings are full-time and lifelong. They have much deeper roots in the place and more intense and varied contacts with each other. No matter how long we work and talk together, almost anyone my age or older will know more than I do about every member of the Hochstetler clan, every inch of ground or Polled Hereford we might come across. Their relations in the neighborhood are given, born rather than moved into or chosen. They rely on subtle proprieties, allegiances that overshadow individual tastes, to organize all those claims on their attention.

Here most sorts of events seem to draw participants from well-scribed circles that are almost concentric: nuclear family, extended family, intimates since childhood, friends or business associates, and then sundry others. Together they comprise a "neighborhood" with personal links to a surrounding "community." These groupings are almost never discussed, but everyone seems to know and trust how they work. Of course, there are a host of complicating conditionals. Women have more responsibility for maintaining the circles and (along with hired men) constraint on their trespassing, while clergy and bankers have less. Divides between rich and poor or Catholic and Protestant can become barriers at almost any time. But for the most part, simple proprieties do work.

For example, one family can assemble (say, for volleyball on Sunday after church) and drive right by another dear family without feeling the need to explain why invitations were limited. Volleyball is for close kin, maybe a few others, but that is all. No offense will be taken. As you pass, just wave nonchalantly. Or better yet, maintain speed, eyes straight ahead, and casually raise the index finger of one hand on the steering wheel. That is the ordinary back-roads greeting. If, however, the function is larger (for example, a hog roast to celebrate the completion of a new machine shed or a high school graduation), you had better invite everyone within the friends-since-childhood circle. Even changes in the way the circles apply are likely to leave the boundaries in place. For example, in the 1960s routine labor and equipment swaps moved from an outer to an inner orb. People put extra effort into the exchanges that circles cannot easily encompass (e.g., rental agreements for cropland or responses to animal and crop disease) because for that very reason relations can be fragile. The point is that our family could not easily fit anywhere in the scheme, even with its regular allowance for exceptions.

As a result, I have had unusually good opportunities to trespass. People from different families have trusted me with fields, crops, animals, and equipment that are worth more than I could afford to buy in a lifetime, a privilege they rarely extend to each other. They have also trusted me with stories that ordinarily circulate only with measured self-restraint. I am flattered and indebted to them for opening up so much to me. On the other hand, it is worth remembering that the stories were still restrained. Even more important, the very conditions that allowed me to trespass also render my experiences idiosyncratic. I have been following serendipitous connections, surfing an unreal social net.

I can easily recall, for example, an episode when a neighbor, part of a father-son team, shook me out of lingering, gone-native fantasy. It was just a normal summer day when I aimed to reciprocate a favor. His dad was helping me drill a bolt out of an engine block, an especially generous act, given a midafternoon prime for making hay. He had mowed some alfalfa, I knew, yesterday and raked it late morning. It would be just about perfect to bale, right when I was a distraction.

When his son happened by in a pickup, I told him, "I hope I can help you guys bale."

I thought it a typical "neighborly" exchange, especially because we had helped each other out in the past. I had even helped him make hay a couple of summers before, when he was pressed by another neighbor to take care of a field that they were working on shares. Timing can make a big difference in the quality of hay, how much protein will remain, how hard it is on the baler, and how well it will keep. The peak period is always shorter than the one required for the job; so there is no such thing as too many hands.

But the son responded, "Oh, no."

I probably should have left it at that, but (trespasser that I am) I insisted: "Why? You know, I owe you. I don't want to be a mooch. Besides, Roger and Phil will vouch for me. I may still have a hard time keeping up with those guys, but I do okay. You know."

He responded, "Yeah, I know. But that's the difference between Roger and me. He'd hire a guy like you."

No offense was meant or taken. It was just a statement of fact to cover a situation short on precedent. Understandings about how you ally and what you do together are supposed to be "self-evident," so obvious that workaday life can proceed without much discussion. They help gild silence. Especially in this part of the world, the less you have to say, the more knowledgeable and connected you seem. But my insistence on this bolt-pulling/baling exchange defied the norm. With generosity signaled in a hush, people usually put up with my trespassing, but what can they do when talk—potentially embarrassing in and of itself—is requested? The no-win quality of the situation suggests, I think, why this anecdote remains so memorable and why on that midafternoon a perfunctory explanation would do just right.

One farmer found a way to distinguish himself from another in his relation to me without broaching the structure of allegiance. After all, these farmers occupy the same circle and therefore "should be" similarly related to me. What set our relationships apart and thereby skewed my farm impressions was less a matter of social propriety than individual style. One guy enjoys oddballs, even university types; the other does not, at least in the same way. Again, I am grateful for the reality check. This little episode is a dramatic reminder to do some checking of my own, to be sure that my picture of farm life is serviceable beyond one *simpatico* quadrant on a few gravel roads between Hills and Sharon Center, Iowa.

In addition to comparing impressions with friends and less *simpáticos* friends of friends, I decided to try surfing the other net, the Internet. Back in 1992, when I first turned to cyberspace to broaden horizons, it was still a pretty novel strategy. It was before the World Wide Web or even Gopher sites were "user-friendly," before plug-'n'-play or point-'n'-click. Going anywhere on the Net required learning that a good source of digitized information existed, its exact name (beware of capitalization!), its location (down to the brand of remote computer and sub-subdirectory), a log-in procedure, password, compatible terminal emulator, configuration, and so on, plus—no simple matter—an account on a decent server. Academics in the humanities and social sciences often were proud of themselves if they could just manage electronic mail. These were the days before state legislators and Vice President Al Gore popularized the "information superhighway," when computer networks were rare outside the halls of America's governmental, industrial, and scientific establishment. But given the concern for agriculture in those very halls, I decided to give it a try.

I was quickly overwhelmed. Alone, the United States Department of Agriculture, its extension offices across the country, and spin-off land-grant university operations comprised a massive Net presence. Every imaginable group of relevant specialists had its list of yet more specialized bulletin boards, data banks, news and discussion groups in animal and plant sciences, veterinary medicine, agricultural or rural economics, history, public policy, sociology, and the like. The gnarl was dense enough, even in 1992, to impress the spiders that work feedshed rafters. Since many of these lists required lurkers to subscribe, my computer "mailbox" has for years been jammed full of incongruous posts. One person wants to know about the similarity of mycotoxins in Scottish and Filipino hog feed; another about the ways to calibrate a goat treadmill or the latest legal and ethical ramifications of splicing genes in sorghum or injecting high-tech *schmutz* into the eyeball of a steer. Expert advice on everything from soybean futures to sow scours was mere keystrokes away. With the passage of every month it got easier both to learn more and to find yet more to learn.

Cyber-swinology may seem an unlikely field technique, but it worked, if anything, to excess. It was clear that nearly all of these people on the Net had more than "virtual" agricultural connections. Farmers and farm stock were among the key suppliers and recipients of their information. Moreover, an impressive proportion of the posts referred to on-the-farm experiences and had a jawing-over-the-fence quality. Emphatically pragmatic, the tone superficially resembles a sillier sort that people favor whenever opinions can be linked to ag-centered economies. For example, when Midwesterners—only a small minority of whom actually farm—get to talking about state programs, they stress their supposed effects on "the farmer," as if reality wore chore boots, one size fits all. Claims adjusters in Beloit condos get down, ag style. They sound like Ethel Mermen doing James Brown or vice versa. There is a gap between realism and reality that earnestness alone cannot fill.[4]

But on the Net that I surfed, concrete and precise terms of reference seemed to fill it. Such circumstantial detail would be very hard to fake, and everyone seemed to take the exchange of reliable information as a responsibility that is serious but also impressively familiar. Lives and life's savings may be at stake, but so is your sanity and sense of humor.

At one point, for example, posters on "Swine-L," my favorite list, wondered if veterinarians should bear in mind the limited jurisdiction of their licensure when responding to out-of-state farmers on-line. The strand ended when a USDA Extension swine specialist, previously a hog farmer in Saskatchewan, asked rhetorically: "If two swine producers meet for coffee at the local elevator and fall to discussing a disease problem on one of the farms, are they practicing without a license? Why should an Internet discussion group be any different?"[5]

The presumed answer: "No, it isn't. We are talking 'here' about ordinary events in genuine communication among flesh-and-blood people." Discussion closed.

Even the *schmutz* injectors and futures traders mixed technical advice with recollections of "Mom and Dad's home place" or "a feedlot I visited last month." But those common allusions tended to be offered as asides in addressing subjects that were anything but common. Gaining their attention and knowing what to do with it required learning enough of their home terrain and blazes to trek where our trails might meet.

In hindsight I know that I squandered some early offers of help. Before they were focused, our exchanges were pleasant enough but too far from garden-variety reality. As an interloper from the liberal arts, albeit with farm experience, I was greeted a bit as you might a kindly Martian. Of course, I was glad that they cut me some slack. I was apparently a reminder of the "general education" that most had endured, at best as an amusing pastime, before hopping aboard a college major on career track. They played along even with the dumbest of questions. But I eventually recognized that, in order to engage their expertise as well as their patience and generosity, I would have to come closer to meeting them halfway.

In particular, two sorts of focus were required. It was obvious that I could find common ground among people who worked with economy or technology, the more "hands-on" ends of agriculture, only if we stuck to a particular commodity. So pigs it is. And I would not get very far with the medical or natural scientists unless I specified a malady. Given nearly hysterical concern for infectious disease, particularly sexually transmitted diseases and HIV in America at the time, I opted for viruses; if pressed, "corona viruses;" if pressed further, TGEV (transmissible gastroenteritis virus), the bug that I had seen hit Phil's farrowing house. In fact, nearly everybody knew about TGE.

During the years that I followed cyber-swinology, particular problems would come to dominate and then fade from fashionable discussion. A parade of maladies, the syndrome or disease of the month, claimed attention: hog cholera, erysipelas, rhino (atrophic rhinitis), pneumonia, rota and parvo (rotavirus and parvovirus), then—as I joined in—TGE, pseudorabies (PRV, or Aujeszky's Disease), ASF (African swine fever), PSS (porcine stress syndrome), PRRS (porcine reproductive and respiratory syndrome), E. coli (*Escherichia coli*), salmonella, and so on. But even as TGE fell off the cutting edge, it left its mark. Whenever someone calls out for help, especially from the farrowing house, TGE leaps to mind.

In Chuck Yeager style, a producer reports, "I've got some death loss, eight of ten in one litter even though they were fine yesterday. Another litter showed scours this morning, but the other sows have yet to pig. Standard vaccination protocols. Standard antibiotic. Advice appreciated."

Among the first: "Eliminate the possibility that what you're dealing with is classic TGE." There is something "classic" about it, a well-blazed trail worth tracking.

So whenever Internet links with food animals or their diseases came to my attention, I would solicit testimony. Over the course of a couple of years, I posted queries on about thirty sites. Each request was slightly customized, but they all invited a description of encounters with TGE or the closest thing to it. Responses trickled in and came up in archive searches, enough in the end so that printing them out required two reams of paper. Of course, none of this represents precisely comparable or adequately sampled opinion. At best it suggests some outer limits, the range of experiences that could be expected among a mysteriously select but interested crowd, one that patently favors both agricultural and computer literacy. If other people had or heard about experiences that were radically different from my own, I would have a decent chance of finding out.

Even though most of the respondents had American addresses and male names, their daily lives were very different from mine. Some of them spent nine to five charting genetics or working pipettes; others much longer hours scraping under pens. But none of them split their time between being a hired hand and a liberal arts professor. None of them was from eastern Iowa, much less my little quadrant. Their experiences with TGE, however, did not range very widely at all.

The vocabulary varied a bit. For example, veterinarians tended to distinguish the "epizootic" vs. "enzootic" varieties; farmers, more often, "getting hit hard" vs. "not too bad." Scientists tended to distinguish "morbidity" and "mortality"; producers, the "sick" and the "dead." Clearly, each occupation had its favored TGEV imagery, ranging from microscopy or blood test printout to an unforgettable stench. But just as clearly, they were all speaking of the same thing, and they spoke with comparable dread. Even those who chose to share experiences with something merely "like TGE" brought up other diseases that wiped out whole herds and credit reserves. Animal scientists commonly told me that members of their family or they themselves left farming because one of those swine pathogens "hit hard." The verb choice is striking, purposeful and violent, especially when you remember how small viruses are. Though more delicate than dust they land with a smack, like the backhand of an angry god.

Stories would often begin, "I got TGE . . ."—as if the farmer rather than his or her stock were infected. These people well knew, of course, that TGEV poses no known health risk to humans, but the expression appropriately captures the sense of responsibility and interdependence that is so intense in agriculture. An animal, even one destined for slaughter, and its keeper are bound not only materially but also spiritually. The quality of life or specter of death for one can—just short of literally—"infect" the other.

But such beliefs are not confined to agriculture. Talk of contagion, character, illness, and justice seem to blend with cruel ease in modern America. In AIDS communities, for example, people often talk about how they feel blamed for the HIV they live with, for "getting it," as if they sought it or as if they should have known they would be so smote for some transgression.

Warnings to "at-risk groups" about their "behaviors" (use condoms; don't share needles) include sound advice. But they can also be abused to stigmatize and shun those for whom the advice has been insufficient. If you are sick, you must have done something to deserve it. And since it can be "given" by transgressors and "gotten" by those who are not, you better steer clear. Viruses cut a wide swath, like Old Testament angels of death.

Of course, none of the TGE posts were so hyperbolic. They talked about pig farming, straightforward, simple. It was I who was so struck by the implications of their accounts, but such implications are rarely far from common sense, once you have witnessed a "hit." A little stirring brings weighty material to the surface.

In December of 1992, for example, I turned to "Ag-exp-L," a discussion group popular with USDA Extension Service agents and their ilk.[6] Since they ordinarily pass information between experts (policymakers, ag-school profs) and the public, they could be well versed in TGE lore. I asked:

> *Have you had experience with TGE (transmissible gastroenteritis, a very common diarrheal disease in swine herds)? If so, I'd appreciate hearing from you. I am conducting research on people's TGE encounters, human experiences and stories about the disease (e.g., typical or unusual outbreaks, folk remedies, technical or emotional responses, etc.) Thank you.*
>
> *—Rich Horwitz*

In came a response from someone using the handle "RGGROSS" on an educational server in New York:

> Yes, sorry to say I have had "experience" with TGE. It can be devastating to both the herd and the pocketbook.
>
> Raised hogs for 25 years in NW Missouri before returning to school and studying ag/rural policy. Also worked for National Pork Producers [Council] for two years; so have had lots of contact with other producers, nationwide. Could give you names.
>
> Let me know what you need.

I stir a little:

> *Thanks. I wish we could just chat face to face (my more usual research mode). Barring that possibility, I'll take you up on your offer, trusting that you'll do only as much as you like.*
>
> *I think I understand the technical side of the disease pretty well [sic], but I want to check out how the experience that I had or heard about from neighboring operators might compare with others'. You could greatly help, then, by just telling me about any particu-*

lar experience with TGE that leaps to mind—whether it's how you first heard about it, the first outbreak, a particularly horrid or easy one, differences you had with other operators or vets. . . . Whatever. Just have fun with it, knowing that anything you share, no matter how banal it may seem to you, will be helpful to me. I'm just trying to be sure how TGE fits in the feel of farm life.–Rich

The reply:

Rich, I agree, this would be a lot more fun to talk about "in the flesh." I meant to ask, though, what your major is at U of I. I'm quite sure it isn't Ag or Animal Science or you would be at Ames [home of Iowa State University, the more ag-tech school]. Were you raised on a farm? Or how else did you get interested in TGE? I suspect by line of questioning that you are looking more at the human side of the issue and its devastating impact on morale. It would help me if I knew more about what you are looking for.

Since I am going to be in the Des Moines/Ames area around Christmas and New Year's, maybe we can get together then.– Robin

Stir one more time:

Well, you've got me pegged. No, I didn't grow up on a farm nor am I in ag science. I am a professor of American Studies, but also been a pretty regular part of things (2000+ acres, 1000+ cattle, 1000+ hogs, farrow-to-finish) for more than a decade, part-time. It started off as a way to keep in touch with neighbors, to have a break from desk work, to stay in shape (well, sort of) – to make sure I don't turn into some lard-butt egg head, if possible. It's a serious occupational hazard.

Anyway, farm life has now become part of my research. I want to learn enough to tell people interested in American culture (people like college students or book readers) what farm life is really like. Needless to say, I think Thomas Jefferson, "Little House on the Prairie" and Big Foot [herbicide, TV] commercials don't quite prepare these folks for the real world – not necessarily the real world of becoming a farmer (few people will) but for living with them. The work I do is somewhat akin to what anthropologists usually do in more exotic places. I am interested in TGE as potentially part of a much larger story, an article hoping to become a book.

Anyway, it would be good to compare notes on TGE experiences. I've only seen a couple, but they sure did affect me. Yes, they're sad, but they also say a lot about more normal circumstances (e.g., how to "manage" or cope with something like fate, how responsibility

can be affected under such circumstances, about money and care for animals, etc.).

But I don't want to get too much into the sense I make of this stuff, both because I may be wrong and because I want to make sure you don't feel bound by my interests or ignorance. It would be good just to trade tales. If you'll let me know when you're free around New Year's, I'll try to meet you.—Rich

She agreed:

Yes, let's get together in Iowa.

It appears that your life and mine have been exact opposites. I was born and raised on a farm, married a farmer, participated as a full partner—complete with "his 'n' hers" tractors, combines, and grain trucks. After 25 years, I returned to school to get a BS from Iowa State in Ag Economics, Political Science, and Rural Sociology in May 1992, with Rural Sociology my real love. Then I moved to continue my schooling at the Maxwell School of Citizenship and Public Affairs, Syracuse University. Quite a change from the farm and Iowa, but I have thoroughly enjoyed living here and find the academic challenge very rewarding.

My primary emphasis is still the interrelationship between ag policy and rural policy. I find it frustrating that government has traditionally tried to address the two issues separately, and as you are discovering, it is very difficult to sort out the economic issues from those that impact families, personal attitudes and satisfaction, and community life.

Your idea of using TGE as a format for looking at these types of issues is very good. It definitely impacts more than just the financial side of things. There are strong psychological issues—personal and family-oriented—as well as community impacts.

I have known firsthand the feeling that I wasn't welcome at the local coffee shop because the word was out that we had TGE. And there was great fear of contamination from boots, etc. It also impacts communities, because usually more than one farm operation in an area is hit at the same time, creating a "community disaster." At more than one high school basketball game, TGE has been the prime topic of conversation: who has it, how severe, what vet they are using, etc.

Will talk more.—Robin

SIXTEEN

A CHAT WITH ROBIN

WE MET AT A TRUCK STOP IN WEST DES MOINES, not far from the headquarters of the Iowa and National Pork Producers. Robin had stopped there on her way up from Missouri in order to catch up with old friends and to learn more about jobs that might use both her farm and her academic training. We chatted over lunch.[1]

* * *

I have been farming nearly all my life. And there's not very many women—particularly there weren't twenty-five years ago—actually out there, hands-on farming. I had to know the tractor, know the combine. Did the grain hauling. Castrated more pigs than I want to think about.

There wasn't organizations where I could share that background and knowledge with other women. I found myself just, like . . . the Pork Producers. I mean, I was one of the few women that had ever been an officer in the Pork Producers Association.

What about women's organizations among pork producers? I know some states at least used to have "Porkettes," but I gather they've been changing their names or folding into the main groups. Didn't Missouri have a separate women's organization?

Yes, at one time they did, although I never found myself particularly comfortable in that. And I don't mean this as a putdown, but they were mostly interested in sharing recipes with consumers, sewing with pigskin, working with school nutrition-type programs. That sort of thing.

And there's nothing wrong with it. I think it is very important. I did work quite a bit with the medical doctors in trying to disseminate the latest nutrition information. I was president of the Missouri Pork Council Women. But what the men were talking about in their meetings was much more interesting to me, because they were dealing with issues that I was having to deal with on a daily basis.

You see, I was born and raised on a farm, the oldest of three children. I've got two younger brothers. One of them is in investment banking, a comptroller for a big savings and loan. He left the farm and went to college. The other ended up in farming—first back home (it was too small to make a go of it), then working for a large farmer up around Osceola, Iowa, and then back to a large operation in Missouri that belongs to someone else. But he was born seventeen years after me. Neither of them were involved in the farm operation as I was.

I started driving a tractor in the hayfields the summer before I started school. I was born in 1948; so it must have been back in 1953 or '54. I remember we had a tractor with a hand clutch that I could just reach. We were using a sled to pick up those little round bales, like they used to have. Dad would put the tractor in super-low for me, and I'd steer between the rows. He'd ride behind, jumping off to throw the bales on and to help me turn around at the end of a row.

In fact, the only time I ever tore up a fence with a tractor was at the end of my first day in the field. I was convinced that I could drive out through the hayfield gate, while Dad kept the cows out. I did make it through without hitting a post, but—as a six-year-old—I was so intent on getting through and then making the sharp left turn required right after, that I cranked and cranked on the steering wheel. Before I could straighten out or grab the hand clutch to stop, I ended up making a full U-turn, right back through the standing fence next to the gate! Needless to say, Dad was wishing he had done the driving. It must have made quite an impression on me, too, since—forty-some years later—the memory of panic, reaching for the hand clutch and watching the tractor rip through fence, is still so clear.

Dad also milked until I was eight or nine years old, and I remember going to the milk barn every morning with him.

I always laughed and said, "I was Dad's oldest son."

And when did you start working with hogs?

Dad always had hogs, but I don't really remember doing much with them. He ran sixty to eighty sows all the time. And he ran purebreds, Durocs: "the best." (At least he had me convinced they were. Of course, anymore, even your purebred breeders are breeding three or four different breeds, and all your market hogs are crossbred.) But as I grew up I was always told, "You're not going to the hog lot by yourself. You can go with me, but not by yourself." That's good precaution.

But by the time I was really getting big enough to be good help, he pretty well sold out all the livestock. He became postmaster of the local post office. So then I didn't hardly do any farming at all from that time (I was twelve) until I got married.

I was not quite eighteen, married right out of high school, 1965. So there were about six years when I was busy doing other things. But I married a farmer with a family that had a lot going. All of a sudden I went from this little 120-acre farm—which in the last six years had no livestock—into this big operation. My husband's family also had three feed stores.

What, if I can ask, was their family like? Their background, say, compared to yours? Ethnicity? That kind of thing.

German. Ronnie's family was German. Mine primarily was Swiss or English, but his was very strongly German. His had been one of those old-line families. He was a fifth-generation farmer on that particular ground. When our children were born and grew up, we had four generations up and down a country road and a half-mile apart. It was a separate community from mine, a separate school district, but it was only seven miles from where I had spent all my life.

They had hogs, and they had cattle. So I jumped right back into a farm operation. And we had our first child fairly soon.

We faced the conscious decision: "Do you want to go work somewhere, or do you want to work here on the farm?"

I loved farming. I mean, I had been raised in that environment. Immediately, from the time I married into the family, I started doing book work. I spent about half my time in town, doing accounting work for the feed stores, and the other half out at the farm. Initially I didn't do much of the hands-on stuff. But by the time I'd had my second child—we'd been married about five years—I just kept getting more involved. His brother had left for Vietnam in '69 (the same year Julie was born); and so he was really needing help, somebody to replace him. By that time I was going out to do chores every day.

We had tractors with cabs, and we had an infant seat in each of the tractors. There was an infant seat in the pickup. There was one in the combine. We'd get up in the morning, and it wasn't any different than going off to a job. I'd take the oldest one into preschool, and the youngest one went with us. And Ronnie would go out and start up the pickup or the tractor and get it warmed up, if it was in the wintertime. I usually went in the pickup to do chores and took the baby with me. I mean, she was born in August, but I can remember that winter, packing her everywhere we went.

We had a hired man, and the hired man and I pretty well did all the hog operation. I enjoyed working with hogs. Really! It was just something I had a good time with. At the time we were using a confined farrowing house, and we had one crate that we never used for a sow. We had heat lamps hanging over it, and when Julie was small, she'd sit in the infant seat underneath the heat lamps in the farrowing crate while I worked pigs. When she got a little bigger—well, we got a piece of indoor-outdoor carpeting and covered the bottom of the farrowing crate with that. And that was her playpen, with her toys out there. Ha! That kid grew up in a hog house, and now she's married to a hog farmer.

Ronnie very rarely . . . He rarely worked with the small pigs. He just didn't enjoy working with hogs, and he had a lot of other things to keep himself occupied. And so primarily it was the hired man and I. I was pretty much responsible for everything from the time they start giving birth until we got them weaned.

I did all the veterinary-type work, all the "midwiving," I guess you'd say. It was me that got up once every hour or every couple of hours, all night long in the wintertime, to go check sows, just to be sure everything was okay. And we farrowed six times a year; so, with that and the book work, it kept me pretty busy, tied up.

We're talking about the early '70s. Let me see: By 1969 I had two children. I had one in '65; one in '69; and then one in '73. So the two youngest ones just went with me wherever I went. And, of course, by the time the third one was born, the oldest one was in school, and the second one went off to a kind of preschool two or three days a week. The youngest one just did the same thing in the farrowing crate. Ha!

I loved what I was doing. It allowed me to be with the kids. I could have a real flexible schedule. If one of them had something, if I needed to stop and go to school (I was a room mother for seventeen years), if I had to do something with that, well, schedules could be juggled to where I could do that.

We started with about seventy-five sows in the early days. That was where it was at when I married into the family. Over a period of fifteen years or so, it kept creeping up, up to a hundred and thirty. That's the most we ever ran. So we just made larger groups, built another small confinement farrowing house, so that we could handle more of them. And in the early '80s, I guess, we switched and went out to the A-types.

Really? So you switched from farrowing indoors to outdoors, from a pretty controlled environment to those little buildings out on a hog lot? I gather that most of the change since the 1960s has been in the opposite direction, to more controlled environments. But you went the other way, right? Do you remember how that happened?

Because I had pneumonia three years in a row. That third winter with pneumonia, the doctors just said, "If you don't get out of the hog dust, you're going to keep fighting this. Now you either got to quit raising hogs, or you got to get out of this stuff."

The two little farrowing houses we were using at that point were beginning to show a fair amount of wear and tear. Rather than replace them with something with maybe a better ventilation system, decrease the dust and all that, we decided that I'd be healthier and we'd be better off to go to the other system. And so we just did away with those confined farrowing houses and went to the A-types, where I wasn't inside a hog building. That'd be about ten years ago, and they're still using that system.

And how long were you doing that? Somewhere in there you must have gone to Iowa State, right? So you work working with hogs till . . . ?

Till '89.

Actually, I started college January '89, at Missouri Western State in Saint Joseph, about thirty miles from the farm. I continued to live on the farm, do chores, run the hog operation, and do all the financial and management work, plus, of course, be mom to Ryan, the youngest. He was in high school at the time. I started off with two days a week of classes and gradually backed off from full-time chores. But I still did the office work for the entire operation.

You see, in the fall of 1988, after twenty-five years of marriage, Ronnie asked for a divorce. After a couple of months of total personal chaos, it became clear that I had two choices: enter into a messy settlement that could mean financial disaster for the farm, with no one continuing an operation that I loved, or move on, hoping to maintain some base for my son to use in the future, if he decided he wanted to farm. Besides, to fight it out, ultimately I'd have to give everything to lawyers and air the family's dirty laundry in public. The last thing I wanted was to risk more humiliation and embarrassment for my children, who are, after all, carriers of the family name.

Anyway, the divorce became final on October 24, 1989, and by then I knew I had to get away from the farm. It just wasn't emotionally a healthy situation—to still physically be living where your old life was and yet needing to regroup, to begin over again, do the emotional "letting go" that has to be part of the healing process.

By winter I decided to move on and had been accepted into the honors program at Iowa State University, College of Agriculture. I arranged for Ryan to move in with my parents during the week, finish his senior year of high school, and then attend a community college while I went to school in Ames. I still had an Asgrow Seed business that I kept as a part of the divorce settlement, since I was the one who had been doing most of the work with customers. I had built up a good customer base, and they were supportive of my schoolwork and weekend commutes home to tend to business. Plus it was a good part-time income while I was in school. So I maintained the family home at the farm and still do.

I went home every weekend, never missed a one. Part of the time it was just to be Mom, regroup with Ryan. And part of the time it was to contact Asgrow customers, deliver seed, and whatever else needed to be done. I was also working for the National Pork Producers as a public policy project director.

I finished Iowa State in '92. Getting selected one of the eighty national Truman Scholars not only provided a scholarship but also opened doors to first-class graduate schools. I decided to go to Syracuse University, the Maxwell School of Citizenship and Government Affairs. The academic challenge and total change in environment have been great.

Quite a jump for this "plain old hog farmer" from Northwest Missouri. I quit doing hog chores in the spring of '89.

* * *

I think I explained to you: I'm interested, for the moment in TGE. For a lot of reasons—mainly at this point just trying to figure out how it matters to people. Do you remember first finding out about the disease?

Yeah, ha! You don't forget it. I can still smell it today.

Walked out and. . . . We were still in the confinement farrowing houses. So, it would have been '75 or '76, the first time we were hit with it. I think it was '75, judging this based on how old my youngest kid was. (How big was he in the farrowing crate, you know? Ha. He was born in '73, and I think he was a couple years old.) Of course, it was in the wintertime. (Always is, when TGE hits.) Yes, December. I'm sure, because it was before Christmas. I can remember us talking about: "Boy, this sure ruins Christmas." Mid-December, 1975.

We went out to do chores one morning. Cold. Snow on the ground.

I bundled up the two-year-old. I'd sent his two older sisters off to school.

The farrowing houses were just out in back of the house. I mean, I didn't have to walk two hundred yards.

This was before the A-frames, right? Did you have bulk bins? I'm trying to get a clear picture.

Yeah, yeah. Before the A-frames. These were the buildings that I was having the pneumonia problems in. The bins were right close by. I'd just gravity-flow feed out of the bulk bins into buckets, and then I'd string feed in the troughs.

See—the kind of farrowing houses that we had—they had ten pens with little individual doors on the outside. Twice a day we'd let the sows out on the lot. So, you'd go out in the morning, let the sows out, get their feed, put it in troughs for them. Of course, there was water for them out there, too. And while they were outside eating, getting a little exercise and so forth, then I would go inside and clean up any pens that need to be cleaned; work with any pigs that need to be worked. By doing it while they were outside, even if you were working a litter, you didn't upset mama. And then you repeated this whole thing in the evening. While they were out, morning and evening, was when I worked pigs. Of course, in between, if you had anything farrowing, you were in there just about every hour. Run out and check, be sure everything is okay.

So I got Ryan, bundled him up, got his toys, and we headed to the barn. He was walking alongside me. We climbed over the fence, through the lot, to where those individual doors are. I started to open that door to let the first sow out . . .

The smell was atrocious! It was a sickening-sweet, scoury smell—certainly not the normal odor you get when you're at that building for the first time in the morning. I was real suspicious when I smelled that.

I started to let them out: first one, then two. . . . Some of them seemed to be normal. But two or three just stood around, their ears kind of drooping, tail drooping, moving pretty slow. And about that time I noticed a couple of them had a bowel movement that just shot about three foot out behind. Terrible diarrhea! And then I noticed a couple of them vomiting. I knew instantly what we had.

Do you remember how you knew?

Well, I knew what the symptoms of TGE were. I'd heard other farmers talk about it, how they had been hit with it before. You know how news travels through a community, in the café, at basketball games, at church. It'd be the topic of conversation: "So-and-so has got TGE."

There were a few farmers who had it but didn't want to talk about it. You'd have some—and I could name names, if I wanted to—who I know dang good and well have had some TGE, but they would never admit it. I think they thought it was a reflection on their management ability, a pride thing: "I'm such a good manager, I don't have TGE." They were the same with a bad crop year or an outbreak of cutworms. They didn't want to acknowledge that they had their problems.

But we had a veterinary that we were very close to, and he had even warned us that winter: "Hey, there's some TGE in the community. Kind of be watching for things." I can't remember specific names, but I knew there was some TGE around.

It was just so different from any other odor. The odor is the thing I remember. And then, when you see them vomit . . . *you know.*

I went back to the house and called the veterinary and said, "Hey, we broke with TGE this morning."

I remember him saying, "How do you know it?"

And I says, "Listen. This is a smell that's different than anything I've ever dealt with. They're vomiting, terrible diarrhea, three or four out of ten. And they were okay last night." (Although I can remember Ronnie saying he had noticed a couple didn't eat real well the night before. So I suppose they were coming down with it, maybe feverish the night before. But by morning they were really sick.)

And he says, "Well, I'll be out, middle of the morning, and we'll check this out.

"I agree," he says, "sounds like TGE."

Then I went on into the farrowing house. I had two or three litters where the pigs looked dehydrated. You'd raise up their tail, and there was this almost watery diarrhea. And, like I said, the smell was just terrible. I still had no dead pigs at that point. It was the early stages.

The vet came over that morning, and I called Ronnie. He was off tending the cattle. We had CB radios in all of our vehicles, but I remember thinking, I don't want to announce this over the CB radio. So I just said something like, "I think you'd better get up here to the barns. We've got some problems, and I think I need some help."

When he got up there, he agreed: "Hey, yes. It really looks like TGE."

And when the vet came, he said, "Yeah, this is what you're dealing with." He said, "I'm not terrible surprised. There's quite a bit of it around."

He took temperatures of some of the sows, and they were real high.

And he says, "You might as well wean all the sows, even the ones that aren't sick, because they're going to be, within twenty-four or thirty-six hours. Get them out of here."

Well, you can imagine. I mean, these pigs . . . some of them were two days old, five days old. They were young!

He says, "You might as well wean them. With a fever like this, their milk is going to dry up. They're not going to have anything, anyway. Those pigs would be better off without the sows there than they would be with them. All they're going to do is just spread it more." We moved the sows to another farm about three miles away.

"You're going to need to get extra heat lamps in here." (We were using catalytic gas heaters in there as well as heat lamps.) And he says, "You raise the temperature of this barn until it's about eighty degrees, because pigs are going to chill. They've *got* to have the warmth."

He gave us an electrolyte mixture that we were using in the little portable nipple waterers. And he gave us a good milk replacer. And he says, "I want fresh milk replacer and fresh electrolyte out here every hour. Don't wait!"

He explained that the pigs nurse in about fifty-five-minute cycles. (If you notice, like with an old sow, about every hour she starts talking to her pigs, and they all get up and nurse again.) And he says, "That's just the natural, biorhythmical cycle. If you'll come in here about once an hour and give them fresh water, fresh electrolyte, fresh milk . . . If you make sure they get up—stand around and make sure . . . If you do that, you'll have a fair chance of saving at least a portion of them."

The ones that were real sick also got a transfusion of electrolyte, glucose [sugar water], with some neomycin [antibiotic]. We had used this before, like for E. coli [bacterium infection of the gut], scouring pigs. You used a fairly long needle, held a pig upside down, stuck it in their belly, up above their navel into the belly cavity. He had me doing that once a day.

We ended up . . . Oh, we lost the pigs that were less than three days old. We couldn't save them. We lost every one of them. But the pigs that were five to ten days old—we managed to save 75 to 80 percent of them.

I wore out a blender.

In order to get the milk replacer to feed down through those nipple waterers, it had to be mixed in a blender. It was strawberry-flavored and colored. Pink! It looked like I was making pink milkshakes every hour. Since we didn't have a water heater out there, I took buckets of warm water and the blender to the barn with me every time. It was so cold that if I'd made it up in the house, by the time I got out to the barn, it'd cool down. And they ate it better, it stayed in suspension better, when it was warm.

That was for the pigs, right?

Yes, for the baby pigs.

But what about the sows? Did you figure they could weather it?

Well, the sows . . . No. We ended up losing two sows from it.

It was a real cold spell. They were running this high fever. They had diarrhea; they were vomiting; they were chilling, really weak, rundown; and then, in the cold, they caught pneumonia. So we lost two sows out of the deal.

Now, the second building, where we had another ten sows, never was hit quite as hard as the first building was. It had older pigs in it. Only one or two of those sows came down with it full-fledged, after we had weaned them. And we lost a few pigs out of that bunch. We always said that it was because the pigs were a little older and could withstand it better. But when this all happened we weaned all twenty sows, both buildings, right then and there. Besides the two we lost, we had a couple of sows that wouldn't breed back after that. They lived, but we never got anything out of them.

And you had been saving gilts? You know, keeping seed stock?

Yes. The gilts were ours. We were buying boars, on a three-way rotation of boars with Hamps and Chesters and Durocs. And about every third cycle, you'd bring about thirty new gilts in. But we were always saving gilts and breeding back. So, it was really devastating at the time.

The workload really jumped, and it was so frustrating. You'd feel like you was doing everything you could to save these babies. Financially, I knew, we *needed* to save them. Plus, it just was heartrending. You'd go out and see a healthy pig; a few hours later you'd go out there, and the little thing couldn't even get up.

By the same token, there's a sense of satisfaction. By going out there every hour, round the clock, and doing all that we did, we managed to save a big percentage of them. I don't know if treating TGE has changed in the past couple of years, but then it was common to lose everything. So we feel, "Well, you did something right." Apparently the veterinary was right: "Let's get those sows away from there and just do intensive care for pigs and do the best you can." Wore out that blender.

If we'd have had a better facility to take the sows to—where it might have had some heat in it—I don't know if we'd have lost those two sows. But with moving them from this heated place to an unheated barn, sick. . . . You know, we tried to prevent it, tried to keep them dry, hit them with antibiotics, but. . . .

And what about the vet, after that first morning? Did he come back?

Oh, yeah. He came back two or three times, when we couldn't get those sows to get over it.

He really doctored sows more than he did pigs. He'd ask me, "How the pigs in the barn doing?" and I'd tell him: "Well, this is where we're at."

I don't remember him going in the barn after that initial visit. He really wanted to stay away from it as much as he could. He said, "Well, if you're not losing a bunch—and it sounds like everything is going well—no need of me being down there." I think he just didn't want to risk the exposure, maybe him picking it up, taking it somewhere else.

That suited me fine. I'd just as soon he not, you know, even though he disinfected boots, changed coveralls and all, coming and going. I didn't insist on it, but I'd just as soon he stayed away. You didn't know who else's barn he'd been in and so forth. By the same token, I didn't go onto other people's lots. It is just "good management." We never allowed anybody in our hog lots in the wintertime. Period!

We always prided ourselves on the fact that we did our own grinding and so forth. So you didn't have somebody from town coming out, grinding feed, and distributing it, because who knows what lot they'd been into before? Tires dragging stuff in? People getting out and doing things? We were real careful. And it was the threat of TGE that concerned us. At least the perception of the threat was very real.

It was commonly understood that, if you went to visit a fellow farmer, at least in the winter, and he was out in the hog lot, you didn't walk out there to talk to him. He didn't want you out there, and you didn't want to be out there. You waited for him to get up to the house, scrub boots, and so on.

But the vet did go on the farm?

He went on the farm. He went out to the sows. Maybe it didn't make a whole lot of sense, but I don't think he ever went into the barn again.

And you didn't do anything, as far as trying to protect one building from another, to keep it from spreading?

No. The two buildings were a hundred foot apart. We felt like . . . The sows had already been together eating, the night before. We'd tracked back and forth between the buildings before we'd noticed anything. I tracked back and forth at ten o'clock, the night it broke. It was probably inevitable that both farrowing houses were going to be contaminated.

But we had another group of sows that were going to pig again in about six weeks. Now, that group we were really concerned about. So we intentionally exposed them. We went out and used a vaccine, modified live virus and all that.

Where did you get it from?

From the vet. He said it would be fast-acting, the most efficient for us to use. He said, "The vets in Iowa are having good luck with it." At that time it wasn't legal in Missouri, but he'd made a contact with a veterinary friend up there. So he called; I came up over the line into Iowa and got it. It was a dry powder that you reactivated, injecting into sterile water. And then I mixed it with powdered milk and grain and fed it in a gruel consistency to the sows. Three doses—three weeks, two weeks, and one week prefarrowing, as I remember.

Do you remember talking about other ways to handle it, ways to avoid getting hit hard with the next group to farrow?

Yes, there were farmers at that time that were grinding up pigs that had died of TGE and feeding that to their sows. But, well, number one: it's not a very pleasant process to have to do. Plus, the veterinary felt like this modified live virus mixed with milk was basically doing the same thing. You were basically giving a mild case of the disease to the sows, hoping they'd build immunity to it before they gave birth, and they'd pass that immunity along to their offspring. So we went that route.

Other farmers, I'd heard, were doing the infected-bedding-type thing. And the vet did encourage us—as we cleaned the farrowing house—to take the bedding out to the lot where those pregnant sows were. He says, "It can't hurt anything. We want to be sure they've been exposed. Hopefully it won't make them too sick. They'll get over it and be ready by the time they pig."

And another instruction: Even though we had been in the habit of letting the sows out for an hour or so, he cautioned me not to let the pigs be away from mama more than thirty to forty-five minutes. He wanted to be sure that those pigs never missed a nursing cycle, because it was the colostrum in the milk that was going to help protect the pigs. So we had to be real careful about this.

And we never did break with TGE again that winter. That one batch was all that had it. We broke the cycle.

The next year or two after that, we were pretty good about doctoring everything, making sure that everything got vaccinated, modified live virus. But the vaccine was expensive (something like three or four dollars a dose), and it wasn't 100 percent foolproof.

And did it cross your mind, you know, that—even with all the shots, the milk replacement, the blender, the electrolytes—that you weren't going to get much out of those pigs, anyway?

Oh, yeah. That crosses your mind. I can remember thinking: "Well, will they ever be worth anything? Are they all going to just die?"

What made me think that was: We'd had an outbreak of salmonella in baby calves a few years earlier. It just . . . . We lost 70-some.

We were buying Holstein calves at three days of age and feeding them out to 1,300 pounds. (A lot of corn goes into a Holstein steer!) And we had them in crates in the barn until we got them weaned. But in a matter of days we lost 70-some out of 117 in that last batch we bought, the one with salmonella in it.

It's far quicker-acting than TGE is. I mean, you'd go out and feed one of those little calves. His little tail would be wagging. He'd be just happy as a lark and eating his bottle. And you'd go back out at noon to check calves, and he's dead in his crate. That one was *really* overwhelming. The calves that we saved out of that batch never did absorb their feed and grow out right.

Plus, the vet said that this type of salmonella is contagious to human beings. And the kids were little. The youngest hadn't even been born yet. And we used to have the kids out there in the barn. We had hot water and wash tubs out there for the calf bottles, and we kept one of those drained. Julie would play in it while I was doing calves.

But the vet said it could be potentially deathly, especially for the kids:

Tell your doctor.
Don't take the kids to the barn until this thing is brought to an end.
Be real careful that you wash your hands well when you come back; again before you do any food preparation.
Keep those boots and soiled coveralls out of the house, where the kids might crawl around them.

Our doctor said the same thing—potentially a fair amount of danger to it. So, they were more concerned about what it was to do to us.

And naturally, there were financial concerns, a major impact on the bottom line. All of a sudden, you had 150 or 200 less pigs you were going to sell. And I knew—because I did the book work, paid the bills, thought of those things more—I knew there'd be major impact. We were relatively young at farming at that point, operating on a lot of borrowed money in terms of major investments in equipment and land. We used hog money to pay off those loans.

In the end we didn't have to refinance or anything. It didn't take that big of a hit. But you couldn't know. You knew you were going to take a pretty large financial loss, but how much? And it is just before Christmas. So, you can't feel quite so joyous about it, about spending for presents and all, even if that money would not have been yours to spend for another six months. So, like we were saying: "There goes Christmas."

And we had two more bunches of sows that were going to pig, one in six weeks and one three months out. Well, maybe we would also get wiped out on those. Could we get it stopped? If you didn't get it stopped, the next time they all farrowed, you'd be facing the same thing again. It was the unknown. And it seemed like, whenever anyone got TGE into the farm fairly early in the winter, they just had problems with it *all* winter.

So, because we'd had that experience with the calves and there were all of these concerns, we really questioned the vet: "What's these pigs going to be like if we do save them?" And he kept trying to assure us, "Oh, they may be a little bit stunted, but if you save them, they'll be okay." In farming, you're just so imbued with that kind of thing, anyway.

But, sure. It affects you. Even the kids. When TGE hit, the kids were little. It really bothered them that you'd gather up all these dead pigs in a feed sack and haul them out. Then you'd haul more of them out the next day.

I can remember Ryan, just a toddler: "Piggies dead! Piggies dead!"

No kid likes to see an animal dead. They love to play with them.

And I'm sure that it got talked about at school: "Piggies dead!"

A couple of other years there was a major amount of TGE that went through the community. (We were vaccinating for it and never did get hit with it.) But it was almost like a community outbreak of disease. In the wintertime, when you'd go to ball games, when you'd go to church, when you'd go to social events, it was a topic of conversation. You know: "We hear there's some TGE. Have you got hit with it yet? How bad is it hitting? Are you losing lots of pigs? What strain? Is it hot?" You know, sometimes you might just lose a few; other times it'd just wipe you out—death loss of 70 percent or so, and nothing you could do to stop it.

Yes, I am interested in what that does to people, what it brings out in them. I was thinking, one of the things about TGE is the way it might bring out the best in people or the worst. You know, the things they do or feel when it hits?

Well, hmm. I'm just reflecting on what I remember Ronnie and I were feeling and people I've known closely that have been hit with this.

There's a certain amount of fear. You don't know how bad it's going to be, what's it going to do to you, how long it's going to last, how much you're going to spend fighting it. How many are you going to lose?

Nobody likes to lose livestock. When you care for those things on a daily basis? I don't think you ever get too big to feel it in a livestock operation. If you're a good animal husbandry person, you care about those animals. I'll never forget the feeling I had, just a sinking feeling in the pit of my stomach when I opened that door. It's one of those things I won't ever forget: Just, "Oh, my God!"

And there's a certain amount of anger: "Why did it have to hit me?" Any time you're facing a loss—and I just knew we were facing a loss, even if I didn't know how much—there's anger.

You knew that there would be an awful lot of work. All of sudden you wouldn't be doing anything but dealing with this thing, until you got over it. My God, I was going to the barn every hour round the clock. Especially for the first few weeks, it was just incredibly intense. You'd get so tired.

But I guess that is the same as with any other kind of emergency, pretty much. But I can see a difference between dealing with a TGE outbreak and, for instance, a drought. (And down in my country, devastating drought is a fairly regular occurrence.) You just stand there and watch your cornfield burn up. That's such a hopeless feeling. There's not a damned thing you can do. You can't make it rain.

At least with TGE, you could take a certain amount of pride. Boy, you were giving it every last thing. You were fighting a good battle with it, and you could see some results. You can take pride in the fact that you didn't just walk away and let it totally wipe you out.

You know people that did that? Who just walked away from it? That just said, "I can't hack it?"

Well, I can't really think of any. But at the time we had a hired man that was working for us who had some hogs of his own. He'd had TGE a couple years earlier, and I can remember him saying, "I don't know that it's even worth fighting. You lose them, anyway." Of course, maybe that's the difference between being a hired man and an owner.

It's really more a matter of some people working harder at saving things than others. That's just a difference in people. Some are going to respond: "Oh well. I knew it was going to happen sooner or later," and not really do a whole heck of a lot about it. Others are just going to do whatever they can.

And how do people decide what can or can't be done? Say, in your case or in the community in general, when TGE hit? There's a lot of potential advice out there, right?

People were always kind of comparing notes of what was working for one guy and what wasn't. You'd go to church or school or whatever, and you'd hear: "That's not what my veterinary said." Or I'd go off to a Pork Producers meeting in Columbia, and you'd hear: "There's pockets of TGE here or there. Well, this is what they're doing for it." And it might be quite different than what we'd been told. And when you know there's major dollars at stake here, who do you believe? It was a little bit unnerving.

At that point there were different vaccines with different claims. There were about three different veterinaries in the community, and one veterinary was telling them one thing, and one was saying, "No, do something else." One was very strongly for this grinding-up and feeding-back pigs or spreading the infected bedding and manure; ours wasn't gung-ho on that at all. For him, it was the modified live vaccine. And there were other controversies: Do you inject the vaccine or mix it with the milk and ground feed? One way might be a lot easier or less expensive, but then was it less effective? You'd hear comparisons made in terms of confinement or non-confinement. Probably, we believed, if we had heated buildings, where you're not dealing with the cold, we would have been able to fend off the effects of TGE and wouldn't have lost sows. We just didn't have as good facilities to try to handle sick animals.

And then some would just depopulate. If you didn't have to use the farrowing house for a month, if you could leave everything empty, supposedly you could clean it out. I understand that the virus doesn't last all that long in an empty building. But we were continuous farrowing. Turning ten sows in with a boar every few weeks, you couldn't be sure of the exact farrowing date. They'd be at least a little spread out. A month empty just wouldn't work, so we had to do something else.

And it was a lot of hassle: Mixing up all that milk replacer, hauling buckets of water, injecting electrolytes, out there every hour, night and day, mixing and feeding gruel. And you had to fix troughs so they didn't leak. And you had to watch out for a boss sow that would try to eat it all and wouldn't let the others.

And then you wonder: Maybe we just got lucky, you know, that the cycle broke. You wonder: If we had done nothing, what would have been the outcome?

But we had a good amount of faith in our veterinary, a good veterinarian. He hadn't been out of school a long while. He liked hogs. We'd just follow his instructions and hope for the best. I'm not entirely up on what they're doing for treatments, but it seemed to me like there was an awful lot of trial and error.

Yes, I am real interested in that: How people know what's going on or at least what sources they trust. When you're farming and there is all of this advice—a lot of it contradictory, coming from all over the place—and you have to make a decision, it's pretty hard to figure out. It doesn't seem to be a very perfect science, but it's not just guesswork, either. You have to trust experience, but you know that might not be enough. You know? Maybe you can tell me some more about how you got a handle on that. How did you learn what's reliable, or how might I? You know, there's different vets, extension, associations, magazines. . . .

Well, the magazines are full of health-related issues. That's where I got a lot of my information.

As for the church, I can't remember a sermon on it or anything like that. No prayers.

It was addressed extensively at Farm Bureau meetings or seminars.

Primarily, my information, sharing knowledge, sharing concerns, just crying on each other's shoulders, so to speak, came at Pork Producers meetings. At that point I was involved at the state level, going to Columbia for meetings. There would be university animal science people there: "This is the latest; this is what we're doing." In fact, there was a time or two that I would come home and share with my veterinary what I'd heard at the meeting. We always considered that one of the primary benefits of belonging to such an organization. It gave you the opportunity to have really cutting-edge information about things.

I had gone to a feminine farrowing school at the University of Missouri in 1973.

What?!

Feminine farrowing school. Missouri was the first university that ever did this. It was strictly for women that were working in the hog industry. They made an announcement that they were going to hold this. They'd take thirty women. And within two days they had the class filled. We had a couple gals from Canada that came down to that. And then the next year they had an advanced one, with a fair amount hands-on. That's where I learned to castrate, and that's the primary reason I went.

Ronnie had the habit of waiting until they were 150 pounds to get around to castrating. And I always thought that was cruel and inhumane and caused weight loss. I was just tired of it.

Yes, the wrestling, the blood, the screaming.

Right. I didn't like it. I thought it was stupid.

He says, "Well, if you don't want me to do it, then you do it."

I says, "Okay, I'll do it, but I'm going to do it when they're little."

So in that school we got into quite a bit of animal health discussions, even did a fair amount of posting [dissections, postmortem examinations] of animals, learning some of the key things you look for. And we had an opportunity to meet with a lot of the researchers.

How about other veterinarians, like the ones who work for the feed companies or USDA Extension?

You have to understand. I set on a county extension board and then a state extension board; so I was always fairly pro-extension, particularly its youth programs. And the state extension people—like the animal scientists at the university who were researchers and had an extension position part-time—I felt like I got pretty good information from them. But I'll be the first to admit that, when it comes to livestock information, at least at the county level in Missouri, forget it. I mean, you just bypassed them. Whatever they told you was five years old. I just never felt like they were really up to snuff.

I'd get better information from my feed company, who had a veterinarian on board. We bought all our feed through Farmland Industries. They would bring in some of their staff veterinaries for meetings in the winter—one livestock meeting for hogs, one for cattle, and then a crop meeting. Personally, I didn't ever have a whole lot of problems trusting the veterinarians that were on staff even with the pharmaceutical companies, if the advice was about swine diseases and swine management in general. Of course, if they were trying to sell their product, telling me that their feed or whatever is a whole lot better than somebody else's, then you'd view what they were saying with a fair amount of skepticism. But if we were talking about overall herd health, I would at least listen to what they had to say.

But, because I had developed those contacts through the state Pork Producers organization, it didn't bother me at all to pick up the phone and call one of the fellows down at the university and say, "Hey, this is what I'm seeing." But I had those contacts, and a lot of producers didn't. So I'm certainly not your typical producer. But what I see happening in the bigger producers today is that they're just like I was then. These bigger operators don't Mickey Mouse around with somebody out at their local country office, feed companies, and so forth. They'll pick up a phone and go to the top.

And I know you'd get a good response if you just contacted the Pork Producers, National or Iowa. In fact, there's an annual meeting coming up pretty soon, and disease is always a topic of conversation. So it might be a good opportunity. I could give you a bunch of names, but there'll be several thousand pork producers right there at one time. You could set up a little table or something, and I'm sure they'd talk to you. Always the second week in January, I think.

Gee, middle of TGE season!

Yes, the middle of TGE season. In fact, one year that I was president of Missouri Pork Producers, I can remember Ronnie didn't come but for one day, because we had TGE all around us. And he was afraid that we would break with TGE with neither one of us there. We didn't want to leave the hired man with it and us gone.

Yes, I remember losing a sow in the farrowing house—just heat stress and old age, I think—last summer when everyone else was away for a couple of days, and I was more or less in charge. I tried to keep her cool, but. . . wouldn't you know it? It was unbelievably hot everyday with no relief even at night. And, of course, she was in a crate about as far as possible from the door. I had to get two neighbors and use chains and a loader tractor just to get her out of there. Basic field work and choring is no problem, and actually I like having to improvise solutions to the things that ordinarily go wrong. But when something serious happens, I want someone else there. I just don't know enough, not enough experience with the technical side of things, especially disease. There may actually be nothing you can do, but it gets expensive and scary fast.

You really should talk to Dr. Beth Lauter, the veterinarian for the National Pork Producers. In fact, I'm going to see her as soon as I get through talking to you. She was in private practice up in northwest Iowa for seven or eight years and left to come to NPPC, to conduct producer seminars and work with federal regulations on health issues. She talks to animal scientists at universities all over the country and knows what they're doing. Or there's David Meeker, Beth's boss, who is in charge of research and has a Ph.D. in animal science. They spend a lot of dollars fighting disease and doing research on it. He's been a farmer and knows pork producers all over this country; addresses animal science groups, works with the magazines. He's on a first-name basis with the editors of *National Hog Farmer* and *Pork Management,* those kinds of things, and professors in the vet schools. In fact, I was just visiting with him, and he says, "Where are you going?" And I explained to him that I was visiting with you. So he won't be terribly surprised to hear from you.

Yeah, for sure he's on my list. In fact—he may not remember—but we talked a little on the phone before. He's great. But, with other people, even farmers I know pretty well, trying to talk about problems like TGE, it's often hard to know when they will open up and when they won't. And I don't think it's just something about me. Despite the way that people talk about the intimacy of family farming, community, trust, pulling together, being practical, down-to-earth, and all—or maybe because of it—people can be awful secretive, even judgmental, too.

Yes. As I was saying, there's some that would never admit they had TGE, because they felt like it was a personal reflection on their management, a measure of it.

Yes, but I thought you were saying that in a real forgiving way. Like: "They shouldn't feel that way. It's not their fault they got TGE. Disease shouldn't be a measure of people." On the other hand, you were also saying that there was a side of the TGE experience—like your being able to save so many pigs—that you can take credit for, like an accomplishment, a favorable measure.

Absolutely. Getting TGE doesn't mean you did something wrong from a management perspective. It's only an indication of poor management if you throw up your hands and walk away: "There's nothing I can do."

I've always operated under the theory that, if you've (quote-unquote) "done everything you can" to do a good job—put your crop in, take care of your livestock—you still have drought; you still have disease. You might walk crops a lot when they're little, check for infestations of insects, cutworms or whatever. But in the case of a drought, what good's it do? It's not going to make it rain. There's some things you can't control. And then you just have to live with it. When it happens, you do the best you can.

With TGE, there is a certain amount you can do to prevent it in the first place. But that doesn't mean you're going to stop it.

And I think you will find farmers, as a whole, pretty judgmental in that area.

More so, though, on everything other than TGE. There a certain amount of it that you can't help and they accept it. But, boy, if they hear of a farmer that's had an outbreak of erysipelas, for instance, I think you hear the attitude: "Well, why in the hell wasn't he vaccinating? Don't feel too sorry for him!"

Yes. As you were saying, there's only so much you can do, but farmers can be very harsh. They can be very hard on people, and it can be very hard on you, if you cross that line—you know, when they think doing "the best you can" is not enough. Like, remember those murders in Lone Tree, Iowa? The farmer, Dale Burr, who murdered his wife and the banker and a neighbor and then killed himself?[2]

Yes, I remember that incident.

Well, that happened near where I live. Our son went to school with some kids in his family, and the president of the bank was murdered only a few hundred yards from where our kid was sitting in school that day. The news talked about "depression in the ag economy," cash flow, collateral and all. But he actually wasn't that bad off, just had to renegotiate some debt, same as everybody else. But his family was just supposed to be, you know, one of "the best." His dad had built the reputation in the township, and he was the only son. . . .

Yes. Tried to uphold it, and he couldn't. It was just more than he could cope with.

You know, ever since we started talking, I've been trying to think of the name of a guy at the University of Nebraska who has done extensive work with farm families in their communication problems and coping with stress. (Ron Hansen is his name! He's in my address book. I'll e-mail it to you later.) He teaches there and has a farm background. He grew up on a farm in Illinois, a farm where the family was totally divided at the death of his grandfather. An estate hassle. That always haunted him. And they were not allowed to talk to an uncle. You know, this kind of bitter animosity that divides a family.

Yes, despite all the old Jeffersonian line, it seems terribly routine.

Terribly routine. Absolutely!

And I lived in that situation, where we had four generations in a half mile. I know firsthand the stress of living with in-laws and Grandpa's in-laws and meeting their expectations and . . . There's just a whole gamut. It's a whole other subject that's just incredibly complex and that can at times be really painful.

But Ron Hansen knows a lot about this. He has spent the last twenty-five years teaching farm management and running classes for farm organizations, having farm kids, students who went home to farm with Dad. Three or five years later they'd be back in his office in tears, because they simply could not get along. Dad was mad, and they weren't speaking. There were so many interpersonal conflicts. The son was starting to look for other jobs. Or you'd bring a daughter-in-law into the situation—one that maybe had not been raised on a farm—and she's expected to do things like Mom did, like Grandma did. It's a whole mass of things, incredibly complex.

I could listen to him for hours. Sometimes he could bring me to tears; at others I'd just roll laughing. He is so much on target with what it really is out there.

Yes, like I remember talking to a guy I just met at an appreciation dinner put on by one of the seed companies. We got to talking about that multiple-murder/suicide in Lone Tree, and he said, "Well, they docked his beans. What do you expect?" Straight-faced, serious. Amazing! You know, he was late in getting the combine out in the soybeans, so all hell should break loose?!

Yes, I found this fascinating. I wrote a paper on it, "the agrarian myth," for a research methods class in sociology at Iowa State.

I realize I am a bit cynical about this, after coming through a divorce, after twenty-five years out there. But there is this attitude that:

"Because you farm, you are just a little better and a little holier, more blessed with moral fiber.

"Because you're rural people or part of a farm community, you're just a little bit better."

EXCUSE ME! When you look at the statistics of all different kinds of things that you use to measure moral living—drugs, alcohol, incest, violence—it just don't begin to hold up. And then, when you know, from firsthand experience as well as from a lot of friends, the conflict within families . . . the silence is deafening.

Yes, that's among the reasons I don't think talking about this is "another subject." It's part of what I was curious about in asking, you know, how TGE might bring out the best or the worst in people.

Yes. It's one extreme or the other. They may pretend that there is nothing going on or they really share a lot: "Well, did you hear this?" But then sometimes it's more a form of gossip: "Did you hear that so-and-so got TGE?"

Right. I'll hear about somebody else, but then I'll be with that same person we just talked about, and they won't mention it. So I wonder, you know, what's going on.

No, no, no! That would be typical. When I say they're "sharing things," it's usually more in the form of gossip while someone else is trying to hide it.

But I thought it was stupid to keep quiet. If nothing else, you needed to forewarn you neighbors: "Hey, we broke with it. Beware. Don't let your dog get over here." These kinds of things.

But there was an awful lot that didn't take that approach. Always a few.

I was always of the opinion that you sought out information, who was doing what, so that maybe you could learn from it. But that's kind of my personal nature. There's also an awful lot of people that went to the other extreme, and it was more like gossiping. There always was an insinuation that maybe, because they had TGE, they'd done something wrong. It's a real dichotomy.

SEVENTEEN

FARM RHYTHMS, HARMONY, AND PROPHECY

Robin helped assure me that the TGE trail, out from my Iowa circle into the wider world, was reasonably true. The signs that she used in tracking her experience resemble those of other farmers and point, albeit obliquely, to other well-worn paths in American culture. At least the ones that track through the rural Midwest seem truer than those commonly glimpsed from aloft.[1]

Her experiences are certainly familiar to Swine Belt contemporaries. The reliance on borrowed money for land and equipment and the combination of grain, cattle, and especially hogs to pay it back; the division of attention between family farmhouse and feedlot; the gradually increasing size of crossbred herds; the rotation of saved and purchased stock; the use of confinement in continuous farrowing or finishing and open air for breeding and gestation; the mixture of improvised and expert veterinary care; the complex array of sources of agricultural information and the effort expended to keep up with the "latest"—these have been typical for successful farmers (the ones who continue to earn the bulk of their income from agriculture) in the Midwest for most of the second half of the twentieth century.[2] Robin represents them well.

Granted, in other respects she was far from typical. As she says, for example, no matter how the actual labor is divided, the people who identify themselves as "farmers" have been and are much more often men. When she jokingly recalls how proud her father was of her involvement as a kid, the image entails a gender bend: "I was Dad's oldest son." And there are ways that she speaks of farming that strike me as more common to women than men on the farm, such as her ease in analogizing between human and animal reproduction, the pride in hands-on care, and assertiveness when it comes to animal welfare (as in her concern for "upsetting mamma" about her "babies"). These talents are among the allegedly "feminine virtues" that men have often, even if ungratefully, relied on wives and daughters to supply. Trade magazine ads for farm equipment feature these talents through photographs of women at work in the farrowing house, an endorsement that appeals to producers in general.

Although significantly gendered, intense regard for livestock is certainly not a "uniquely female" trait. As Robin emphasizes, it is nearly impossible to tend animals on a daily basis without coming to feel for them in your gut. In some respects, she was actually much tougher than Roger, my boss on the farm who gave up farrowing when TGE first hit. He simply could not bear witnessing the illness with which she, in the end, proudly wrestled.

There are parts of her story that have a distinctly rural Missouri twang, as when she shortens "Doctor of Veterinary Medicine" to "the veterinary" or recalls running the border for TGE vaccine. But it is also obvious that she is way too intelligent and assertive to fit any "hick" stereotype. She has been very active in state and national organizations and early gained confidence, as she says, to bypass "Mickey Mouse" consultants in the countryside and "go right to the top" for information. At the same time, authorities be damned, if her own good sense should recommend it. If confinement buildings make you sick, get out of them; if men cut pigs too late, do it earlier yourself. It is hard to beat the image of thirty women, Robin among them, rushing to "feminine farrowing school" in the early 1970s, polishing their castration technique.

But even in this respect, I think she well represents other farmers who have long prized formal as well as practical education. About the only hayseeds to be found in the heartland are on *Hee Haw* reruns. Farming requires tremendous technical knowledge and commitment to lifelong learning. It is appropriately labeled "agricultural science" in land-grant universities, where the boys who are going back to Dad's farm have their own national fraternity, Alpha Gamma Rho. In fact, the vast majority of people who work on hog farms, including 60 percent of the hired hands, have had some college or vocational training. Robin's life and mine have been "exact opposites" only in the order in which we have moved from farms to books or universities to learn what is most important.[3]

Still, while American pundits since Jefferson have found noble lessons—democracy, dignity, self-reliance—in the lives of sturdy yeomen, modern farm men and women mark time with more tangible details. Of course, they care about grand issues in national life, but they are generally preoccupied with the simple bodily functions that many urbanites (and eighteenth-century slaveholders) consider unmentionable. A lot of those functions are dirty and dangerous and smell entirely ignoble. Contrary to urban lore, no one thinks that hog lots "smell like money." They smell like shit, and so does every exposed pore in your body after a spell of "working hogs."

Whatever might or might not gild their character, farmers—especially if they tend livestock—have little time to contemplate it. They are busy preparing meals, measured by the ton, a couple of times a day for hundreds of bovine and porcine dependents and keeping an eye on "that baldy with the red ear tag" that was listless last night. Depending on the locale and season, they are hassled to keep stock cool or warm, bedded and growing. Daily routines include manning the barricades against disease with the matter-of-fact resignation of a suburbanite checking the backdoor lock. And they are accustomed to strides that hit, not with the smack of sensible shoes, but the squoosh of boots in steamy manure. If they are zooming down the road, they are apt to be headed for town to get bearings, roll pins, or a half ton of mineral block. Or they are off to mend fence, to pick up some medicine from the vet or the kids from 4-H, to meet with the Soil Conservation Service, or to gossip, with greasy hands gripped round Styrofoam cups by the pellet mill's percolator.

The time markers in Robin's memory are also the usual ones. They are "biorhythmic" things, the frequency with which mamma sow talks to her pigs and the time of day for each chore. For most farmers "choring"—just routine animal care—fills about two-thirds of the day, and stock set their watches by it. Alter that tempo the least bit, and you are likely to hear bellowing and face an outbreak of indigestion. Hence, feed stores sell sodium bicarbonate by the fifty-pound bag (like a giant Alka-Seltzer to mix with stock rations) and Tums by the cash drawer (for the person who does the mixing). The remaining hours of every day are reserved for surprises—receiving an unscheduled shipment, working around a bad transmission or frozen hydraulics, "pulling a problem calf." The emergencies are so predictable that the counterpoint is written in the score.

Longer periods, months and years, also have a steady beat. The date of a photo can be identified by the bloodline of the boars in the background, or the pathogen of the month by the age of the kids in front. Robin, like most producers in the Swine Belt, kept three or four groups of sows, which bred and went through 114-day, staggered gestations before farrowing, weaning, and starting the cycle anew, two cycles per year. Cow and sow estrus could even substitute for months on the community calendar. Posters at ag stores usually include gestation schedules—long columns of dates matching breeding and birth for each species—but only an ignoramus would need to consult them.

While the scale and technology vary and change, the rhythm remains, and it exists for every crop and animal and the people responsible for them. The fuguelike quality of it all is so much a feature of the culture that most farm people are apt to know with a glance, say, when they drive by on November 23, what you will be doing the first week in March. If their prediction proves false, it would be neighborly, late March, to ask why. Something probably went wrong.

When the smell of TGEV-induced diarrhea wafted out the door, Robin and Ronnie could instantly extrapolate to accounts and farrowing cycles six months down the road and then back to Christmas gifts for the children next week. In a way, their lives flashed before them. The smell of that shit assaulted dreams and memories as much as olfactories. In fact, the smell was "sweeter" than usual, like poopy diapers when a child is on breast milk rather than solid food. But in the relatively sweet smell of TGE, Robin also sensed hundreds of deaths, heartrending visions, children's tears, fatigue, and the bank on their heels. Just: "Oh, God! There goes Christmas."

All that in a smell. When it hits, "*You know.*"

Well, you know, *if* you farm. That requirement is part of what makes farming a distinct way of life, one that, every beat from birth to death, is supposed to bind a person to work and workplace, person to animal, kin to kin, and future to past. The wish for such a "natural" alignment is far from extraordinary in America. People often express such an ideal and blame its alleged loss on a favorite villain. But family farmers are among the very few who are supposed to be invincible, to make the wish come true simply by virtue of their occupation.

Of course, it often does not. For over a century, more people have been leaving American agriculture than joining it, and domestic violence and divorce are as common in the heartland as they are on the coasts. For the past decade, increasing numbers of farmers have actively discouraged their children from staying on to work the home place. Groups that focus on rural youngsters, like the Future Farmers of America (FFA), have changed their mission accordingly. They shifted from vo-tech training for kids who might be kept down on the farm to public promotion of agribusiness. In 1988 to 1989, sixty years after the birth of FFA, delegates voted to remove the words "Future Farmers" from the name of the organization and its publications. It is now called the "National FFA Organization" (NFFAO), with no prescribed function for the letters "FFA" beyond their residual allusions.

But for those who remain in agriculture, no less (maybe even more) than those who wax quotably about it, farming is supposed to be different. Farming is supposed to be a practical and moral universe that you "grow up in" or (more perilously) "marry into." There is unrelenting, soft-spoken pressure to follow the righteous path of your (or your husband's) ancestors. This ethic of continuity is immediately visible in a pecking order that also reflects common American ideals.

"Everyone knows" that the "old-line" families light the way. In this part of the world, they are more likely to be ethnically Swiss or German and Protestant than their Bohemian and Catholic neighbors, but such contrasting fortunes are taken to indicate other virtues. For generations, the "best families" like Ronnie's (or Roger's or even Burr's, before all hell broke loose) have been models for their communities. Their success—the well-painted buildings with vigorous stock, the large (but not too large) holdings of fertile ground, the shiny (but not too shiny) equipment, their names on the roll of church and school donors—are taken as signs that they "must have done something right." They lend credence to the promise that acts of individual character attract material reward, a crucial ingredient in American dreams.[4]

These families are the ones whose farms are getting bigger as they rent or buy ground from neighbors and as they enter long-term agreements with the corporate giants that corner food markets. Much of their success may be credited to well-moved capital and to federal programs that favor expansion (also known disparagingly as "speculation" and "the government"), but they are unlikely to say so. Along with and on behalf of admirers, they will insist that their success came in doing what everyone ought to do. They worked hard. They kept the faith in diligence, ambition and moderation, practicality and experimentalism, each in appropriate measure. And God saw that it was good. Being part of an old-line farm family means you got what you deserve.

By "insinuation," if you are less successful, you also got what you deserve. An outbreak of erysipelas is less a prompt for sympathy than a reminder of how the Lord treats those who fail to vaccinate. And if you harvested beans when they were too dry to pass whole through the combine . . . ? Who knows what might befall you? With dispensation, you might get by with refinancing; without it, your noble rights and duties are revoked. You take a job in town or go "work for someone else," which is to become fundamentally transformed, even if the bulk of what you do everyday remains the same. In Robin's account as in most others', farmers turned hands have, in effect, fallen from grace. The hired man is nearly a breed apart. If he really knew what to do, if he really were convinced of the agricultural covenant, he would be an owner rather than a hired man. Right? At least that is the "insinuation." From top to bottom, the system of farm-family prestige is associated with a holy meritocracy.

Even within farm families at a given rank, there is a clear hierarchy of generation and gender. When son becomes dad, he gains veto power over farm decisions, and he will not cede that power lightly. Sons are usually thc ones raised to imagine inheriting the home place; daughters to "help," at best like honorary sons, or to "marry into" the in-law's operation. Everyone is busy in valuable ways, but usually men are the ones working where more of the family's credit is visibly invested, with larger animals and equipment. Women are more likely to know the books and to "do the midwiving, so to speak" with calves and hogs, and to manage care for kids, schools, church, and other community institutions.[5]

The closer you are to the top of this hierarchy, within and among families, the more tightly good management, good family, good character, prosperity, and God's grace are supposed to entwine. God help you if they don't.

These are among the reasons that, if you listen carefully and know the code, especially when there are silences, farm talk sounds emotionally dense, almost impacted. With the least magnification minute feedlot observations or memories of mere instants in and around it can incite laughter or tears.

It cannot be surprising that Robin speaks of farming as something she "loves." Most farmers do. To stay in the business, you have to love it—including birth and death as well as boredom, high finance, fear, and anger. Dad, the best families, Thomas Jefferson, and a vengeful God are looking over your shoulder. Have you done everything right? Followed the righteous path? If you do not love farming enough to also love all of that, just about any other line of work would be preferable. No wonder, whatever the macroeconomics, farmers' numbers are dwindling. I enjoy farming, no doubt, largely because I have been spared those pressures. At least I was until TGE hit, and I have felt that I understood farming better as a result ever since.

Part of that understanding comes in seeing how Euro-American culture burdens the business. Being "good" at farming entails confirming well-pedigreed fables about how individuals, generations, genders, families, and communities are supposed to fare, if given half a chance. And that is what European emigrants and their descendants have for centuries claimed in America.[6] Alibis or sympathy for failure might leap to mind if the sufferers were born in a ghetto or with a "birth defect," if they were too young, old, or recently immigrant. But for ordinary folk working the Heartland, the American errand is less forgiving.

Of course, some allowance may be made for random occurrences, "acts of God"—flood, drought, a sudden storm in the sky or "the economy." In farming as in the insurance industry, these are matters of "chance" rather than "risk." They are distinct kinds of things precisely because they cannot be subjected to human will. They "just happen," even to "the best people." People do the best they can, but God presumably moves weather fronts and stock exchanges with a roll of the dice. Nothing is overpromised (except that the dice are fair), and nobody is at fault if you lose. The house wins; that is all. But if other sorts of things—things that are supposed to be avoidable—can be implicated in misfortune, then you have violated a central promise of the society, one that elites regularly rekindle as a beacon of hope for the world.

Failure of any sort is, of course, painful, but it is eased when it is at least according to plan. Good is something you make, and bad, a mistake from which you learn or an act of God that you nobly endure. But what about diseases like TGE? They seem to fall neatly on neither side of the divide. They are neither entirely subject to nor free from human control. They defy the insinuated choice for heroics: struggle against that which is changeable or humbly accept that which is not. You can share in neighbors' pride when their crop is healthy and their grief when it just refuses to rain. But sickness, especially contagious disease, seems to pose a tougher challenge to modern hearts.

Part of what makes disease so revealing, then, is its location, so to speak, atop a pressure point. Plagues like TGE defy the usual calculus for explaining what is or is not supposed to happen. Americans seem ill-prepared for such mysteries. They can easily be interpreted in the manner of Chillingworth or Mather on the *maleficium* trail. They become evidence of witchcraft, like a strange mark on a woman's body, a milch cow gone dry, or a sudden disquiet in the neighborhood.[7] No one can be entirely sure how the evidence got there, but it may be a sign of something powerfully wrong, the sort of thing that is left when a child of the Lord has struggled with evil and lost. If you were to see such a thing or hear about it, you would not want to sit next to it in a café, either. Next time you contemplated a bleacher at a high school ball game or a feed truck rumbling toward your lane, you, too, might want to check local rumors for help in maintaining safe distance.

Of course, such a dramatic interpretation risks overstating superstition in the countryside. In fact, disease hysteria seems more evident to me away from the farm (for example, in hearing people talk about HIV or Ebola) than near it. With only her first diploma, Robin better understood the workings of contagion and immunity than most of the university graduates I know. But even with their relative sophistication, maybe in part because of the calling attributed to their craft, farmers still treat things like TGE as if they lay at a spiritual crossroads.

When it hits—whatever your understanding of "it" is—how should you feel? Which way should you turn? To find out, I decided that I had better turn to other sorts of folks who specialize in such questions. I headed to the halls of science.

PART V

DISEASE, BODY AND SOUL

Deep hemorrhagic infarcts—the phrase began fastening its hooks in my head. I had assumed that there could be nothing much wrong with a pig during the months it was being groomed for murder; my confidence in the essential health and endurance of pigs had been strong and deep, particularly in the health of pigs that belonged to me and that were part of my proud scheme. The awakening had been violent and I minded it all the more because I knew that what could be true of my pig could be true also of the rest of my tidy world.

—E. B. White, *"The Death of a Pig"*

Existence is precarious for the animals on the farm, as it is materially for many people, and as it is existentially for all of us, whether we recognize it or not. We can try to avoid this recognition with illusions of "agency," fantasies of staying young forever, and the distractions of "self-improvement," but it only lies in wait for us.

—Susan Bordo, *Twilight Zones*

The culture of democratic societies requires not only the civic virtues of participation, tolerance, openness, mutual respect and mobility, but also dramatic struggles with the two major culprits—disease and death—that defeat and cut off the joys of democratic citizenship. Such citizenship must not be so preoccupied—or obsessed—with possibility that it conceals or represses the ultimate facts of the human predicament.

—Cornel West, *Keeping Faith*

The single most important thing to know about Americans—the attitude which truly *distinguishes them from the British, and explains much superficially odd behavior—is that* Americans think death is optional.

—Jane Walmsley, *Brit-Think, Ameri-Think*

I learned this in epidemiology school: life is a sexually transmitted disease, and it always ends in death.

—USDA scientist Scott Hurd

EIGHTEEN

SOME SWINE DISEASE SCIENCE

ALTHOUGH I CAN MAP HOG-WAR FRONTS from press coverage and gain a decent feel for farming from personal experience, fieldwork, and the Internet, disease is another matter. In the last decades of the twentieth century, talk about it has been ubiquitous. Contagion of all sorts—best exemplified by mysteriously "emergent" sorts, such as AIDS—has been among the most pressing matters of public policy, media commentary, and individual concern. For people suffering with disease much of the attention has been too discordant, too little, too late. But amidst the cacophony have been a few tones that are so steady and harmonious that they often are attributed to nature itself.

Whatever ails you (at least if it is not something "in your head," something you were "born with" or "damaged" in an accident—a neurosis, birth defect, or injury), people will suspect that you "got a disease" and that it came from "germs." Every disease supposedly has its own distinctive "bug." In ordinary depiction, these germs or bugs are innately aggressive, swarming things, like microscopic street gangs. They lurk invisibly in dirt or ply damp, dark alleys and orifices of the body, especially the bodies of people who are careless about such things or unfortunate enough to have been born in their midst. When some germs (the pathogenic ones) "invade" the body, they "make" you sick. You have "picked up a bug," likely one that "is going around." If you are "strong" enough, your "immune system," a circulatory infantry, will eventually "fight it off." But in the meantime, you are the pathogen's "host." Especially, again, if you are careless or disadvantaged, you provide those little muggers a cozy hangout, opportunity to multiply and transit to other victims. Just walking around, even if you cover your mouth when you sneeze, you are apt to "spread" germs as effortlessly as you cast a shadow. If a pathogen is too strong or your "system" too weak, you need advance protection (say, vaccinations or regular doses of Echinacea) or reinforcements (antibiotics and chicken soup). If such a fortified system is still too "weak," you get very sick or even die. If only because this whole process is too small to see, people privilege the advice of natural and medical scientists with the instruments and expertise to observe and manipulate it. In general, the tinier the details of a disease story—the more it deals with molecules or at least cells in addition to organisms and populations—the more it is said to be "understood."

Of course, variations on these themes are significant. People who differ in their gender and sexuality, in education, ethnicity, age, income, religion, and locale, also tend to differ in their health expectations and experience. The combat lingo that colors so much disease talk is pervasive but hardly universal. Moreover, at the very same time that contagion has contributed to the authority of science and health professions, resentment also has been on the rise. There is growing suspicion that science is overrated and physicians overpaid. Nevertheless, since the 1970s nearly every newspaper and half-hour of local news has added its "Health Beat." Digital monitors are now integrated in fitness gear, and cookbooks contain passages that read like physiology or organic chemistry. Even people "into wellness" and "alternative medicine" are encouraged when mainstream scientists endorse their crystals, aromas, pressure points, or herbal regimens. And when your child is spiking a fever, trust in traditional medicine seems to spike, as well.[1]

Faith in veterinary medicine and associated science seems to be even more resolute. Among the American public, as best I can tell, vets are considered distinctly diligent and dedicated but otherwise plain folk. Since their version of medical school is about as difficult as any and the graduation perks relatively slim, they "must be" genuinely devoted to the field and the animals that are their patients. There is much the same response to folks who get a Ph.D. in quasi-medical fields like biochemistry or neuroanatomy. People rapidly calculate the requirements ("Gee, you could have been a doctor!") and credit researchers for electing a less cushy (and thereby presumably more principled) profession.

Through such understandings of disease and expertise, at least in agricultural lore, the organization of knowledge and the social order of professions are presumed to be intrinsically aligned. As Robin explains, farmers tend to prize relations with vets who are "up on the latest." The latest is the understanding of disease that is developed in universities, hospitals, and research labs. Cordial, regular contact with such outfits is a key qualification. As "good" veterinarians bridge barn and gown (say, by calling Ph.D.'ed pals, when stumped by your sickly sow), they keep cutting-edge science and practical husbandry in sync. Farmers count on the integrity of these relations, especially when the technicalities that are the substance of exchange—the science—so often remain alien. Know the right people (who know the right people . . .), and the science takes care of itself. Natural scientists produce knowledge that medical or animal scientists adapt, extension agents promote, veterinary clinicians transport, and farmers apply. If anything does not apply, the news feeds back through the same chain, one link at a time. Since ordinary intercourse forges those links, no comprehensive planning, monitoring, or supervision is required. At least that is the way it is supposed to work.[2]

Almost every field has its relatively "basic" ("pure," "theoretical" or "experimental") vs. "applied" subspecialties. At one "end," lab-coated ascetics foster findings with precise but purely abstract connection to ordinary circumstance. At the other end are rough-and-ready adventurers. Their explorations are more improvisational, more soiled by shit happening at the moment, but they are also more obviously related to the world as we know it. For example, a "bench-oriented" epidemiologist might test if a disease is airborne by placing a lab animal in a biosecure box, blowing into the box a stream of air that contains only respiratory requirements plus a known concentration of pathogen, and monitoring what happens. (This experiment would be replicated and contrasted with results under sundry "controls," etc.) A more "applied" or "field-oriented" epidemiologist might instead plot on a map actual outbreaks of the disease that have been observed and see if they followed prevailing winds more than other conceivable paths of contagion.[3] Of course, whole fields (such as microbiology vs. public health), even with their internal diversity, can be considered much more experimental (or, as rivals complain, "exotic") than others.

So, ideally each of the fields and subfields has its place in a grand continuum. Foundational, at least in the case of animal disease, is science that is both "pure" (lab-tested) and microscopic. At bottom, "understanding" rests on carefully controlled and perfectly replicated manipulations of the smallest possible pieces of pathogen or host—say, one piece of a compound that has been synthesized or extracted from one cell cloned from a standardized source (an in vitro sample). Then, if all goes well, that finding can be reproduced in experiments on a compound taken from a cell of an actual laboratory animal (an in vivo sample). Once passing both in vivo and in vitro experimental muster, findings may be referred, say, from biochemistry or microbiology on to the pure and then the applied end of the next "larger" field, such as virology or immunology. There samples might be whole cells or even organs, while other fields (e.g., epidemiology) might deal with several animals, species, or even global populations and clinicians with those sows that are coughing in your crates. In this manner, it is presumed, the "latest" deserves faith because it has been ratcheted as fully as possible through the basic to applied ends of fields that are well arrayed to trace understanding from the far corners of the living universe right down to individual atoms.

Scientists themselves, though, well know that the division of labor within and among disciplines is nowhere near that tidy. People trained in one discipline often end up working in another or joining a hybrid team. Analysis does not necessarily begin in any particular field or proceed in any particular direction, except maybe the one that at the moment is easiest to fund.

During economic upturns, basic (or "esoteric") science becomes less of a luxury. In downturns, more goal-oriented (or "practical") inquiry gains favor, as do particular maladies, when outbreaks or wonder cures make headlines. Around World War II, for example, research money was easier to come by if you were interested in antibiotics and vaccines; after the war, in cancer. More recently, also largely in response to AIDS, viruses and immunity have gained research priority. Some specialties scurry to reorient or repackage their strengths in response, while other fields—particularly microbiology in the past decade—are ready and able to jump to the head of the pack. Given such variability and change, at any given moment "understandings" are very unevenly developed. "The latest" may include thorough knowledge of a couple of chemical bonds that are theoretically implicated in pathology but next to nothing about how to alleviate the suffering of millions of creatures. Similarly, there are some terrific cures for maladies (such as aspirin for diffuse aches and pains) that remain pretty mysterious at the cellular and chemical level.[4]

Chatting with dozens of swine-oriented scientists and clinicians in the United States and Europe, 1993 to 1997, I encountered myriad tales of this ironic sort—half-bemused, half-hostile recollections of breaks in the basic-to-applied and micro-to-macro chain. It is certainly a pervasive and, I think, revealing body of lore. It bespeaks both an extraordinary commitment to the chain as a whole and concern for weak links. The tales often come up when talk turns from interdependence among diverse, far-flung associates to public or funding-source favoritism. Right after testimony of heartfelt solidarity, I hear about a "they" who have it so easy and a "we" who get no—or least not enough—respect. Lore that surrounds social boundaries of all sorts often has this quality, but the boundary between each medical and scientific discipline seems to have its distinctive variant.

For example, I asked John Butler, an immunodiagnostics consultant and med school microbiologist (whose experimental media include swine tissue), to compare his work with what I had seen at the National Animal Disease Center (NADC) and the College of Veterinary Medicine of Iowa State University in Ames. He answered:

Well, most of the people there are concerned really more with studying the microorganisms that are responsible for the disease with the intention that they will make vaccines from them. They're more interested in those subjects, okay? Now, there are exceptions, a few people in Ames—both at the university end and at the NADC—that are also interested in immune systems. But the funding of agriculture research, the political pressure, is such that a lot of the budget goes to people coming up with vaccines. So, that's why you see so much of it going on. It's much more difficult to convince a congressman that they ought to pay for studying an immune system than it is to convince him that they ought to pay somebody to make a vaccine.

Now, long-term, I think they would have perhaps been better off to spend some money to understand the immune system, okay? But, you know, Congress always wants a quick fix to every solution. They don't want to wait ten years for something to be solved. They want it to be solved next year.

Can you explain what that "long-term" picture is like, how it relates to a "quick fix" or a "solution"?

Well, I'll give you a good example, and that's coming from mouse immunology. You know, the NIH [the National Institute of Health] funded basic research in immunology and continues to do so. Through the 1970s, at least, it resulted in the identification of all of these cell-surface markers on lymphocytes, okay? Just because it was part of basic science, one wanted to be able to identify one lymphocyte as being different from another: how one kind of T cell differs from another kind of T cell, such as a CD-4 T cell from a CD-8 T cell. Well, it paid off a lot, because suddenly the retrovirus HIV came along, okay? And had it not been for some of the work that had been done in basic immunology—identifying CD-4, which is one of the main receptors for HIV—it may have taken another ten years. It was only a couple of years after the discovery of HIV until they hooked it up to the fact that the receptor was a well-known protein. So the investment in basic research paid off almost immediately. Many times things pay off that way. That's just one really good example.

Now, I could tell you another very good example in agriculture, but you know there are much fewer examples like that because agriculture has been very reluctant to fund that kind of basic research. They want a quick fix for everything rather than a long-term solution.

But I will give you a good example in agriculture, one that was done right over at the NADC in Ames, okay? There's this adhesion receptor on bovine neutrophils. . . .[5]

Or consider memories tossed in the opposite direction, from the more messy, "big-picture" applied sciences back toward the basic sciences. This one comes by way of a conversation with Jeff Zimmerman, a veterinary epidemiologist at Iowa State:

Yes, I remember hearing about ways (like with rhinitis) that, due to viral interactions, a vaccine might be less effective than lab trials would lead you to believe. But do you know other, sort of "classic" epidemiological cautionary tales? You know: How you think you have a problem solved but there's some complication?

Right, about how you figure you've got it "fixed," but . . . Right? Yeah.

Yeah, if I remember right, I got this story from Wayne Rowley. He's a neat guy, an entomologist over in the Department of Zoology. It's the funniest story . . . about DDT and malaria, back around World War II when DDT first came on the market. I'm not sure I'll get it right.

Anyway, there's malaria, and "We've got to wipe it out," right? They've got DDT and U.S. troops to go around and spray everywhere; kill all the mosquitoes. DDT's "wonderful," they think, because it will actually stick to surfaces and remain active. It will kill the bugs that land on those surfaces much later on. It's cheap to produce, a "wonderful" thing, a "wonderful insecticide."

So, they went around this village and sprayed all the huts. The huts are made of twigs interwoven tight enough to keep the rain out. Well, in those twigs live these lizards that eat the insects; so they won't bother people. The lizards are eating the insects; everybody's happy.

But then, when they spray these huts, of course, they poison all the bugs. The lizards eat the poisoned bugs, and then the lizards start getting sick. They wander around drunkenly in the open. And then a cat would come along and eat those stupid, drunken lizards, and the cats would die. Well, that meant there was no more check on the rodent population, on rats in particular. So the rat population proliferates. And because the rats are out of control, the rats started dying, and of course, the fleas start leaving those dead rats and go onto the people. Pretty soon bubonic plague gets in there.

So instead of people being sick with malaria, they're dying from the plague.[6]

These anecdotes are among the huge inventory that scientists maintain about ways that nature belies expertise. Someone—often in another discipline vying for funds—thinks too big or too small, too long- or short-term; they know a lot but do not know what in the world to do with it, or the world tricks them into thinking they understand, when they really don't.

Such perplexity in the sociology and lore of expertise can test your faith in the singular wisdom of "the latest" and in the ultimate coherence of science as an institution. In barnyard light, the difference between micro and macro or basic and applied research (in vivo vs. in vitro samples, immunization via injection vs. ingestion, etc.) seems too slight to render one "closer" to reality than the other. They are all reports from Mars. Between workaday science and Monday-morning chores stand piles of literal and figurative shit, of social hierarchy and happenstance stuccoed with wishful thought. But when pathogens hit the farrowing house and animals start to groan and die, who are you going to call? Whose advice are you going to trust? As concocted and wind-driven as the whole business may be, for farmers (like the rest of us, I bet) it is the doctor who is up on "the latest," whatever it may be.

TGE is actually quite unusual in that much of the disease story—basically everything except its moral—has been quite clear and familiar for many years. Ask just about anyone with a relevant doctorate, and you will be introduced to the same protagonists, settings, and potential plots. The canonical précis appears in the holy scripture of hog science, *Diseases of Swine*, now in its seventh edition. Abbreviated versions begin nearly every technical article and training manual, every pamphlet, World Wide Web page, and flier on the subject produced in the past couple of decades:

> Transmissible Gastroenteritis (TGE) is a highly contagious, enteric viral disease of swine characterized by vomiting, severe diarrhea, and a high mortality (often 100 percent) in piglets under two weeks of age. . . . In the densely swine-populated areas of the midwestern United States, TGE is recognized as one of the major causes of sickness and death in piglets. Swine producers are especially apprehensive about this disease because (1) mortality is high in newborn pigs; (2) there is no effective, practical treatment; (3) entrance of the virus into a herd in winter months is difficult to prevent because of the probable role of birds, especially starlings; and (4) the commercial vaccines available are of limited effectiveness.[7]

Just about every pig person knows that the culprit is a virus, TGEV, and scientists know a good deal more than that. It belongs to the family Coronoavirdae and the genus *Coronavirus*. The "corona" prefix makes immediate sense, given its physical form, its morphology. In electron micrography a slice of TGEV looks exactly like a crown, a circle with jeweled stubs ("peplomers") projecting around. In three dimensions, then, I imagine it resembles one of those harbor mines shown in World War II movies, an orb covered with ball-tipped spikes. The orb, though, is tiny, only 60 to 160 nanometers (billionths of a meter) in diameter, and each spike is only 12 to 25 nm long. Nevertheless, researchers have been able to dissect and analyze the virus, down to the exact glycoproteins in those spikes and the structure and function of just about every piece of the huge RNA strand that the orb contains.

Under common conditions, the virus is also quite fragile. It is essentially eternal when frozen, but it is very vulnerable to temperatures near or above those that are normal indoors or outdoors, come summer. Hence there is a bit of good sense in common impressions that the bug is around only in winter. Even on a cool spring day, just a few hours of sunlight render TGEV harmless, as it is in animals other than swine.

Viruses and bacteria are curious that way. Under normal circumstances, a germ can be innocuous in one species or even one region of the body—functionally equivalent to the healthy host itself—yet wreak havoc in another. It is this relationship, then, between specific sorts of germs and hosts, as conditioned by their environment, that makes for a "pathogen," not something essentially "good" or "bad" about the bug itself. Researchers, for example, have identified only about a hundred or so "zoonoses," infectious or parasitic diseases that can pass from livestock to humans (or vice versa), even by convoluted means. Improvements in food-processing sanitation, especially the pasteurization of milk, and efforts to eradicate certain livestock diseases (such as brucellosis and bovine tuberculosis) have succeeded in making actual incidents of zoonosis extremely rare in the United States. The ones that remain will be difficult to eradicate, short of scorching the earth.[8]

For example, one of the really tricky problems in dealing with salmonellosis (the cattle infection that worried Robin as both farmer and mother) is its ubiquity and variety. More than 2,400 serotypes of salmonella have been identified in animals and the environment. Only about half of the ten that are most common in U.S. swine herds actually make hogs sick. Four of that top ten are also on the list of culprits in human disease identified by the Centers for Disease Control and Prevention. Unfortunately, the three lists—the most frequently isolated salmonella serotypes in swine, the most frequently pathogenic ones in swine, and the most frequently pathogenic ones in humans—are not the same. Only three serotypes (*S. agona, S. typhimurium,* and *S. heidelberg*) are on all three top-ten lists. Since these bacteria (undifferentiated by serotype) are just about everywhere, and since the particular serotypes that harm humans are common enough in healthy hogs easily to escape attention, it is hard to imagine eradicating the disease.[9] Similarly, TGEV (or a virus awfully similar to it) is quite common in dogs, cats, foxes, and starlings who then shed it all over the place, but it does not seem to cause them any harm.

It sure does bother pigs, though. Once ingested, the virus well tolerates the warm, dark, acidic, and enzyme-rich trail that food travels. When it reaches the small intestine of a susceptible pig, TGEV rapidly invades the lining (the mucosa), particularly the villous epithelial cells of the jejunum (the middle section of the small intestine). These cells normally comprise a sort of shag carpet that soaks up nutrients and electrolytes and passes them on for distribution throughout the body. It is a challenging environment under any circumstances, and villous epithelial cells need regular replacement. But TGEV alters or destroys them too rapidly. Soon the villi—those shag threads—begin to atrophy. As they shrink, so, too, does the capacity for sodium transport and thereby the transfer of electrolytes and water. Among the first functions to go is an ability to digest milk ("hydrolyze lactose"), which means just about everything for newborns, and the concentration of undigested lactose itself compounds distress. The cumulative result is "an acute malabsorption syndrome." In effect, a fuzzy, spongy vessel becomes a sewerage pipe. Everything that comes in curdles in sour juices and then blows right on out. That process is the source of the distinctively foul odor of the disease. Piglets are especially vulnerable because they so depend on a steady supply of milk and because their ability to replace epithelial cells is so limited. If infected before they reach three weeks of age, pigs will nearly always die of dehydration, metabolic acidosis, and then heart failure. In susceptible animals, the disease runs its entire course from exposure to death in 18 to 72 hours.

This is the course, though, of only one variety of TGEV infection. It is the much-dreaded "epizootic" sort that "hits" so hard. It is the kind that Phil, Roger, Robin, and I vividly remember, because everything gets so sick and death among piglets so surely follows. It happens only to herds that are "naive." They have not been previously exposed and therefore have no immunity specifically for TGEV.

Alternatively, infection strikes a herd in which all or at least most of the breeding stock has survived an earlier bout, say, the prior winter. Such disease-seasoned sows have lifelong active immunity that they share via colostrum with their young while they nurse. If TGEV is reactivated or reintroduced under those circumstances, the outbreak is "enzootic," a milder "hit." Typically, the sows will not even appear sick. Suckling or recently weaned pigs will likely develop diarrhea but, depending on their age and protection from other challenges, only one or two in ten will die.

Since piglet "scours" (diarrhea) and death are fairly common, anyway, and since those symptoms have many possible origins, enzootic TGE often goes undiagnosed. Farmers could easily think that they just so happen to have a little more problem with scours on their place than others. What animal scientists term "chronic enzootic infection" farmers might consider a bit of "bum luck." As long as they "continuously farrow" (keep mixing new, naive stock with immune ones that also shed the virus), they will be perpetuating TGEV infection and immunity to it. Since in so doing they are also likely to be spared the infamous, horrific symptoms of "full-blown" TGE, they could easily believe they "never got it," at least not a "bad strain."[10]

If they mix naive and immune animals intermittently and are relatively isolated from sources of contagion, herd immunity (and hence significantly reduced vulnerability to epizootic TGE) will last about as long as the breeding life of once-infected sows. So, if worse comes to worst, some farmers figure, you should just expect that TGE (or some other damned thing) will wipe you out every six years or so. There is no point in worrying about it. You would best think of it as a sad fact of life, like drought or a rainy day that you save up for rather than avoid:

> Some diseases take time to reach the "flash point." . . . If you're adding gilts, or boars, or both without isolating them from your herd for a time, you're headed for trouble. Think of it as adding sticks to a fire. . . . The sow herd is your "fire." . . . "It doesn't give off much heat. But you bring in 20 gilts. Two weeks later, you add 20 more. Two weeks later still, you need 40 gilts. Ka-boom!! It doesn't matter if it's PRRS, influenza, mycoplasma or whatever—any of those health issues are potentially capable of causing a health explosion."[11]

From this perspective, the question is not whether but when catastrophic infection will occur. Washing germs out of the hog house is about as promising as pissing in the timber to prevent forest fires.

Likewise, even producers who know and credit the scientific story can consider well-placed filth essential for animal care. Enzootic infection might be a relatively small price to pay for protection from the epizootic variety. With that reasoning, some old-timey farmers regularly, purposefully spread TGEV (say, by bringing in bedding from an infected herd well before farrowing) to be sure that immunity is established, that pigs are not "naive" at an inopportune moment. This is the sort of folk vaccination procedure that Robin considered and that is still common among small-scale producers.[12]

The only other well-credited option is the more high-tech route: biosecurity—all-in/all-out, shower-in/shower-out, segregated sites, positive air pressure, air washes, and all—to maintain a physical barrier between naive hogs and sources of contagion. God help you if it breaks.[13]

These tales of disease and coping options—ranging from ignorance-is-bliss or raise-the-white-flag to bunker mentality—are the essential, uncontroversial ingredients in TGE lore that just about every scientist and clinician knows. But there is much more to it than that. "The latest" has been accumulating quite a while.

TGEV was first identified back in the mid-1940s, about the same time and within the same scientific circles as human pathogens that were the target of celebrated breakthroughs. But for its unresolved plot, the TGE story would be just as inspiring as that of polio and its protagonists as heroic as Jonas Salk. Mindful of such Cold War-era precedent, Americans expect pure science and prevention to advance together. Hence it is especially galling that TGE research has enjoyed so little practical payoff. Of course, effort on behalf of pigs and pork profit are less concerted than they are for people, but you would still think that, after fifty years, someone would have come up with a reliable treatment or vaccine. In this respect, the frustration for hog farmers and embarrassment for swine science is comparable to the feelings of people who care about AIDS. One can only hope—and I certainly do—that yet more dedicated care, including support for research, will eventually reduce the suffering. But TGE suggests the horrifying possibility that it might not, at least for an excruciatingly long time.[14]

In the meantime, however, the relevant basic science has advanced tremendously. Fortunately, in the case of swine disease, that science might yet yield payoffs that are unanticipated and that in significance dwarf hog interests. In their difference from humans, for example, pigs may be extraordinarily helpful in building understanding of life with disease of all sorts for all species. Sitting amidst the flurry of activity in his microbiology lab, John Butler explains:

> That's what we're interested in here, because the swine is a unique model in the field of developmental immunity. Mice (which are the immunologist's favorite) are like humans in that the maternal antibodies as well as a whole variety of regulatory factors—like interleukins, cytokines, and lupokines, these kinds of things [molecules that are produced by white blood cells and that help coordinate their disease-fighting activity]—are transmitted in utero from the mother to the fetus. And—let's be fair to the people who have used the mouse—it is economical. The mouse has a very short gestation (twenty-two days); it reproduces many times in a year. People therefore have been able to develop all of these pure, genetic strains of mice, which are essential for research. It would be tough and expensive to do that with large animals. Europe has gone for the mouse,

too. Now, the National Institute of Health—which of course is interested in human health—will essentially only fund research with animals if they are mice. Otherwise it's "agricultural."

But in many ways, the immune system of a pig—IgA and IgG antibodies, the MHC antigens of cells—these have a very high degree of homology with humans. And, even more important, pigs—farm animals in general, ruminants like swine, the cow, the horse—do not transmit any of these factors in utero. As I always write down in my little reviews, herdsmen 2,000 years ago—before "science" ever came into being—knew that a calf or a piglet would die if it didn't suckle its mother, while a human newborn could be fed cow's milk. It may not be the best thing to do, but it's not going to die, okay?

So a piglet puts us in a good position experimentally, because we can study all of the factors that drive immune system development—passive protection, regulatory effects. . . . We can follow exactly how the mother is affecting the development of the B cells. In a human or a mouse it happens before birth, but in a piglet you can regulate and observe the whole process.

Like in that place in the picture on the wall there, in Hannover, Germany, or in Ames, Iowa, we have neonate-biotech, germ-free facilities. Piglets are taken by cesarean section and then moved through a sterile transfer device. It's a long plastic tunnel that goes from the site of the C-section into an isolator, a germ-free animal incubator. Those animals are actually germ-free—not just SPF [selected pathogen free] but germ-free. They have not come into contact with *any* bacteria. We have gone at least eight weeks with a pig raised in an isolator, and it responds to antigen, to immunization, as a newborn piglet. There has been no development of their immune system. We think it sits there on idle, okay? We can feed them purified antibodies; we can feed them all kinds of things, and it is all computer operated —- how much and when they eat, a snooze cycle, washing their trays and weighing everything—so we can know exactly how these things affect immune system development.

Among the things we're thinking is that there are two things that drive that development. One is the maternal factors that are transmitted through the colostrum, including antibodies and all the other things that haven't been identified. And then the other are materials released from the gut flora.

You mean just normal bacteria, like E. coli, strep, the kind of stuff that is everywhere?

Yes. You see, mostly what we do here is work on immunoglobulin genes [the ones that control the production of proteins of which antibodies are made]. All these guys you see around here are gene cloners—okay?—because we're really interested in the development of the system that makes the antibodies, what we call "the B cells." The B cells are the type of lymphocytes [a variety of white blood cell] that gives rise to everything that makes antibodies, and we're interested in everything that affects the B cell's development. Like: What about the microbial environment in which the animal lives, the "normal flora"? These microorganisms live in the gut as commensal organisms, and they release various things through activity or their death, which also quite clearly stimulate the development of the immune system.

It's a very tough area, very complicated with thousands of organisms. You know, the gut is a very, very, *very* complex microenvironment. And our interest is, not identifying all the organisms that live there, but understanding how that very complicated microenvironment has an important role in the development of the immune system, specifically in the development of the B cells, which give rise to the antibody-producing cells.

Like endotoxins—not "*exo*toxins," which are produced and secreted by bacteria cells (if the bacteria is a pathogen, some of them will kill you)—but "*endo*toxins," which are inside the bacteria and released only when they are destroyed. The endotoxin is a lot more benign substance, but it's present in huge quantities. And it is the material in the endotoxin that causes the release of various lymphokines and neutrokines [messenger molecules] from the enterocytes [intestinal cells], like the epithelial cells in the gut. And then these factors affect the lymphocyte population, start it developing in the normal manner.

That's why there was a substantial change, at least in academic medicine in the 1960s or 1970s about the use of antibiotics. Antibiotics often kill off the "normal flora." And the normal flora, you realize, also occupy the surface of the intestine. It is attached to all these cells. So, if you kill the flora with antibiotics, then you leave those cells open for the pathogens to take up residence; you actually make the organism even more susceptible. The best examples of that are in human urinary tract infections, you know, where continual use of antibiotics especially in women, kills all the E. coli that are the normal ones and produces worse problems. All these pathogens take over.

So, to put it crudely, a certain amount of germs or "dirt" is essential? And too much "cleanliness" is just unhealthy?

Well, maybe you should change your categories. Its really more a case of whether the floral environment is a normal commensal flora or it's a pathogen.

We're continually surrounded by these commensal organisms, one living on the other. It's just a normal part of the ecosystem, okay? And if you want to call that "dirty," that's *real* "dirty"! But there's nothing wrong with that. It's how the system works.

You know, I often think about how, biologically speaking, ruminant animals are extremely healthy creatures. Like, with a cow . . . it's rumen changes everything, this whole swarm of life and digestive system. It makes you wonder if we might be better off to have a ruminant that could eat all of the grasses and benefit from all those byproducts that must be made in the rumen. It probably has a lot to do with the health of the animal.

It's when there's that one oddball that gets in there. It is a pathogen because it has some virulence factor, some enzyme, some toxin it releases, something that alters or destroys a cell in the body.

But it's a very subtle thing. As you know, like in bacteria, genes can move from one organism to another by things called "plasmids" [small DNA rings, accessory genes]. We do that all the time in the lab. It's just a tool for gene manipulation, but it's a normal process in life, too. E. coli can transfer plasmids from/to each other and just might transfer a plasmid which encodes genes from an organism which is pathogenic. So you might have normal and pathogenic material in the same E. coli, see? Very subtle.

So how does your immune system know the difference?

That's a good question. You know, there is a phenomenon in the immune system we call "tolerance." That is something that develops when the organism is still young, and in general that tolerance has to be maintained. In other words, there are examples where continuous exposure and tolerance are good for life. The organism, being exposed to normal bacterial flora, this gut flora, at an early age begins to recognize those as, not foreign, but as "self." That's why—unless you're suffering from an autoimmune disease—your immune system is not attacking you. It sees the normal flora and the rest of your cell-surface chemistry as "self," okay? It's when one of them mutates or when a new bacterium comes along (it doesn't have to be a pathogen) that the immune system will recognize it as foreign and develop a response that could eliminate

> it over time. If it stays around, as long as it doesn't really harm the host, another type of tolerance is induced and maintained. But if it is new or mutates and becomes destructive—giving off new proteins, peptides, carbohydrates—as far as the immune system is concerned, it will be recognized as foreign, and it will respond to eliminate it, once that line is crossed. But it's a fine line between foreign and self, and there's a lot of unknowns in where that line is. You can imagine![15]

There are a number of lessons to be drawn from these stories of modern biological science. First is the potential benefit of looking in unlikely, even silly places for help with serious problems. Seemingly esoteric experiments, such as those on mice with designer genes or pigs in neonate bubbles, in fact can cut years off the time needed to develop tests and treatments for disease. Even if only in happenstance and hindsight, just about anything could fit the doctor's order. Hence, although frequently typecast as bean-counting compulsives, scientists credit imagination no less than certainty in the pursuit of knowledge.

Second is the attraction of considering all sorts of organisms models for each other. Swine in particular are pretty decent as model people. Relations between the two species have, for example, made hogs attractive as human-tissue donors (e.g., for heart valves and skin grafts) and laboratory animals.[16] Butler points out a fundamental kinship in their immune systems that makes their developmental difference amazingly convenient for research. In such ways, it is a particular *kind* of "likeness"—the quality of similarity between two species rather than the sheer quantity—that allows one to speak to another.

Third is the value of recognizing both diversity and continuity among God's creations. At one level or another—from the vantage of a single cell in the gut or of an immigration official at La Guardia—but for a few "oddballs," we are all "commensal" organisms. The distinction between "self" and "foreigner" is itself unclear and unstable, a concocted, necessarily dynamic, heuristic fiction. What stands between any two of us—say, between a healthy organism and a variety of E. coli—is less some fixed, essential difference *in* us than a chemical and cellular process *between* us, in effect, an ever-under-appeal judgment about the history of our interaction: How well we have been getting along lately? Nearly all of the time it goes amazingly well. Hence, functionally speaking, nearly all of the vast hordes of microbes within you *are* you.

Fourth and finally, there are reasons to appreciate an elegant, almost anarchic messiness in these relations. Our bodies—whether human or porcine—are probably healthier to the extent that they have borne an incalculably intricate history of relations with alien beings, welcomed those that act neighborly and defended themselves against the few that prove troublesome. Even those defenses are stronger to the extent that they are impure, dynamic accretions. Gearing up for "oddball" foes has less to do with hunkering down and warding off than accepting nurturance and the "whole swarm of life"—viruses, bacteria, their corpses, and byproducts—that are in and around us. It is such happening shit that keeps us alive.

This interpretation of scientific lore has its romantic appeal, one well-suited to current turbulence. When elaborated just a bit, it prophesies divinity (or at least a brighter "on the other hand") amongst disorder, diaspora, and plague. Modern Americans distressed by the multi-, post-, and polyglot moment can take heart. Maybe what appears to be dangerous here and now is just a small, essential part of a heavenly plan.

But let's not get carried away. It is wise to remember, as scientific lore also emphasizes, some microbes are dedicatedly nasty neighbors. Insofar as they bear prophecy, it is more of the Old than the New Testament variety. A bug like HIV or TGEV does not seem to me the sort of gift that God grants some of His flock for the sake of their tender care; He gives it to them to scare the bejesus out of the rest of us. Relatively rare though they may be, they cause next to nothing but agony and death among their hosts. When the hosts are commercial livestock—only more plainly than when they are human—these pathogens also devour lots of dollars, the lymphocytes of the body politic.

NINETEEN

CONTEMPLATE "ERADICATE"

To REDUCE UNCERTAINTY, pain, and public expense, the state hires scientists and clinicians to manage a list of "regulated," "reportable," or "quarantinable" diseases, the epidemiological equivalent of an FBI Most Wanted List. Once a bug has been so designated, its fate in livestock is less dependent on the faith, philosophy, and resources of individual farmers and their vets. It is a public enemy. Sundry officials aim to track it down and then "control," incarcerate, or exterminate it.

For more than a century, national governments have embarked on such germicidal crusades. It was, in fact, a swine disease—hog cholera—that inspired one of the earliest and most dramatic of U.S. successes "eradicating" a pathogen. Ask just about anyone in public health about the practicality of eradication, and they will remind you, "We haven't seen a case of hog cholera or even vaccinated against it since 1978."[1] A standard text explains:

> Eradication of a major livestock disease is expensive, requires large amounts of manpower, and may continue many years without success [as in the case of BTB, bovine tuberculosis, which persists in pockets despite effort since World War I]. Even considering these things, eradication programs that have been successful have proven profitable in all respects. Hog cholera was eradicated from the United States in fifteen years at a cost of about $140 million. Prior to that time, hog cholera cost swine producers nearly $50 million a year in swine deaths and control measures. In addition, eradication of hog cholera opened numerous foreign markets for breeding stock that had previously been closed.[2]

When compared with other ways of responding to disease, eradication is a final solution. For its advocates, the name of the game is not just easing the suffering of an animal or even protecting whole herds from an occasional outbreak; it is getting rid of the bug itself, once and for all: "To eradicate a disease, the causative agent must be eliminated from a defined area. The area may be as small as a farm or a ranch, or as large as a continent. Eradication is the only perfect means of disease control."[3]

Hence, much of the new technology in pork production (early weaning, biosecurity, all in/all out, shower in/shower out, etc.) can be considered instruments of eradication on a small scale. More commonly, though, the area eradicators have in mind is as large as a nation or a cluster of trading partners. Through a program that combines testing and vaccination with quarantines (restrictions on the movement of animals and associated products), agents of the state erect barriers around every source of contagion. Animals that have been exposed to infection must be perfectly identified and segregated. Then every one of them must die. The strategy goes by the name "depopulation." In effect, it is a house-by-house search-and-destroy mission. The "causative agent" must be denied safe haven, even if every hideout in the hemisphere must be burned to the ground.

Of course, the alternative—never-ending rounds of vaccination or recurrent infection and the suffering and cash it costs—is hardly very attractive, either. But even the word "eradication" has Draconian overtones, like Indian "removal," Westmoreland's "pacification" of Vietnam, or Sherman's "March to the Sea." In *Hog Cholera and Its Eradication: A Review of U.S. Experience*, the USDA explains how, in its extremity, such a response to disease bears spiritual as well as financial rewards:

> During the years of the hog cholera campaign, hundreds of federal and state veterinarians and other field personnel had direct experience with the rigorous measures necessary in a large scale stamping out program. The logistical problems of safe disposition of large numbers of condemned animals, of organizing strict quarantine and inspection over large areas, of quickly establishing field laboratories, and of meeting the many concerns of affected livestock owners had become a part of their career backgrounds. . . . This extensive practical knowledge can be quickly and directly applied to combat future introduction of any foreign animal disease.[4]

According to its veterans, the hog cholera campaign taught public health officials to be combat-tough.

I was pleased, however, to find that one of the many heroes of U.S. eradication efforts, a retired veterinarian whom I will call "Paul Reimer," was both as committed to eradication as ever and about as thoughtful and sensitive a gentleman as you can find.[5] I was "out East," mainly around Washington, D.C., chatting with sundry USDA officials and pork industry lobbyists, when Dr. Reimer agreed to meet me at the Amtrak stop near his home. Since a thunderstorm knocked out the power in the whole city just before my train rolled in, he had to walk down eight flights of stairs to get to his car and pick me up. For the next several hours we talked nonstop, and it was clear that Reimer was anything but a megalomaniac. Despite an unlikely beginning, his career put him in the middle of nearly every national and hemispheric effort to eradicate livestock disease in the twentieth century, a total of nearly a dozen different campaigns.

> My father left and then my mom died when I was very young; so my brother and I had to pretty much raise each other. Initially, I was a high school dropout. I went to work for this vet, $3 a day, holding cows while he tested them for brucellosis and tuberculosis. Of course, I'd never seen anybody like this before, because I was born in the city. Everything impressed me, you know? Going out onto the farms and talking to these farmers, you could tell that he had a respect for them and they believed in him. And I decided that's what I want to do.
>
> *So you still remember him clearly?*
>
> Yeah, sure. He told me about his career and how foot-and-mouth disease had got into the United States and how he was sent off to help eradicate it, what the problems were associated with it, how bad the virus was and all that. All that stuff impressed me. He was so committed, so dedicated. He was the first one I ever ran into in my life who was trying to do something that made me feel: "That's what I want to do."
>
> So I had to go back to high school and finish, even though I'd been out two years. And I did. I just really pushed. I kept saying, "I'm going to be a veterinarian."
>
> I ended up at this small college. This would be around 1938. I was in a boardinghouse up there with nineteen Jewish boys from New York. They'd been trying to get into med school, and they couldn't; so they were up there. A lot of our teachers were displaced from Europe, extremely capable people who had fled Hitler. But then, after a year up there, I found out they weren't accredited. So, God, here I am lost for money, and I need to get into vet school. I went down to Washington, to the BAI [the Bureau of Animal Industry] to see what I could do. They offered me that same job, holding cows, $3 a day: "You go out down to

Blair, Texas, and work for a year, and then you can go to the state university as a resident."

I suppose at that date, it wasn't bad money, but I had worked my way up to foreman on the late shift, up from $19 to $34 a week at a factory. But I was committed. And when I got to Texas, they knew it, and they were kind of supporting me. Boy, that was quite an experience.

A break came when they needed somebody in the laboratory in U. City, where they were collecting all these blood samples, sending them off to laboratories for brucellosis tests. So, they transferred me to it. Same pay, but my job was to keep the lab clean. For security reasons, I could sleep in a room off the lab. So I could get money for tuition, washing test tubes in hours when I didn't have classes, study and sleep right there.

After five years, I graduated. And because they had been so good to me, I decided—rather than go into private practice—I needed to stay with the government and at least pay them back something for what they'd done for me. They transferred me up to Rhode Island, but now as head vet of the brucellosis laboratory there. But it wasn't a big deal, because it is a small state. There was just an assistant and me. In Texas there were at least a half-dozen lab technicians just reading samples. But in Rhode Island, in addition to running the lab and reading the samples and hiring some practitioners part-time, I had to go out there and test cows.

I remember, there was a guy—I had been out to his farm two or three times—and I was out there testing his cows. "I think you're a pretty good vet," he said, "but I don't think that guy reading these tests is worth a shit." I just laughed, and didn't say anything more about it.[6]

These words come from just one person selecting memories that can speak to a younger stranger. But this elder also happens to be among the Americans responsible for an alliance of science and the state in its take-no-prisoners approach to disease. He spent his professional life tackling one plague after another—testing cattle and swine hither and yon; wandering jungles on horseback and tundra by rail; helping inoculate and destroy millions of livestock; working with acrimonious, acronymic public health agencies of Europe, the Americas, and the Caribbean; coaxing pharmaceutical company executives, aboriginal peasants, government officials, professional colleagues, and anyone else who might doubt that eradication has its virtues. And these memories show that such crusades need not be motivated by lust for combat or glory. The attraction has much in common with that of other vocations: a chance to model oneself after a mentor, to extend a tradition, to build a life of service that deserves public trust.[7]

But in addition to the pull of such generic psychic and social reward, there is the push of experience with disease itself. Here, for example, he describes a breakdown in standard disease-control measures (including the requirement—a statutory legacy of earlier campaigns—that garbage be cooked to destroy bacteria and parasites before it can be fed to hogs) that led to the spread of African swine fever to the Americas in the late 1970s:

> Every year our people, along with all the head vets of 160 countries, meet over in Europe, the Office of International Epizootics in Paris. And they knew African swine fever was running wild in the Iberian peninsula, in Portugal and Spain. But we hadn't alerted our people at the ports to be concerned about it until it was too late. I could have crucified our people. The disease spread through food products off airplanes coming out of Portugal. Here again, the garbage feeders had spread it, getting garbage at the airports and feeding it [uncooked] to their hogs. When it broke in the Dominican Republic it spread like wildfire.
>
> In looking at it, it's just like the most severe case of hog cholera you ever saw. It affects a pig's immune system almost like AIDS. Every organ of the body has hemorrhages throughout it, ulcers in the intestinal tract. And they die. That's one disease—everywhere it's broke, Spain, Portugal, the Dominican Republic, Haiti, Brazil—hogs have died so fast that they're everywhere, dead hogs floating in the rivers.[8]

Nevertheless, the shift from routine defense to full-throttle offense against disease requires determination. Even with its clinical name, "depopulation" (the wholesale sacrifice of carriers for the sake of future generations) is not only technically difficult but also disturbing to initiate, witness, and recall:

> The fact of killing those animals—I hated that part of the whole thing. To see those people lose them—awful. But you've got to have a frame of mind that—geez, it's awful hard to kill all these animals, but in the long run they're going to produce healthier ones that aren't going to have to live with these kinds of diseases.[9]

His story of each campaign is peopled mainly with dedicated staff and grateful beneficiaries but also with a few obstructionists who must be enlightened, bought off, or outsmarted. There are impoverished people who must be swayed to alter routines that, he worries, keep them just barely alive. What ties together the memories of germicidal combat is less any soldier-of-fortune bravado or hostility toward pathogens than compassion for their victims.

A half century later, for example, he recalls going to Newport News, Virginia, the gathering point for livestock that the United Nationals Relief Association acquired from all over the United States. Massive motley herds awaited shipment to Europe, where they were desperately needed in the wake of World War II. It was a horrific vision:

> Geez, all kinds of things broke. When I got there I could see almost every major horse disease out in those pens. Normally, you couldn't see all this stuff in a lifetime. Remember, too, that this was before antibiotics or vaccines, right? Sulfa was just coming. So we didn't have much we could do. The losses were very high. It impressed upon me what happens when you have exposed stressed animals and the consequence of it.
>
> They used to teach us in school, you know, when you get hired by the farmer, you've got to make your decision based on the economics of it. If his animals are going to die, then you get him to sell them as quick as he can, because otherwise he's going to lose everything on the farm. But when I went to Newport News and saw all these animals sick from this exposure to animals being shipped like that and saw the consequences, I changed my mind. This "great idea"—move them if they're sick—is just giving the problem to someone else.
>
> Of course it's tough to stop. Ever since Mexico [the campaign against foot-and-mouth disease, 1947 to 1952], we've had these arguments. I've been through this eradication/control thing all me life. People would get upset: "What is the definition of eradication?"
>
> It means to pluck it out by its roots! So, it takes a lot of time and effort and people and money to find those roots.
>
> When you have one of these outbreaks, people go through three phases. First, panic: "Do anything!" Next is cooperation—when everybody tries to get together to eradicate. And third is apathy—when you're trying to get the last remnants.
>
> It's like looking for a needle in a haystack. And people don't want to do that. "Hell, I want to get back to my business. I'm tired. . . ." And they try to walk away from it. But you have to keep fighting until you've got every last thing out of there. That is the difference between control and eradication.[10]

Of course, not every disease can be eradicated. Some viruses, such as HIV or influenza, are so various in strain, so genetically variable and volatile, that it is hard to imagine corralling them once, much less for all. A candidate for eradication must be stable and "understood," at least well enough for scientists to set sights.[11] Bill Mengeling, a veterinarian and virologist at the National Animal Disease Center, explains just a bit of how hard that is:

> We know that during the acute stage of illness most infectious agents can be transmitted directly or indirectly among animals kept in the same general environment. However, we know less about how some of the same agents are disseminated over long distances, or how they survive between epidemics. One of our major limitations is that (simply because of low probabilities) we can't always reproduce natural, farm conditions. For example, we have available for our use ten isolation rooms, space to hold about 100 young pigs. But there are about 90 million pigs slaughtered each year in federally licensed plants in the United States—90 million "experimental units" raised on hundreds of different farms. Imagine a starling stepping in the feces of a pig that has diarrhea because of TGE, and then flying to another farm (perhaps miles away and until now free of TGE) and carrying the causative virus with it. Well, you can see the potential for transmission without any clear evidence for its cause. It may only happen once in 100,000 times. But it may happen a lot with 90 million pigs on hundreds of farms without ever being confirmed by experimental data.[12]

Nevertheless, before even imagining eradication, people have to know how the disease spreads. There must be reliable, affordable tests of exposure, quarantine protocols, and if possible, a single, standard vaccine—one that stimulates the production of immunofactors that both well target the pathogen and remain distinguishable from those stimulated by the pathogen itself. Ideally, then, once the disease has been "understood" and "brought under control," naive animals will be protected through vaccination. Continued testing can determine if the pathogen as well as its clinical consequence have been eliminated. After a few years of negative tests, vaccination also can be phased out. Should tests reveal a surprise return, just "knock it out"—quarantine and depopulate the site of infection and vaccinate around its perimeter. The key to a "large-scale stamping out program" is both the initial blitzkrieg and the hair-trigger sentinels permanently stationed on its flanks.[13]

Theoretically it may be possible to take such an approach privately, one farm at a time, but no amount of effort will suffice if neighbors, suppliers, truckers, and packers with whom you trade are not allied. Everything you barricade out the front door can just slip back through the cracks. Even fastidious hog buildings, for example, are vulnerable to dust- or airborne infection when they are close to farm-to-market roads. It hardly matters that the place is scrubbed and strung with barbed wire, if tractor-trailer loads of pathogen shedders are bouncing by upwind.[14] Hence, eradication efforts tend to require, not only scientific understanding and reliable technology, but also the united support of the livestock industry and the resources of national and local government. At issue are the bounds, purity, and permeability of the state. Which forms of life does the body politic ignore or include, and which are its enemy aliens?

For better or worse, in American experience with food animal pathogens, that fundamentally political question—should we tolerate or should we eradicate?—has, as Reimer explains, been driven by "the economics of it":

> Unless it's a matter of public health—people, not animal—the first thing in this country is the economic significance. What it comes down to is: Eliminate the disease to produce food at minimal cost.
>
> The U.S. probably has had an advantage in that we have so many animals that depopulating—getting rid of an infected herd—is less traumatic to us than it would be in other countries. But to control disease you have to spend, like we did with hog cholera. Before we eradicated hog cholera, the sales of vaccine—not the cost of vaccinating, just the vaccine itself—was $25 million per year. There hasn't been a hog vaccinated for hog cholera since 1978. So, when you look at the economics of eradication, that's where you get you payback.
>
> I look at the vaccine as an operating cost. If you can eradicate, then you eliminate it. That's the benefit to producers. Plus, of course, some countries won't allow imports unless you can prove that your herds are pathogen free. So, we're fortunate to have producers, particularly certain leaders, realize: If we can get rid of some of these diseases, let's get rid of them.[15]

I must admit, as much as I find Reimer charming, as much as I respect his compassion, generosity, and resolve, I remain skeptical. I am inclined at least to quibble with the bookkeeping. If the question is the promise of eradication as an objective, why balance only profitable accounts? What if the cost of less successful campaigns (e.g., against BTB) were figured into the "payback" for winners like hog cholera? I do not think "the economics of it" would look so good. This is not to say that money has been squandered; only that the financial benefits of eradication are less dramatic than its signal successes suggest.

To be honest, though, my skepticism is less calculated. "Eradication" just sounds so aggressive and violent to me. Of course, I agree that it takes a cold heart or appalling cowardice simply to accept the devastation that pathogens can wreak. Thank God, humans have managed to rid the earth of some of His most hideous creations. Along with eradication advocates, I cannot help but fantasize about nuking not only "new" bugs like Ebola or HIV but also old ones like trypanosome, the protozoan blood parasite that is transmitted by tsetse flies to cattle, which then fall victim to African sleeping sickness (nagana). A standard textbook laments:

> Vast grassland areas of Africa could produce more than enough protein and fiber for the human population of the continent, but where the tsetse fly prevails, nagana prevents the development of anything but inefficient nomadic animal husbandry. Effective immunization against nagana has not been devised, and attempts to eradicate the tsetse fly have failed. Meanwhile, a large part of the human population of Africa exists in poverty, and with minimal or insufficient protein.[16]

Hence, maybe I remain skeptical of eradication only because I am writing from the comfort of my American home with its well-stocked fridge. Surely plague and starvation add luster to alternatives. Or maybe my reluctance can be traced to the ways that history conditions our allusions.

> *I'm interested in this because so much of what you're talking about comes in the postwar period, and the word "eradication" sounds so "military," you know? And I wonder . . . like these parasites, bacteria, and viruses are living things, too. They sort of hang out in the world, and the idea that you can somehow remove them from the face of the earth—it puzzles me.*

> Well, you keep right on asking me. See, that's my whole career. I've always worried about how we're going to do it. Could we do it? Like in dealing with foot-and-mouth disease in Mexico [1947 to 1952], it *was* a "war"—a war to eradicate the disease.[17]

For Reimer and his contemporaries, I suspect the word "war" mainly evokes memories of the "good one," World War II, followed by Korea and the Cold War. Targeting a pathogen resembles fighting a fascist or containing a Stalinist aggressor. For me, though, the wars of reference are "bad ones," like Vietnam and Nicaragua. In my lifetime, the United States has mainly fought revolutionary patriots. Killing livestock to eradicate disease sounds an awful lot like napalming villages "to save them."

In order to be sure that that my reluctance is not so easily chalked up to privilege or period effect, I ask:

> *So why haven't people decided to eradicate TGE? It's awful. It's been around for a long time. It's well understood. It's as common and costly as things like pseudorabies, which have been targeted. People are talking now about eradicating "new" things like PRRS and salmonella. So why not TGE?*
>
> Even though TGE is very severe in piglets and this type of thing, my feeling is that it hasn't reached most producers in the States to where they're screaming that something needs to be done about it. And unfortunately, that's our system. It's driven by complaints.[18]

To gain a clearer picture of that "system," I interviewed some of the people whom government agencies currently employ to draft, interpret, and enforce the regulation of animal and plant health. I chatted with administrators working for the USDA in Maryland and Washington, D.C., for the Animal and Plant Health Inspection Service in Colorado, their counterparts in the Netherlands and Denmark (the leading pork producers in Europe), and their representatives to the European Union in Brussels. I followed up with visits to their "people out in the field" and "lab techs" scattered about the United States. From top to bottom, nearly every one of them was trained in biological or medical science and was determined, above all, to improve the monitoring and control of contagious disease. They wanted to help the political process better reflect good science.[19]

What should not have surprised me is their nearly universal complaint that science suffers expedient abuse. Trade agreements such as NAFTA or GATT, which sanction medically sound public health relations, are hijacked to afford one set of producers advantage over another. Officials of some import countries, for example, trump up foreign disease threats to justify nontariff trade barriers, such as absurdly elaborate monitoring requirements or embargoes against exporters whose herd health and biosecurity are, in fact, superior. Disease monitors find themselves beating swords, not into plowshares, but into pseudo-medical instruments of trade war.[20]

The closer you get to the local level, the harder it gets to detect any singular idea, much less scientific consensus. A legislature, for example, enacts codes aimed to protect public health as animals ply public roads and rails. Before livestock can be shipped, someone must certify their health. If they carry a certain disease (one of the "reportable" ones) or if they visit a place where exposure is likely (a "terminal site," such as a stockyard or exposition where animals of diverse origin are mixed), they must be quarantined or killed. Hence, with some medical justification, by law in ordinary, actual circumstances a hog like Wilbur in *Charlotte's Web* would never return from victory at the county fair to Zuckerman's farm; he would be delivered to a packer and butchered within twenty-four hours.[21] As plainly, painfully good-science as this procedure may sound, the showring-to-slaughter routine bespeaks other interests. It also affords 4-H'ers cash reward for their effort (maybe even a small profit) and confirmation that their animal met market as well as club standards of agricultural performance. No doubt, some of the support for this routine—the public acceptance of regulation and the cooperation of packers—comes from such "extraneous" concerns.[22]

The purpose or function of any regulation is in this way understandably "impure," but there are enough vagaries in the regulatory apparatus as a whole to frustrate a rational mind. It is, after all, political. Scientists work for the government, which responds, albeit grudgingly, to a multiform electorate, and influence is tough to move upstream. Reasonable people disagree about the best approach to a particular disease, the reliability of the attending science, the level of tolerable risk, and the efficacy of restraints. Furthermore, within public-health agencies themselves there is more at stake than health alone. Each sector has its precedents and routines, each varyingly vulnerable to the influence of "interests" that are staggeringly diverse. In a pinch different levels and branches of government credit the expertise that preserves jurisdictional prerogative. Even strong commitment to principle—say, to eradication—may or may not be associated with the will and capacity to effect it, if only because principle, funding, licensure, inspection, and enforcement tend to be responsibilities of different bureaus.

TGE and pseudorabies, for example, stand right next to each other in my own state's Most Wanted List of hog diseases. According to administrative code, testing, reporting, and shipping requirements similarly apply to the two.[23] But everyone knows they are elaborated and enforced for one (pseudorabies) and not at all for the other.[24] You would think that TGE were missing from the list. This was among the curiosities that led me to chat with Walter Felker, the state veterinarian in the Bureau of Animal Industry of the Iowa Department of Agriculture and Land Stewardship:

> By legislative code, being an agency of government, the Department of Agriculture has the authority to write the administrative code. It has the authority to declare a disease to be reportable and quarantinable.
>
> *Yes, and part of what I'm interested in is how TGE is treated so different than pseudorabies, even though the two look pretty similar in the code.*

Well, TGE should not be in there. It should be removed.

Yeah, but it is *there in the code. See, it's written right here. So, how did it get in there, then?*

Oh, back when TGE made its appearance in the United States. I remember the first case I saw, when I was in [veterinary] practice. It came from Indiana, this direction, I suppose, in the early to mid-1960s.

When it first hit there was a big wave, an attitude in the industry: "Well, maybe we'd better report this and eradicate it," or something of that nature. It was kind of a novelty at the time.

They wanted to restrict the movement out of those [infected] herds, until they learned it was kind of like the common cold. You know, there's no sense in quarantining everybody that's got a cold, or there'd be no way to conduct commerce.

And since TGE is such a widespread disease, with very little commitment toward eradication or control beyond what you manage on your own, it really should not be in there. It doesn't cause an awful lot of concern. If somebody's got TGE, we don't react to it. The farmer knows how to handle it.

We could remove it from the list. But things traditionally don't get removed from rules very easily, you see? But it would be no problem if they removed that tomorrow. There's no public-health aspect to it.

Well, I wonder if it is more or less a problem for different folks—small vs. large producers, packers, or whatever. Like, when you talk to producers who want eradication programs, is there a difference in the way those producers influence policy, some generalization about how that works?

Oh, yeah. Any eradication thing, anything the government does—let's face it—it's influenced largely by complaint, you know? Many things are, you might think, almost totally "complaint-driven." And the very small producer can be a very vocal complainer. They call half the people who influence a decision, who make the policy. *We* don't get money for eradicating, you know; *they* make it, very effectively lobbying pro or con on an issue.

The reason pseudorabies is moving ahead is that the political muscle of the swine industry in Iowa is putting the funds out in the feedyard to eradicate it. It is the purebred producers, the breeding stock suppliers, who have probably been "carrying the water," as they say.[25]

Since our conversation was interrupted by phone calls about every two or three minutes, I knew that I had the right person to tell me about complaints and disease control. Some of the complaints came from people who wanted better enforcement; some from people who wanted to evade it. In either case, they were questioning a particular instance or strategy of disease control rather than the desirability of control itself. No one was ringing the phone off the wall on behalf of free-range pathogens.[26]

I did, though, meet a fair number of experts who were in the business of drumming up public support for germicide in general. Not surprisingly, representatives of the companies that manufacture and sell vaccine were prominent among them. For example, Mark Welter, vice president of research for Ambico, Inc. (the main licensee for TGE vaccine in the United States) was extremely generous with his time. He patiently explained how he, too, had misgivings about unnatural procedures, painful injections, and early-weaning technologies. He felt compelled to respect farmers' and consumers' desires, economic realities, background flora, and the wonders of a normal, healthy immune system. But he also felt that with foresight and determination it was possible and desirable through vaccination to eliminate TGEV once and for all. It had been done with hog cholera, thank God, so why not TGE? For that matter, why not other ubiquitous pathogens? He even helped me appreciate his vision of producing yet more "natural" vaccination through "recombinant plants." Common foods could be genetically engineered to include protein that is a harmless version of the one that pathogens ordinarily produce. Eat the fruit and you stimulate immunity to the truly harmful variant. Get the right hepatitis-A, cell-surface proteins to grow in tomatoes, for example, and then all you need for vaccination is some catsup on your French fries. Eventually, by denying pathogens vulnerable hosts, you eliminate them. Thanks to the vision and vaccine that his father, the founder and president of Ambico, C. Joseph Welter developed, TGEVicide is just a start.[27]

Unfortunately, independent advocates of a TGE vaccine are hard to find. Of course, there are some, but they are decidedly in the minority among researchers and clinicians. Once you get out of Dallas Center, where Ambico is headquartered, just about everyone claims to know that the vaccine pales before the disease itself in inducing protection from future infection. As shit happens, so to speak, the vaccine just does not seem to work very well.[28]

Other scientists—in particular, many of those "county extension" types whom Robin as others malign—envision eradication as a potential reward of more germicidal, state-of-the-art "management." As the trade magazines well document, enmity toward disease is part of the cutting edge. While old-timey farmers put up with chronic infection (such as enzootic

TGE), high-tech big guys nip the whole business in the bud. By segregating early-weaned pigs from their SPF (selected pathogen free) parents, they achieve "high health status." Less feed is "wasted" coping with infection or vaccine and building immunity to it.[29]

Since I resist this boast, I was pleased to hear some of the people who first developed these technologies endorse my skepticism. At least the innovators whom I met promise much less than their promoters. Although they aim to minimize disease of all sorts, they also suspect that outbreaks are inevitable. Biosecure, multiple-site production, for example, is best considered an insurance policy, a way to spread rather than eliminate risk. When that inevitable outbreak does hit, at least you will not lose everything. After painful experience of precisely this sort in the mid-1990s, trade magazines slowly deflated the eradication promise of early-weaning technologies.[30]

Nevertheless, along the germicidal trail, I also discovered yet more evidence—much of it in line with the latest biological science—that allowing disease to "run its course" may, at least on rare occasion, have its own attraction.

In the Swine Belt of Europe, for example, where the concentration of livestock, manure, and pathogens is even denser than in the United States, some disease threats have waned without the benefit of human intervention or complaint. Just as epidemiologists predict, mutations of pathogens that perchance spare their hosts have an advantage over more lethal competitors. European experience with TGEV itself seems to have taken just such a fortuitous turn.[31] That is among the possibilities that I discussed with TGE expert Dr. A. P. van Nieuwstadt when we met at the Dutch equivalent of the National Animal Disease Center in Lelystad:

So, what sort of work are you doing with TGE?

Well, there is not TGE here in our country. I hope that is not disappointing for you.

Ha! No, that's one of the reasons I came. In the U.S. as here, there's a program to eradicate pseudorabies (what you call "Aujeszky's disease"), but there isn't one for TGE. I gather that there never really was one. But, as you say, even though TGE remains a problem in the U.S., it is no longer one in the Netherlands. I wonder how that happened, how TGE could disappear, apparently without anyone doing anything.

Well, I think TGE has eradicated itself. See, TGE was first described in the U.S.A. in 1946, first recognized in Europe in 1958, and first reported in the Netherlands in '62. In '68 it became possible to make reliable diagnoses by a technique of immunofluorescence [IFT]. . . . Every year since '75, when that technique was introduced at the regional health services, outbreaks of TGE were diagnosed in our swine population—every year, until 1988. During the last years, there was a rapid decrease in the number of outbreaks of TGE and then, in 1988, we had our last outbreak.

The reason for this decrease and the end of outbreaks might be that another virus came up, PRCV (porcine respiratory corona virus), which is antigenically related to TGE virus. We think that PRCV has its origins in TGE virus. (It has also been detected in the United States, but we think that virus might have a separate origin.) It's in all Western European countries, and in all Western European countries we have observed a coincident decrease in the number of outbreaks of TGE.

So, how does PRCV resemble a mutation of TGEV? How so?

Resembles, yes. It looks very much like TGEV. The sequence of nucleotides, the genome can be determined precisely and compiled. In the genome of PRCV, they have detected only a few deletions when compared to TGEV. But it has lost the capacity to multiply in the intestine. So it has what you call "tropism." It grows and multiplies in the epithelium of the upper and lower respiratory tract only. So, while TGE virus multiplies in the intestinal tract (as well as the respiratory tract) and causes diarrhea, PRCV has lost this capacity. PRCV infection has no clinical signs.

But it seems, from circumstantial evidence, that the antibodies induced by PRCV protect from a challenge by the TGE virus. It is very difficult to prove experimentally, but some protection has been found. Similar experiments have been run in the U.S., France, England, and Belgium.

Although benign in itself, PRCV may provide some opportunity for bacterial infection. The combination may cause serious lung disease; so I would be reluctant to use it as a vaccine. But there have been experiments showing that PRCV works very well as a vaccine.

TGE is an example of a disease that has eradicated itself by producing new variants.[32]

So, in this case, a nasty virus spontaneously mutated into a more prolific strain, one quite harmless to its host but via that host deadly to its own progenitors. PRCV is at best an ersatz vaccine, and given its rapid spread, there is reason to worry about its own mutations or interactions with other pathogens in the future. But at the moment, it seems to work better in saving European animals from the TGE scourge than anything humans have been able to concoct in a half century of concerted effort.

In the interim, evidence mounts that biosecurity can—in some cases, over the long haul—be worse than unnecessary. Manifestly beneficent acts of germicide may, as it turns out, have subtly dire side effects. For example, when challenged by PETA people who allege cruelty in hog confinement, I normally counter with two points. First, note that for many generations hogs have been selectively bred to thrive in such environs. Rooting around in pastures would be no more "natural" for them than it would be for naked suburbanites to roam the jungle.[33] Second, note that enclosed, hard-surface, easy-to-clean floors allow farmers to limit swine exposure to worms, which are likely to plague them in a wallow. I stress this point, because such rudimentary sanitation seems so obviously reasonable, like wearing shoes outdoors or washing hands before you eat. Who would want any creature to have to compete with parasites for nutrition?[34]

But there is now reason to believe that sparing creatures that competition may encourage other health problems. For example, gastroenterologists hypothesize that recent increases in the incidence of chronic gut problems such as Crohn's disease in the industrialized world are attributable to success in reducing parasite infections among children. What appears to be human "disease" may be better considered yet another sign that developing immune systems benefit from "dirt," from interaction with a diverse and challenging environment like the one in which the species evolved. Sanitize the environment and you may find yourself more vulnerable to it.[35]

This is how John Butler responded, when I pressed him for a disease-control bottom line, at least in the case of TGE and market hogs:

> You know, we look at it maybe differently than the USDA or farmers might, because many times what agriculture looks for is a foolproof way to vaccinate animals, okay? You have to realize that part of the problem in raising animals—and I came from a farm, too, right?—is just animal husbandry, just common sense in management. I mean, all the great vaccines in the world aren't going to save your animals from death if you've got a horrible environment, mud and cold, for them to be in. You know, you can't make the world safe for fools. But if you put the animals in a normal, healthy swine facility, if the dams are immunized properly beforehand . . . it's been shown time and time again that there's adequate

immunity transmitted to the piglets. If they suckle properly, that's it. It works.

And what about eradication, you know? Yeah, you can control the effects on an individual through decent facilities and immunization, by vaccination, feedback, or whatever. But why not shoot for eradication—total biosecurity with herds that are naive and won't shed the virus—and you wipe out the disease altogether?

Well, it hasn't worked very well in the human situation.

We thought we got rid of some diseases and—because people stopped vaccinating—suddenly we have them back, things like polio and smallpox. People thought, "We could just have a population that is disease-free. Then, theoretically, it's not possible for anybody to contract the disease. Right?"

Except it never works that way, because there's always new people coming into the population that you can't really, totally control. In the case of the farming industry, somebody might inadvertently move from a hog farm in southern Italy and have dirty boots and bring them to Iowa, or something like that. I mean, unless you're going to live in a totally pathogen-free facility (like what we use in experiments), you're always going to be in that situation. I don't think it's practical for farmers, especially when there is so much transfer going on.

You know, it used to be all small, individual farmers that were fairly self-contained operations. Now you take feeder calves or pigs growing up in one environment, gradually developing active immunity there, and move them to another. You take animals that are carriers and put them in an environment where the rest of the animals are naive. And suddenly you've got an epidemic on your hands. Plus, you've stressed the animal out by putting it on a train or a truck, banging down the road all day. It's frightened and stressed out. You know, you probably couldn't do anything that would be worse for a breakdown of the system that keeps disease under control.

So, you know, it's quite clearly the case that the solution with newborn organisms, including humans, isn't to raise them in a totally germ-free environment. They'll never be able to survive in a conventional environment.[36]

Such experience counsels the troubling conclusion that sanitation and eradication, the effort to achieve "high health status" through barricades and germicide, may be exceeding the point of diminishing returns. This is among the impressions that I shared with Eldon Uhlenhopp, director of Veterinary Field Services at Iowa State University.

Is there something unscientific or just plain dumb about my fears of biosecurity, "naive" stock, and "pathogen-free" environments? I know that neither one of us would wish disease on anyone, but might illness over the short run be better in the long run? Is there reason to doubt that pigs (or people, for that matter) are better off building immunity than avoiding infection?

So you want citations for this idea that "high health status" herds are at risk? Is that what you're asking?

Yes. We're on the same wavelength.

It's a traditionally accepted concept.

Oh, okay. That's one of the things I'm concerned about because—to me, anyway—calling an animal "high health" and "susceptible" sounds like a contradiction, but they mean the same thing, right?

Yes. That's why Tim Stahly [a colleague at Iowa State University] has gone with "high and low immune status."

See, "high health status" first evolved to mean a herd that is very healthy. But they're not immune stimulated. They're very susceptible. And so Stahly has approached it from a different viewpoint. Pigs that correspond to a "high health status" herd he calls "low immune status." And it's probably more accurate. But it's hard for a veterinarian to use that adjective "low," because that's what we've striven for all these years. We'd like to think we're shooting "high."

Yes, I've often wondered how the structure of science . . .

Hold your thought, because I've got to interrupt you. See, that's why pig medicine is so critical in looking at human medicine. Because what are human beings? Especially in the U.S. they're "high health status" or "low immune stimulated" populations, generally. So that whole concept is interwoven.

Exactly! I'm glad you said that. And I didn't make you say that!

Yes, unsolicited testimonial!

What the microbial world is telling us is "Okay, you high-health-status people, we're still here. Even though you haven't seen the signs in the populations for decades, we're still here." . . . I think people have to realize that they are a population. More and more, we do the same things with people that we do with pigs. That is, we put them in small areas and commingle populations. The ones who have never had disease before, if they don't recognize the contribution of these little building blocks for their health status, are going to fall victim.[37]

Nevertheless, fantasies of final solutions persist. One of the most prominent of all swine scientists, Stan Curtis, recalls a meeting back in the 1980s where the agenda was particularly germicidal:

I was called out to the University of California at Davis for a two-day symposium. It involved all these different departments, animal science, veterinary medicine, and so on. They were trying to get the state assembly to fund a big grant to "take care of all the ills of animal agriculture in California."

All ills? All species?

All! Everything! They invited all of these politicos in and so on.

Now, you're going to get a charge out of what this man told me. We're talking about the former dean of the veterinary college there. He was at the helm of the conference, sort of the "grand statesman" of the veterinary college.

At the reception on the evening of the first day, it was just the two of us, and he said, "I'll tell you—with these new biotechnology tools that we have in the veterinary research labs, you give me $25 million today; in five years I can have eradicated every major disease in every species of animal in California."

I just looked at him.

And then I said, "Are you kidding?"

He said, "No."

I said, "I want to tell you something, sir. I am not a veterinarian. But I have studied a bit of microbiology, and I think I have a hell of a lot more respect for those little bugs than you do."

> Look: The reason those diseases are what they are is that they're not very well adapted yet. These sort of pathogens that kill their hosts—they are crazy. That's why eradication works. It works for things like cholera and so on—I mean, diseases that are highly contagious and have clear symptoms, 70 percent mortality rate regardless of the kind of exposure, the animal, the environment, and so on.
>
> But most of the real bastard diseases are not that way. It is the chronic ones, the insidious smoldering diseases that are well-adapted, that are tough to fight. They only get a foothold when the environment is wrong. Or maybe they only cause an infection but don't cause a disease unless the animal is stressed. They only bloom when other factors come into play. These "multifactorial production diseases" are the ones that are actually extracting the big toll. They are the ones that no one, not some guy in California nor anyone else, is going to eliminate. They are going to be around forever.
>
> What we have to learn how to do is how to live with these things, how to provide environments that are supportive of the animals and let them do battle with these natural enemies as best they can, and quit kidding ourselves that we're powerful enough to overcome these microorganisms! What we do with these living creatures is just so egotistical and foolish![38]

Complementary cautions come from many others—especially scientists or clinicians who have both farm memories and an academic appointment. They seem most common, almost formulaic, among epidemiologists. When, for example, Jeff Zimmerman and I met in his lab at Iowa State, he began with both an obligatory homage to the hog cholera campaign and an admission that he was "not a big promoter of eradication programs." Although "there are some bugs out there that you're better off just not to have," you probably are wiser to think about ways of living with them than plotting their annihilation.

> The biggest thing seems to be being able to manage the ecology of your own unit, making sure you just don't commingle exposed and unexposed animals.
>
> Like, Al Leman—you've heard of him, right? The editor of the last edition of *Diseases of Swine?* He died last August [1992].
>
> He called himself a "disease agnostic," which I think is an interesting term. I think he changed later on, after he got a pseudorabies outbreak in one of his operations. He wasn't quite as devout an agnostic after that. But he was convinced that the trick was management. And I think he was right.[39]

So, even under extreme duress, the key bearer of the Holy Scripture of swine disease himself remained convinced that disease was no demon. Within at least this sect in the church of science, eradication is the recourse of backsliders.

On a trip to Ohio, I pressed Linda Saif—the scientist who coauthored the gospel of TGE in Leman's Scripture—to see if there were some sort of bottom line:

> *So what is the deal, anyway? Where should the research and medicine go? What can you do about TGE?*
>
> That's part of the problem. You can have outbreaks, even in well-managed operations. Even the most stringent barrier systems break down. You still get TGE. And vaccines are marginally effective. In the face of outbreaks, for most people, their only chance to save some animals is feedback [well-timed exposure to the wild virus], which is what most producers use. Part of the dilemma is that there is no highly effective control.
>
> If eradication were feasible, I would favor and support such efforts. However, it is questionable whether this can be done, not only practically, but also because of economic constraints. If the disease can't be eliminated practically, then I have offered my suggestions for approaches that could be used or developed to control the disease.
>
> *So is that your main direction?*
>
> Yes, that's been our major emphasis. As I said, we know that's possible from the natural-case scenario. Animals exposed to wild-type TGE virus recover and protect their litters and have fairly long-term immunity. We know that.
>
> *Is that the best we can do? Even though they'll still be getting some infection, having some setback, and shedding pathogen?*
>
> There may not *be* 100 percent protection. In other words, the ideal scenario that you want is protection against disease, but not necessarily 100 percent protection against infection.
>
> If you protect against disease, you're protecting against the thing that's most critical in terms of the death of the piglet. But, on the other hand, you'd like to see them at least get a sub-clinical infection during this time, so that they become a serum-positive, actively immune ani-

mal. The next time they see TGE virus, they have active immunity protection against it.

If you had complete passive protection—so that you had absolutely prevented both disease and infection—those animals would still be susceptible later on. They would not develop good active immunity.

I want to be sure I understand, because what's going on in the pork industry—for that matter among humans, I think—is a greater emphasis on avoiding infection of all sorts as much as possible: pathogen-free environments and breeding stock, early weaning, biosecurity. I've heard it said that "surrendering" to infection is old-fashioned inefficiency. Why tolerate pathogens that can be avoided for the moment and thereby, in theory, eventually eradicated? Why "waste feed" on the whole cycle of infection and immunity?

Well, I think if you had that scenario, you'd have what you have with humans with HIV. HIV in humans is a good example of what you have with a compromised immune system.[40]

Here the extremes of purity and danger converge.

When the prospects are so risky—so clear but limited in demonstrable promise and unlimited in imaginable harm—why engineer for "low immune status," for maximal naiveté to infection? Why does eradication remain attractive? Whatever the best explanation, I do not think the reasoning is uniquely swinological. Faced with modern human no less than livestock epidemics, many Americans—I among them—wistfully imagine ridding the world of their "causative agents" once and for all. At the very least, we are exasperated by parents and principals who let sick kids sit next to our kid at school. If only we could seal them off. . . . These are, in effect, eradication and biosecurity fantasies.

On the cultural right, bolstered by homophobia and white racism, many among us are even willing to sacrifice citizens living with contagious disease like AIDS to limit its spread: "Barricade the border; let 'em die in Haiti or some San Francisco hospice, away from 'our' bodies, public hospitals and schools." Such de facto "depopulation" whimsy, though too plainly cruel for polite conversation, is easy to overhear in impolite circles, easier than it is around the USDA. But even on the prissy cultural left, hymns to inclusiveness and biodiversity seem to fade when the verse turns to viruses.

I detect hypocrisy when "nature" or "wilderness" disciples treat pathogens, not as sacred elements of a wondrous whole but as ephemeral instruments of divine warning for man. Blockbuster jeremiads like *The Hot Zone* and *Outbreak* treat newfound germs as if they were alien harbingers: Stay out of tropical forests or expect punishing plague. It is a message that shockingly parallels the Pilgrims' rationalization of aboriginal genocide in North America more than three centuries ago. Why should we continue to believe that disease is a sinner's penance or quarantine and germicide the calling of saints?[41]

Maybe in the human case, where fear strikes so close to the heart and where hosts are ready targets for bigots, we should not be surprised by the contemplation of final solutions. But in the case of food animal science—pork production, for goodness' sake—you would think that less drastic measures would prevail. Since stamping out pathogens and promoting immune-deficient stock are so questionable, why would big-guy investors risk millions of dollars to do just that?

One possibility, of course, is that the germicidal side of high-tech agriculture is only its veil, a convenient cover for power grabs by the big guys. They can afford expensive games and drive little guys out of business by requiring them to play along. A second, more forgiving and less purely porcine possibility is to remember that disease elimination technologies, as questionable as they may be over the long haul, can look awfully good when compared to disease itself. Routine quarantines and biosecurity, antibiotics and immunizations, even the occasional antipathogen blitz deserve much credit for great leaps in the percentage of piglets (and children) that survive infancy and for reductions in suffering thereafter.

In the end, I do not think eradication in itself can be found right or wrong. The same skeptical spirit that makes it seem questionable counsels against such blanket judgment, pro or con, regardless of circumstance. Why consider disease free of circumstance, anyway? What we can see, though, in treating eradication as an idea is an alignment among tenacious and ubiquitous categories at work in American culture: health vs. disease, self vs. alien, purity vs. danger, spirit vs. body, and friend vs. foe.

Arguments about the way we best respond to disease subtly invoke yet more challenging questions about who "we" are and the ultimate purpose of our being. The pragmatics of medicine, politics, farming, and finance gain some of their straightforward quality by presuming and then effacing a hallowed ground. Of course, Americans are not unique in so orienting themselves. In fact, I would be even more alarmed if they did not. I am pleased to live in a society that, pretense to the contrary, still arranges the relations among people, pigs, profits, and pathogens in a spirit of ethical conviction. What seems to me less comforting is the substance of those convictions—the demonization of disease, impatience with suffering, and a penchant for shunning or brutalizing the beings and bugs with which it is associated. Winding through the halls of health science, the TGE trail counsels more reverent regard for the impure, painful aspects of mortality itself.

TWENTY

THE BODY PORCINE

SCIENTIFIC TALES OF DISEASE are far less diabolical than those I heard or imagined around the farm. Hence, I always left interviews in schools of veterinary medicine or research centers feeling better, almost relieved. In particular, I was glad to have put some distance between viruses and the Angel of Death or other *malefecium*. Threats to body and soul still seemed substantial but considerably less daunting.

Yet, too, confusions remain: Why does this TGE virus seem so intractable in the United States? What ought we do about that appearance? What in general *is* the best way to think about the relationship between "pathogens" and other living things? Are plagues or at least sorts of them best considered avoidable? Or, following Rodney King's plea, should we—people, pigs, and viruses, living things large and small—just try to get along? Do we just wait patiently for the worst to pass? Or, moved by the suffering that disease brings here and now, should we pursue final solutions, biosecurity and eradication, once and for all?

It was comforting to have a clearer sense of how these questions figure in expert experience and of how they might ultimately matter, technically, morally, and politically. Even though I still have no sure answers, doctors and researchers inspired confidence that I was a little closer to them, certainly closer than I was the day that TGE broke in Phil's farrowing house. With their help, for example, I could follow a fair share of what passes for sure answers among people who have been trying to find them for years, decades before the questions even occurred to me. The intricacy of the stories, the Latinate expressions, and laboratory paraphernalia helped convince me that my problem, the nagging uncertainty, was mainly a matter of my ignorance. Mainly, but not entirely.

Something is missing in their stories that is important even if short of crucial and hardly intentional. That something is surely not diligence or data or even a sense of humor. These scientists are sharp, dedicated folk. Our conversations easily move from acronyms for protein strands to stories of love for research or medicine or the children at home. Everything they do (except, maybe, fighting for grants) is intimately and concretely connected to life. But still, there is something about the way they talk that seems a bit too concrete or not intimate enough. At least that is how they seemed as the National Animal Disease Center (NADC) faded in my rearview mirror or as field recordings replayed in the hours after work on the farm.

Two particular sorts of exchanges most prompt that sense of omission. One comes when these experts treat me as a source of "real-world" information. For example, a NADC swine virologist asked, "Before we go on, let me ask you: How much do farmers still have trouble with TGE?"[1] Of course, he might have been doing a bit of opinion research of his own. But all I could think was: "Hey, you're asking *me*?! If I know more than you, viruses have nothing to fear. You folks don't get out much, do you?"

Actually, they do not get out much, and for good reason. No farmer is apt to welcome down the lane anyone whose workplace is also a pathogen warehouse, no matter how effective the environmental controls. Hence, as a matter of policy, most research scientists avoid dealing with food animals in ordinary farm environs. Basic biosecurity and public relations require it. Many of them fondly recall Mom and Dad's home place or others that they visited in early years of medical practice, but many of them (especially researchers in colleges of veterinary medicine) are also correct in feeling disconnected from contemporary, workaday agriculture, even in comparison to me. Of course, life around the grain bins is no more authentic or all-encompassing than it is anywhere else, but it is decidedly different there than in a laboratory.

Just how it is different comes out in another sort of exchange, those in which questions that seem "normal" or "realistic" to me simply baffle them. For example, I asked a group of NADC researchers their response to a hypothetical "old-timey" hog farmer, a man like many I know who decides to put as little energy as possible into avoiding TGE. He does not vaccinate or attempt to isolate his animals from the pathogens that cling to his boots or blow in the pasture breeze. Instead, he banks his money, time, and energy for the outbreak (the "epizootic" variety) that is apt to occur, at most, every half-dozen years. A scientist's response is a look of pained amazement: "How could he do such a thing? Doesn't he care about his pigs?" Since the amount of caring does not seem to me relevant (*how* to care seems the issue),

I backpedal and simply assert, "Well, such people exist, doing the best they can." I try again: "Well, what about other breaches in biosecurity that can be expected? Like . . . in most towns there's a farmer—in the Swine Belt probably one or two for every elementary school—who usually invites the second-graders out to the farrowing house to see newborn pigs. What about that sort of thing?" Again, the response is pure incredulity. "That's just inviting disaster!" he gasps.

In such moments what is missing, I gather, is a sense of how messy farm life is. Almost everything is barely, if ever, under control or in its place. Grease, corn dust, manure, blood, mud, snot and spit, the traces of work, motion, birth, sickness, and death are endemic. The smell works its way into the fibers of your coveralls and the calluses of your hands. Routine medical care (such as the equivalent of a toddler vaccination to ward off "shipping fever") appears radically different once you leave the lab. On-the-farm "inoculation protocol" for a tractor-trailer load of feeder cattle means, not surgical gloves and a few hundred disposable syringes (one for each dose for each animal) but a reasonably sharp, stainless-steel-and-glass hypo six-shooter. You reach it through the slats of a manure-spattered sorting chute, poke through the hide of a nervous neck or rump, and squeeze on the trigger of an "Ideal pistol-grip." Since random bits of hair and blood accumulate on the needle, morning and noon you rinse it off at the nearest hydrant, or twist on a new one with the pliers you last used to pick up something too gross to touch. Given such standard procedures, I have not been alone in imagining the potential horror of anything like a livestock equivalent of HIV. There does not appear to be any, but talk of the specter has arisen regularly without any prompting from me. Of course, some places—milking parlors, surgical kits, and the like—are remarkably clean and orderly, but none of them resembles a scientific laboratory. And the disparity is most marked in the everyday world of pigs.

In fact, what seems most lacking in my conversations with swine virologists, epidemiologists, and immunologists is the animal itself, a whole breathing, eating, willful, grunting thing, exuding sounds and fluids, warmth and smells, a body as well as proteins, cells, and sundry systems. Despite concerted effort, the cavern between practicalities and bench science or "applied" and "exotic" research remains as intractable as TGEV, and in-house bridge building seems beside the point or off the map. For example, whatever their proximity to reality in the world of microbiology, in vivo tissue samples seem just as removed from reality on the farm as in vitro ones. It is just very hard to keep the scientific stories about pigs and an actual pig both vividly, simultaneously in mind.[2]

A pharmaceutical advertisement in the trade magazines plays to that very difficulty. A glossy, color photograph of handsome, leather-bound books and well ordered serials spans two pages. Off to the right, sitting on a table or podium in front of the books, a piglet stares out at the viewer, unimpressed. The heading reads: "To really tell tetracyclines apart, you'd have to be a microbiologist . . . or a pig." Should there be a difference of opinion, put your money on the pig. At least that is the logic to which the ad appeals.[3]

Part of what makes a pig unmistakably a pig is its conspicuous difference from science, books and other trappings of civilization. Juxtapose a hog and anything "high," and one of them surely will suffer in the comparison. It is the recipe for a joke (and *traif*). Even when cast only as a walk-on—without lines, through its mere presence—a sow instantly ridicules the affected urbanity, wit, beauty, or wisdom of its costar: Eva Gabor in *Green Acres*, Michael J. Fox in *Doc Hollywood*, Delta Burke in *Designing Women*, or Carlos Riqueline in *The Milagro Beanfield War*. Or the two combine with an outlandish effect in a character like "Miss Piggy," only one of the innumerable, incongruously grandiloquent swine of children's literature and lore. In a culture that sets clean against dirt, mind against matter, and soul against body, swine are dirty matter. They are so purely body that the perception of human resemblance or other soul-to-soul connection seems unimaginable, ridiculous if not blasphemous.[4]

Even in comparison to other animals that "people keep," pigs are remarkably free of soulful mystique. Unlike the family dog or cat or work animals, they were never prized or selectively bred for companionate qualities. By the time they are one month old, they are no longer cuddly, dependable, or subservient, nor are they expected to be. Unlike sheep, llamas, goats, or other critters that could be prized for their fur, hogs are valued almost entirely for the corpses they will become. Of course, other animals (fish, fowl, frogs) are also considered relatively spiritless, purely instinctual or at least "too dumb" ordinarily to anthropomorphize. But when they are anthropomorphized, as they are in children's literature, the effect is less outrageous. A fantasy of romance or Socratic dialogue among salamanders may be novel, but it is not shocking, funny, or frightening in itself as it is, say, for Miss Piggy or for Napoleon and his comrades in George Orwell's *Animal Farm*. The transgression that comes in attributing something like humanity to swine is much more profound.[5]

Feelings might run high because of the supposed intelligence of the species. Just about everyone has seen or heard that hogs are smart, perhaps as smart as people and certainly smarter than the cattle with whom they might otherwise compare. Experience convinces me, too, that cattle are by far the half-wit ("bovine," "beef-brained") of the two, but they seem also to invite intense human associations. In *Little House* or latter-day incarnations, families prize "Daisy" or "Elsie," the household milch cow, even if she is as dumb as a post. On the farm where I work, we can recognize by ordinary demeanor the calf that must have come from an Amish neighbor last March and has yet to figure out that it left. As you walk through a crowded feedlot, cattle jostling to clear a path before you, it is that steer that stays ahead to nuzzle, as if you were one of the Yoders' seven kids. It is hard to imagine (though hardly unimaginable) having such a relationship with a 500-pound boar, even if he, too, were dumb as a post. The difference does not seem to have anything to do with brain power.[6]

More commonly, beef cattle are set apart as props for the whole Western/cowboy shtick: boots, multi-kilo belt buckles, and Stetson hats. For more than a century Americans—especially men—have not only tolerated the cattle bond but also defined themselves in relation to it. For example, the importers of distinctly dairy (vs. beef-and-dairy) cattle had to tolerate severe ridicule in the early 1800s. American breeders of "good-looking" (blocky, thick, square-stanced, heavy-muscled) cattle treated the curvaceous form (bowed ribs, light legs, and bony rump) of Guernseys, Holsteins, and Jerseys as a challenge to manhood itself. Putting one in your pasture was akin to donning a tutu. Even now urban cowboys (as likely as not out-of-work actors from New Jersey) are used to sell tobacco, beer, and other good-ol'-boy accouterments. Through commercials featuring virile carnivores, the National Cattlemen's Association and the Beef Board have styled themselves manly "stewards of an American heritage." In the countryside, the infamous conservatism of cattlemen—as compared to cotton planters or pork or poultry producers—can be blamed on the burden of ropin' and brandin' machismo.[7]

By contrast, hog farmers seem less anxious about their gender. Their self-confidence perhaps is easiest to detect in a distinctly proud tradition of crowning "Pork Queens" at ag fairs across the nation. For example, one "Iowa Boy Columnist" (with an Angus breeder's sarcasm) wrote to the *Des Moines Register* to issue a challenge: "If you surveyed the nation, how many high school girls would you find who would consent to wear a crown and a satin sash . . . to extol the glories of the hog?" At the age of 37—about two decades after her reign as Iowa and then National Pork Queen, 1972 to 1973—Soo (Klingman) Greiman responded:

> Iowa pork people introduced the nation's first "Pork Queen Pageant" in the State Fair hog barn in the summer of 1960. . . . Who first but Iowa would envision combining the image of the hog with the enthusiasm of vibrant, young women in order to promote the pork industry? . . . While the hog is undeniably woven into the very fabric of our state, so, too, is the Pork Queen an important dimension of Iowa's pork industry, rich in energy and visibility. Her position is like a symbol of our belief in sound, solid agriculture along with our most precious resource—our young people.[8]

Judging from his half century of USDA experience dealing with diverse livestock people, Paul Reimer agreed: "People in the swine industry are down-to-earth type of people. They are willing to try new things. They're innovative, where the cattle industry are dominated by their culture, which is kind of: 'We know it all, and we know what's good for us. You all stay away from us.'"[9]

The challenge of raising a "good-looking" hog bears no such burden. When pork producers get together to decide what "I like in a hog," they have very different concerns. This was among the topics I discussed in 1993 with David Meeker, then vice president of research and education for the National Pork Producers Council. His office was in the crisply decorated headquarters, a square brick building, about the size of a paunchy Wausau home in suburban Des Moines, smack in the middle of the Corn and Swine Belt. It is next door to an even more modest building that houses the Iowa Pork Producers Association just off Interstate 35. David is dressed business-casual, trim, youthful, behind a desk buried in stacks of letters, memos, and draft reports. On the wall hang sundry pictures of pigs, pig people and family members, and a diploma for a Ph.D. in Animal Breeding from Iowa State University.

> The Council represents pork producers. Across the U.S. there are about 250,000. They have at least a few pigs, anywhere from a couple to 2 million in some cases, like Murphy Farms in North Carolina. Technically all the individual members belong to the associations in their respective states, and national has forty-five state members. (There are only five states without organizations: Alaska, Massachusetts, New Mexico, Rhode Island, and Vermont.) So technically an individual is a member of a state—say, Iowa, and then Iowa is a member of national. But the Council is a grassroots organization. The Pork Producers elect their leaders democratically, have annual meetings, and so forth.
>
> The funding makes it a little complicated. We are a membership organization, but since 1986 we have been funded through a check-off system. Every hog that's sold in the United States has 35/100ths of 1 percent of its value deducted from the check and sent here.[10] So every hog contributes to a check-off fund, whether or not the sellers consider themselves a member. About 90,000 of them do. And they account for about 90 percent of all the pigs in the country. So there are a lot of very small, part-time producers that don't bother with us, that don't add up to very many pigs. But they all, all 250,000 still have some money taken out of their checks and still benefit from our advertising [e.g., the NPPC campaign for "The Other White Meat"], research, and so forth.[11]

I was wondering if you had—you know, maybe from your research—profiles of the membership. Part of what I have to do is tell people what pig farmers are like, right? I can tell them about what I've seen here in Iowa, but I want to be sure I give them the bigger picture. I don't want it to be just my own weird experience. Have you got a profile, what's typical or a set of types of producers?

Interestingly, we have not. I've got some things that might help. We did an employee-employer survey not long ago about benefits, wages, attitudes. That's the closest thing we've come to studying the people. The bulk of research and education programs, 99.9 percent of it, is research on pigs—research on pork products, on feeding them better, on genetics, disease, equipment, all that sort of thing.

But it sounds as if your experience is about right, pretty accurate at least for Iowa. And something like 70 percent of all the pigs in the U.S. are raised within two hundred miles of where we're sitting. So when you say "Iowa," you might as well include eastern Nebraska, western Illinois, southern Minnesota, northern Missouri, Wisconsin. I mean, it's a circle that's a lot bigger than Iowa, where most of all the hogs are.

It wasn't all that different when I grew up. When I was in high school, my dad—with the help of me and my six brothers—farmed about 800 acres, raised a lot of corn and soybeans, fed some cattle. But our main love and interest was purebred Hampshire pigs. And my dad gave me my first sow of my very own when I was about eight. So I've always been interested and involved in the pig business.

I think the hog industry became a major industry in the United States because it was a much more profitable and mobile way to get grain moved in the marketplace and get some money for it. You know, "If you fed your corn to a hog, at least it would walk to town." You wouldn't have to carry it. And now, you know, most grain in this country is raised at cost or below.

I get calls every week from farmers that don't raise pigs saying, "Look, I'm thinking about putting in some hog buildings, so I can add some value to this grain; I'm going to go broke this way." And you know, the historic profit on pigs has been pretty good. In part, this is because of the hard work producers are willing to do. Not everybody wants to do it. But anyway, this is the way the hog industry grew up and matured, simply as a way to get rid of corn.

But that isn't the way it is in the entire nation, and it's changing.

And it may stand in the way of quality improvement, efficiency, what's necessary to compete with poultry, etc., etc. Now Iowa farmers are realizing—partly by looking at the North Carolina example—that they don't have to waste as much as they used to, that they can use three-site production, all-in/all-out, early weaning, split-sex feeding, better genetics: "Look, if we're going to compete in the world market as a source of food for people, we've got to get rid of some of this fat; we've got to be able to compete at a lower price; got to get our costs down, etc." So even in Iowa the attitude is changing very quickly, changing from "hog feeders" to "food producers." It's an orientation change. They may do the same thing from morning till night, but it's a different mindset, a different way of looking at things, and it gives them more connection with products beyond the packing plant. They take ownership . . . not ownership physically, but responsibility for the meat on somebody's plate, instead of that responsibility ending when the hog walks off the farm.

All trends in all businesses are towards more efficiency and more global competitiveness and all of that. We're doing the right thing, I think, by becoming more efficient and embracing new technology and so forth. They're basically models of the (for lack of a better term) "North Carolina way of doing business"—big, high-tech, and so on. But there are limits, too, some things about the biology of the animal that are so much different from, say, poultry that I'm sure we're not following exactly the same model.

Whereas the poultry industry could take a tray of eggs and incubate them and handle them and do everything by machine, the swine industry has a whole . . . a whole culture in farrowing sows—an actual animal that has to go into labor and have babies. And they have to be warm and dry. And none of this is very suited to mechanical manipulation, as the egg is. So there are other biological differences that will prevent the total mechanization of the industry. But the trends are obvious, and the numbers are easy to get.

You know, as I said, there are about a quarter of a million producers now. Well, as late as 1968 there were a million. So, in the last twenty-five years, 750,000 people who raised pigs don't raise pigs anymore. Those trends are there, and they're in all of agriculture. I think they can continue for a while, and maybe we'll end up in twenty more years with 20,000 producers or 40,000. But at some point it will level off because of (1) the sheer size of the industry and (2) the biology of the animal. It just takes more people and a little more human touch than some other parts of agriculture.

And hog people are different?

Yes. We haven't studied this, but we've talked a lot about it in our circles and with our leaders and so forth: There's very little or, you might say, no glamour in the pig business, you know?

You've worked on a pig farm; so you know. It can be smelly, and it can be dirty, and it's hard work. And different from cattle producers or sheep producers or horse producers is this ominous dark cloud on the horizon of what disease can do to you. Contrast that with a cowboy with shiny spurs and rhinestones riding on the range and glamorized on TV in all the Westerns. That's far from the basic foundation of the people we have. We have to kind of bite and scratch and claw our way to make a living without glamour.

There are a lot of people in the cattle industry and with ranches that have made their money in oil or movies or something. They do cattle because it's classy and it's cool. Now, for some of those reasons, the pork producers are much faster to embrace change, quite a bit faster to adopt new technology, willing to lessen their labor, and so on, whereas the cowboys are clinging onto their image and their boots and their horse, because they love it. They are kind of committed to a way of life. They like it the way it was and never want to change it.

You know, I'm not saying people don't love pigs and love the pig industry. But it's a different kind of commitment. It's a commitment to the hard work and, at the same time, a willingness to change.

Swine have to be valued "merely" for the animals they are. What you see when you "like what you see in a hog" is just an individual and the sow, boar, or market hog that you can imagine helping it become. It is not an icon of intelligence, wholesomeness, cuddliness, glamour, or virility. It is neither a mirror of human vanity nor a natural resource. It is just one living thing that might with care, by the end of its life, be exchanged for something else. Looking good, then, is the quality of a body observed and then projected into the future. It is, then too, an estimate of the prospects for a successful relationship of a sort. One life, the farmer's, is considerably longer than the other's, but they share interdependence and the physicality of it all. Hogs are muscle, bone, and blood made of ("converted from") the food a farmer provides, less the excretions he or she hauls away. A good-looking pig is simply one that will eat, shit, grow, and die, all in good time—good, that is, for the farmer but only for the farmer if also for the hog up to the moment of death. The "value" it gains comes from genetics, credit, the grain bin, the market, patience, diligence and luck, "good management" and "that human touch." But the potential to gain value is supposed to be visible in the pig itself. A good one, you see, will make pork and make money.

PART VI

Say It with Swine

I am thinking of the capacity for intimacy as a virtue, the virtue of friendship, as what Lewis has called Philia, *the emblem of which, he says, is two figures holding hands and gazing at some third object. (This is unlike* Eros, *that love whose emblem is two figures gazing at each other). . . . A swaybacked, diseased horse was used by Cervantes to illustrate what can happen when we seek actually to live a life apotheosizing the divinity of the beauty of horses but only desultorily, getting it almost right. The case of Don Quixote, like the case of Midas, also reminds us that getting it almost right may be the highest (human) possibility, which is another reason art is so dangerous. To love great horses or gold is to know how to renounce them, and how to time it.*

—Vicki Hearne, *Adam's Task*

Babe *illustrates a different kind of success, one in which the "it" (of "Just Do It," "making it," "going for it") is interrogated and challenged. . . .* Babe's *world is the one we live in; heroic moments are temporary and connections with others are finally what sustain us. This is a reality we may be inclined to forget as we try to create personal scenarios that will feel like Olympic triumphs and give us the power and "agency" over our bodies and lives that the commercials promise. But we still feel the emotional tug of abandoned dreams of connection and intimacy and relationships that will feed us in the open-hearted way that the Boss feeds little Babe and Babe's eating feeds* him.

—Susan Bordo, *Twilight Zone*

It is a non sequitur to talk of death or love rationally. Once experienced, both defy description, and without experience, both are nonsense. Though it would be stoutly denied by most rural people, husbandry is a form of love (it is no accident that spouse is synonymous with husband), and the ultimate goal of all husbandry, be it of plant or animal, is to mourn the death of that which we nourished and loved. We call the wake a harvest celebration.

—Karl Schwenke, *In a Pig's Eye*

Pig, sleep well. Gather your dreams
for the day when we become one.

—David Lee, "The Pig Hunt"

TWENTY-ONE

MULTICULTURAL PORCINALIA

IN PRECEDING CHAPTERS I have tried to make hog affairs speak to human, especially American conditions. As I explained early on, I know the urge is not mine alone. Scads of scholarship, tchotchkes, and whole civilizations bare the soul through swine. They have symbolized everything profane among Semites and sacred among New Guineans. There are abundant American popular sources, as well. Nearly everyone in the United States knows about Arnold Ziffel (Eva Gabor's prize Chester White in the hit TV series *Green Acres*), Miss Piggy (the sow diva of Muppet cinema), Napoleon in *Animal Farm* (which became a Cold War master text), and Wilbur in *Charlotte's Web* (probably the single most frequently read and critically acclaimed children's book in America in the second half of the twentieth century). These star swine are readily available for and widely used in evaluating human challenges. More Americans may first contemplate their mortality through imaginative identification with Wilbur the pig than with any other single creature.[1]

With factual and fictional supplements, these connections can be elaborated considerably. It is now a matter of some personal and professional pride to be able to do so. Whatever the topic, I am convinced, you can say it with swine. Just to show how far full-boar analogizing can go, I here elect a tough one.

That choice actually began with a dare. A friend who was organizing a conference in Finland asked me if I might have something useful (a follow-up to a prior essay) to say about multiculturalism in the United States. "Well, I don't think so," I said. "I am awfully busy trying to finish this book, you know, the pig book. I don't think I should take on anything all that new. Unless . . . unless it can be done by way of talking about hogs and disease. What do you think?" I was kidding, of course.

"Sure," he said.

"Really?"

"Yeah, sure," he said. "Just do it."

His acceptance of that condition started me down this course. (Hence Scandinavians can share some of the blame.)[2] I now aim to show only that, in its whimsy, porcinalia can help enlighten a topic that has been just about serioused to death.

For the past decade or so especially in academia, Americans have called upon each other to stand plainly "for" or "against" multiculturalism, and hyperbolic extrapolations proceed rapidly from there. On the "for" side, what is at stake is basic decency and respect for human difference. Anyone who opposes it might as well be a Brown Shirt. On the "against" side, people fancy themselves protectors of America on the verge of going Balkan or Babel. Californians, for example, have trouble imagining a future of civility if officials recognize that citizens speak more than one tongue. In places like departments of English (where they teach one tongue), people voice the opposite concern. Something as subtle as admitting that you once enjoyed reading Norman Mailer might be enough to signal that you are a hell-bent fascist. If you do not "fully" embrace multiculturalism (and swear off Mailer), someone will conclude, you must approve of lynching. I hope that a more restricted use of the term and then some playful, porcine elaboration can help.

By the word "multiculturalism" I mean welcoming experience with differences that social stratification, physiognomy, proud heritage, and bigotry provide. Challenges to and among multiculturalists center on two issues: (1) the ways people categorize themselves as kindred vs. alien—as "us" or "some of us" as opposed to "them"; and (2) the quality of life that encounters with those categories condition. I share with many others professors of American Studies an interest in clarifying the best and worst we can hope for in such encounters.

Limiting conditions seem especially clear when hierarchies are tested and cultural boundaries breached. The boundaries that I here consider are the usual ones: those of gender, social rank, and ethnicity or race. In this case, too, the purpose for tests and breeches are familiar. Those with the capacity to overcome cultural divides are apt to view their aggression (or, for that matter, their isolationism) as a matter of generosity. The supposed beneficiaries of this generosity may welcome it in a crisis, but that is about all. No one likes to remain "wretched refuse" any longer than necessary. In the debates surrounding multiculturalism, whether in Somalia or Newark, we all can hear strains of the rhetoric of liberal developmentalism vying with "Yankee Go Home." And my contribution does not stray far from that familiar theme. Where it does stray is in its particulars. Exemplary cases all deal directly with animals—pigs in particular—and only through interpretive leaps with people.

A good starting point in the world of swine literature is the children's story "Pigs is Pigs," which Ellis Parker Butler penned for readers of *American Magazine* in 1905. Butler interests me not only as a fellow Iowan (he was born in Muscatine in 1869) but also as the son and grandson of two pork packers. Like E. B. White, he had a visceral connection to the animals he processed for fictional consumption. Butler turned to a career in writing only after the meat-processing business failed. The year interests me because it places his work in the middle of the second great immigration in the United States. It was a time when Anglo America was even more challenged than it is today to accommodate vast numbers of legal and spiritual aliens. Nativism and racism were on the rise, and the salve of the "melting pot" would not seize the popular imagination for at least another decade. Butler decided to address this encounter obliquely, as White did the fear of death, through a work addressed to children, through humor and pigs.[3]

The magazine story was a smashing success. The book version of *Pigs is Pigs* came out immediately, sold for just twenty-five cents, and sold out multiple editions. Although Butler published a total of thirty-two books and innumerable magazine stories before his death in 1937, the 1906 edition of *Pigs is Pigs* made his career. And for more than a half century, reprint editions followed. In 1945 the story was among *The Best American Short Stories* assembled for the Modern Library Edition, and it was part of the Golden Book series, which dominated children's literature in the United States through the 1960s. Total sales must number well in the millions. Although I am not a professor of English, I would like to take this opportunity to claim a place for *Pigs is Pigs* in the canon of pig lit.

The story begins with a confrontation between an officious Anglo American, Mr. Morehouse, and a stereotypical Irishman, Mike Flannery, a semi-literate and infuriatingly literal agent for the Interurban Express Company. Mr. Morehouse just wants to ship a pair of guinea pigs, which as pets would cost twenty-five cents each. But Flannery insists on sending them at the higher, farm-animal rate:

> Mr. Morehouse shook his head. "Why, you poor ignorant foreigner, that rule means common pigs, domestic pigs, not guinea pigs!"
>
> Flannery was stubborn.
>
> "Pigs is pigs," he declared firmly. "Guinea pigs or dago pigs or Irish pigs is all the same to the Interurban Express Company an' to Mike Flannery. Th'

nationality of the pig creates no differentially in the rate, Misther Morehouse! 'Twould be the same was they Dutch pigs or Rooshun pigs. Mike Flannery," he added, "is here to tind to the expriss business and not to hould conversation wid dago pigs in sivinteen languages fer to discover be they Chinese or Tipperary by birth an' nativity."[4]

This passage defines the comic situation that the plot must resolve and gives the story its title. And a number of its features are intriguing. First, I suppose, is its directness, which would have been even more extreme had *American Magazine* accepted Butler's original title for the story: "The Dago Pig Episode." But even with its more telegraphic name, *Pigs is Pigs* gives voice to bigotry with startling clarity, certainly for us at the end of the twentieth century but even at the beginning.[5]

When this story was written and first read, hostility among ethnic groups in the United States was simmering. In official circles there was a détente between Anglo and Irish Americans, an uneasy alliance, forged less with each other than against newer immigrants from Southern and Eastern Europe as well as East Asia and Scandinavia and nearly all of African and Native America. But Butler anticipates what might happen when that simmer turns to boil. His characters speak in a language that is anything but politically correct. Morehouse calls Flannery an "ignorant foreigner," and Flannery assumes that the word "guinea" must be an ethnic slur, rather than a matter-of-fact name for a breed of rodent that is actually native to Latin America. I doubt, for example, the story would work anywhere near so well if the animal in question were a Yorkshire terrier or a Jersey cow. I must admit, it took me a moment's pause to follow Butler's play on the word "Guinea."

Of course, contemporary adult readers might well have recalled the way that Italian emigrants came to be slandered through association with the name of an African place. Emigrants from Mediterranean Europe after the Civil War had some difficulty passing for "white," especially in the South, where racial categories were predominantly binary. If your complexion or hair looked even remotely "Negro," you might as well have come from Guinea, even if your family had spent the past couple of centuries around Palermo.[6] This quality of indirection is also part of what draws my attention to the story. It deftly effaces Anglo/Irish, male solidarity and white racism just as it faces other bigotries with comic candor.

Together these aspects of the story—the direct address of xenophobia and the indirect address of patriarchal white racism—show just how well it fits the current dilemma of boundary crossings into the United States. As large numbers of people immigrate from Latin America and the Caribbean, South and

East Asia, Africa, and the Middle East, as many of them and older stock citizens claim distinct American identities, how can they deal with each other? How can the old categories of differences, especially ones that name positions of privilege and oppression—Dago, WASP, Kike, Dyke, and the like—help or hurt?

Butler has set up a comic situation for exploring that very question. In style and story, he follows the lead of Mark Twain, who lived just down the river from Butler's childhood home and who Butler claimed was his key inspiration. The influence may be most obvious in Butler's treatment of vernacular speech. For example, when Flannery calls someone "sir," it finds its way onto the printed page as "sur," as if misspelling were distinctly audible among the prole. Twain was also the author whom his mentors, his aunt, his father, and a high school English teacher, most admired. In crude dialect and sentimental farce, in fare crafted for children, Butler like Twain challenges key categories of American social life.

The story begins with the paralyzing effect of category conflict: Are guinea pigs pets or domesticated animals? Is their "guinea" name relevant or irrelevant to the way they should be treated? Like a straightforward essentialist, Morehouse invokes a biological basis for his categories, and he demands that it apply: They are pets, not pigs; do not let the name fool you. Flannery is more of a rough-hewn social constructionist: If they are called "pigs," treat them as pigs; do not be distracted by some self-serving and ultimately contestable history of the name or the bodies to which it is applied.

How in the story does political economy adjudicate these positions? As we might predict, to the systematic advantage of the Interurban Express Company. Policy requires that, if there is a question about which of two shipping rates apply, the higher rate should be assessed. It is a comically simple solution that Butler invents, but one that exposes profiteering in the pragmatist's solution.

Morehouse is outraged. He stomps out of the office, leaving his two guinea pigs in Flannery's care, and files an appeal with the Interurban Tariff Department. In the midst of a cycle of exasperating memo exchanges, Mr. Morgan, the head of the Tariff Department, approaches the president of the company:

> "What is the rate on pigs and on pets?" he [the president] asked.
>
> "Pigs thirty cents, pets twenty-five," said Morgan.
>
> "Then of course guinea pigs are pigs," said the president.
>
> "Yes," agreed Morgan. "I look at it that way, too. A thing that can come under two rates is naturally due to be classed as the higher. But are guinea pigs, pigs? Aren't they rabbits?"
>
> "Come to think of it," said the president, "I believe they are more like

rabbits. Sort of half-way station between pig and rabbit. I think the question is this—are guinea pigs of the domestic pig family? I'll ask Professor Gordon [the famous zoologist]."[7]

Before anyone could track down Professor Gordon, Flannery faced the sort of horror that contemporary social Darwinist and eugenic tales foreshadow. While the limp-wristed guardians of civilization churn memos, muddle morals and split hairs, their swarthy charges multiply. Certainly those two guinea pigs did. A panic-stricken cable from Flannery to Morgan read:

"About them dago pigs," it said, "what shall I do they are great in family life, no race suicide for them, there are thirty-two now shall I send them do you take this express office for a menagerie, answer quick."[8]

Soon Flannery is tending 800 guinea pigs; then 4,064; then 280 cases of them; then cattle cars full. The cost of daily cabbage rations alone surpasses the return for shipping a troupe of elephants, much less a pair of swine. And so he becomes convinced that he just ought to accept the loss and get them the heck out of his charge. Even Flannery can no longer bear the cost of his stubbornness. He ships the lot of them for fifty cents and draws this lesson:

"Niver will ye catch Flannery wid no more foreign pigs on his hands. No sur! They near was the death o' me. Nixt toime I'll know that pigs of whativer nationality is domestic pets—an' go at the lowest rate."[9]

It would be worth discussing what we are to make of the story's end. Focusing on the interaction of the two protagonists, Morehouse and Flannery, we might respond chiefly to their white, male bond. It is forged through a renewed commitment to practical, present circumstance, their common interest in money and convenience over contestable principle. For the right price, essentialist and constructionist unite. Ethnic slurs like "dago" or "Rooshun" are never explored from the position of the subject. They are trivialized instruments of male bonding and white privilege. Anything that even reminds you of one of those "new" immigrants, "lower races," or reproductive functions is best segregated and banished as soon as possible. According to the story, it may be silly to worry about the purity of language or nationality or "race suicide," but even a stupid Irishman will have to agree that his fortunes lie with Morehouse and getting those foreigners on their way. While, we learn, the boundaries among the variety of Americans can be considered an irrational jumble, two categories remain: "we" who are pragmatic, white, adult, English-speaking and male, and "other," inscrutable, promiscuous foreigners who belong elsewhere. The story congratulates Flannery for his belated recognition of that lesson. And, of course, then we might congratulate ourselves for seeing the trick in the tale, albeit with about a century of hindsight. How incisive of us!

But I want to draw attention to another trick that might not be as obvious and that might make the story seem more useful to us now. The trick I have in mind is the one that leaves uncontested the boundary between species: human, pig, and guinea pig (or, for that matter, TGEV). Of course, I would not expect anyone to confuse the three and would not find you more incisive or politically correct if you did. Pigs *is* pigs, not people. I draw attention to this divide in part because, now as in 1906, it remains so commonsensical. Through examining species boundaries, I am looking for valuable alternatives to the usual menu of multicultural choices: fortify, forgive, or deny our differences.

Even in zoology, the science of such classifications, the lines are far from crisp. Through transgenic technology (for example, the introduction of bits of human DNA into the genes of bacteria, fish, and—yes—pigs), the lines are demonstrably permeable.

As Ernest Mayr has argued, biology itself has advanced through three senses of the term "species," in roughly the same order and time as conceptions of race and culture in the West. Well into the nineteenth century, scientists worked with essentialist categories. Each category represented a set of unique features. Hence, in recognizing that animals belonged to a species, scientists attributed to them a distinctive, unchanging, fundamental character. With such a view, Linnaeans could rank life-forms into taxonomies that might suggest ranks, too, of God's favor. Darwin was a key figure in shifting the sense of the term in a more nominalist, historical direction. A species, from this vantage, was less a pure, distinct type than a group of organisms that have come to share enough common features to warrant being given a common name. As the flexibility and dynamism of the living world have impressed biologists, their notion of species has become even more fluid and avowedly heuristic. For most biologists now, the term indicates less a set of essentials or even a class of organisms than a conjunction of organisms, space, and time. A species is a "community" that over time has come to occupy its present, "reproductively isolated niche." In kind though not degree, the difference between species resembles that between genders and ethnicities. They are "historical entities" that change and that we, as historical actors, might seize, refigure, or reject.[10]

What are we to do, then, when essential differences or hierarchies among species seem to resemble those among humans? Especially when the difference between people and, say, pigs seems so obvious? For most of Western history, the preferred answer has been to use the "obvious" differences between species to rationalize the human ones. Essentialist science in league with the Book of Genesis renders one category of human "naturally superior" to another. But since, outside devoutly Royalist and racist circles, this idea has been losing favor since the Enlightenment, I assume that most Western readers now find it objectionable.

More recently, a diverse collection of reformers has rallied under the banner of animal-rights activism, in effect, to reverse the logic. Since everyone—pigs and people no less than blacks and whites—ought to be considered "essentially" or "naturally" equal, how dare we employ categories that divide us? How, for example, can humans raise fellow creatures simply for slaughter, creatures that—but for accident of history—would be considered one of "us"? I assume that this position has a few more takers in the West now, but I resist it for reasons that I cannot yet justify very well. Suffice it to say, after nearly twenty years of working, albeit part-time, on a large hog, grain, and cattle farm, I have found a way both to care for farm critters and to enjoy meals made of them. Living purposefully through that apparent—and, from my point of view, inescapable—contradiction may well be more defensible than trying to avoid it, say, through vegetarianism. After all, some of us are just greener than others.[11]

Whatever your stance on animal rights, the point is that boundaries between one ethnic group or gender and another—between humans and animals or animals and plants or plants and microorganisms, bacteria or viruses—are at once obvious and contestable. Likewise, the reasons for treating categories differentially are at once obvious and contestable. Bigotry is wrong, less because it uses categories than because it is used by them. Membership in a category is taken as definitive, delimiting, natural, permanent, rather than a more or less useful fiction. Again, I turn to animal stories, this time nonfiction, for a lesson.

The 1950s and 1960s were decades in which the categorical divides were strong. In particular, the division between so-called "higher" animals and microbes became a battle line. Microbes that were classed as pathogens were the enemy. Any country that harbored pathogens was, as a result, potentially an enemy, too. The United States geared up for the fight. The Animal and Plant Health Inspection Service of the U.S. Department of Agriculture (APHIS) organized campaigns to control and eradicate creatures (parasites, bacteria, viruses) that were deemed threats to American food animals and crops. If the name of your country was implicated in the transmission of those creatures (and if the authorities could get away with it), your exports could be embargoed. (Faced with a truckload of Mexican cattle, an APHIS inspector in Brownsville—or more likely, the Mexican exporter, well before loading—consults Washington directives: "Let me see on our list here. . . . Nope. Can't let you in. We've got a record of foot-and-mouth in Mexican cattle.") In fact, until very recently such embargoes were painfully reflective of the whims of officials in more powerful, northern, industrialized nations. In trade as in civil affairs, segregation is generally for the convenience of the privileged, and the effect can be contrary. It is both an assertion of authority and an admission of its injustice.

Next I sketch a couple of cases with more implications for multiculturalism than can be fully developed here. I can, however, demonstrate both that they deserve development and some ways that it might be done. I begin with a story that can be told with two morals, one emphasizing the virtue and the other the danger of policies that, insofar as possible, assert categorical controls. I then turn to another story that also has been told in two modes but that implies a third, one that recently has become persuasive to an influential audience. It allows for more mixed and measured approaches to categories and deserves more consideration among cultural critics who have heretofore paid them little attention.

In 1946 there was an outbreak of foot-and-mouth disease in Mexico that rapidly spread through sixteen of its twenty-eight states and infected sheep, goats, and hogs as well as cattle. Such epidemics amply demonstrate that, at times, some species thrive only when confined with their own kind. The mixture of pathogens with cattle and then uninfected with infected animals was a disaster that threatened to spread north with trade across the Rio Bravo. Fearful ranchers pressed U.S. authorities for protection, and the Department of Agriculture responded. The importation of all susceptible Mexican animals and their byproducts, no matter what their state of origin or health status, was forbidden.[12] Bright-eyed young field staff, like Paul Reimer fresh out of vet school, joined in an attack on the disease that included, not only severe shipping restrictions (strict segregation by nationality), but also thousands of U.S. and Mexican personnel, scattered throughout the Mexican countryside with the power to vaccinate, quarantine, and kill. They monitored millions of animals for seven years (1947 to 1954), completed a series of vaccinations on 16 million, and killed 1 million that were otherwise infected or exposed. There were significant losses, including the lives of about sixty employees. Veterinarians and inspectors from both countries were subject to highway and plane crashes, typhoid and malaria, mutiny, ambush, and stonings. But by nearly all accounts the operation was a success, such a success that its design survives in many forms, including the Peace Corps.[13]

Before chalking one up for the benevolence of nationalism, it should be pointed out, however, that nationalism appears to have precipitated the outbreak in the first place. Although even in retirement he remains one of the foremost champions of eradication programs (what I would call the "strong-categories position"), Reimer also well knows the politics involved. Categories are never pure, and attempts to invent purity are vulnerable to profane manipulation. If rumor is to be believed, he explains, in 1946 the disease was unwittingly brought to Mexico from Brazil through the efforts of U.S. cattle speculators. They admired the tick resistance of Brazil's zebu cattle and aimed to breed that resistance into their stock. If they could get rid of the ticks they also might reduce the incidence of anaplasmosis that the ticks spread. To avoid treatied restrictions on imports of Brazilian stock into North America, these U.S. speculators got Mexican entrepreneurs to ship the zebu to Veracruz, where animal health monitoring was more lax. Veracruz did for foot-and-mouth what Jurassic Park did for dinosaurs.[14] The fact is, entrepreneurs do not much care about boundaries, and pathogens care even less. Genocidal wars of one against the other seldom end in total victory or defeat and usually entail a fair share of collateral damage among noncombatants.

Very similar complexities attend nearly every disease eradication program, whether it is successful or not. Geopolitical entities make a great deal of sense as species—historical entities, cultures, communities in a niche—when it comes to the management of resources or the deployment of troops. For example, in funding public health or corralling an epidemic, the state seems to be a very useful fiction, indeed. Under other circumstances—such as anticipating who will get along with whom or dividing dirty from clean or high from low—they make very little sense. They are closer to cruel lies than useful fictions.

A similar lesson can be drawn from the response to an outbreak of African swine fever in the Caribbean in 1978. An epidemic threatened all of North America. It spread very fast, in just a few months from the Dominican Republic to Haiti and then to Cuba. On the advice of veterans of the Mexican campaign, Reimer agreed to participate in the counterattack. He moved from serving the U.S. government to the InterAmerican Institute for Cooperation on Agriculture. For several years he worked with a team that included the Veterinary Services of the twenty-nine member countries on their "number one priority: the eradication of ASF [African swine fever] in Haiti."[15] It worked, but doing so required killing every pig, every potential ASF host, wild or domesticated, in the country. Coincidentally, a generation of rural children, whose schooling traditionally was underwritten by the timely sale of the family hog, lost years of education. The number of illiterate people rapidly increased, and they were well numbered among the infamous "boat people" who subsequently fled to Florida.[16]

No one has ever proven a direct, cause-effect relation between the eradication program and an increase in illiteracy or refugees, and Reimer doubts that there is any, beyond isolated cases.[17] He is understandably proud of the success of the campaign in agricultural and public-health terms. The spread of ASF was halted; Haitains received millions of dollars for swine that were doomed, anyway; latrines were installed in the countryside, thereby reducing the circulation of trichinae (the parasites of trichinosis fame) between foraging hogs and defecating humans. But he also recalls less fortunate results. Too much time passed between a family's loss of their hog and their receipt of compensation (replacement and technical assistance). Suffering could have been spared if more cash were available for payments on the spot. Moreover, the uncomfortable racial and national preconditions for the program simmered just below the surface. When I invited Reimer to criticize the Haitian operation, he recalled the way those pigs were replaced:

> We had to bring the new pigs in, to repopulate. A consultant group, who I thought was the best in swine production, advised us to bring in hogs from the United States and Canada. I think, if we had to do it over again, we would have had a sociologist in that committee and made sure we had a lot of black pigs. Because that was a mistake. . . . They like anything black. The country's 96 percent black. But, psychologically, it would have been much better. That's the major criticism we got out of eliminating it [the disease].[18]

We may think of the misfortune as more or less than psychological. I certainly do not feel qualified to evaluate disease eradication programs. But I admire the light they shed on potential relations among categories of difference, of species, races, and nationalities. Since the dangers of category pollution are undeniable and the willingness to experiment great, we can see in these programs, maybe even more vividly than in fiction, how different multicultural affairs might be arranged.

It is worth noting, for example, that the USDA is no longer adding to its list of diseases targeted for eradication, not because they welcome epidemics or because they have deconstructed categories of containment, but because the cost of such total warfare has proven prohibitively high. Likewise, in the wake of GATT and NAFTA, those who aim to monitor and control food animal and plant diseases around the world, including APHIS, are in the midst of defining alternatives to national borders organizing their efforts. In a movement that is called "regionalization," officials are trying to chart the world's risks—such as constructive sorts of segregation or integration—according to the way those risks actually are distributed rather than old categories of convenience, such as nationality.[19] The boundaries are moving and shaded rather than crisply drawn. Scott Hurd of APHIS explains:

So they are moving away from the nation as a container to working with regions. A good example is African horse sickness. The region of northern Spain does not have it; the region of southern Spain does have it. So, we have to do a risk assessment next year [1996]. We have horses coming to Atlanta for the Olympics. How do we decide if those horses are at risk for African horse sickness? Do we divide Spain up? Can we legally divide it up? Most of the Olympic horses will come from northern Spain. So, we have to do a regionalized risk assessment. . . .

And we've been working on TB in what's called the "El Paso Milkshed." Up until now that was presumed to be [pointing to a map] here in New Mexico and down there around El Paso, where there's a bunch of dairies. But when I talked to all these people, epidemiologists and whatnot, I'd say, "Well, what about Mexico?" And they'd say, "We can't go to Mexico. They won't let us go to Mexico.". . . That's what we're struggling with. Finally, somehow we got permission to go into Mexico. So, just a couple of weeks ago we used geographic system units that shoot from a satellite to pinpoint the location of these dairies. Because disease knows no boundaries. There is just a muddy little irrigation ditch called "the Rio Grande." And so, when we were finally able to plot those locations, the picture got so much broader and so much more understandable. See, these dairies have a lot of tuberculosis, and it's only 2.8 miles from this dairy on one side of the border to this dairy on the other. . . . But the paradigm, the way people were working, was that our purview, our concern, stopped at this river. And finally we were able to break through and help them understand how disease doesn't care. So let's think about this whole area as a problem.

This has just happened. I mean, we were the first federal officials, U.S. federal officials ever on these Mexican dairies, just three, four weeks ago. . . . And we found out just the other day that (I don't know if this is going to turn into anything or not) there's a lot of bird traffic in this whole area here. So, we're looking for reasons the disease gets from one place to another. . . .

Just to kind of emphasize (I think it's really interesting): You can't let international boundaries stop your thinking.[20]

I suppose it is the bald pragmatism of these animal tales that appeals to me. The outcomes invoke values and feelings, but the stakes are also plain to see: how much categories cost, how they are both defined and applied, who gets sick, who lives, who dies. Yet when considered together, these tales also undercut the crackpot realism of Morehouse and Flannery and their modern analogues—bigots and identity politicians alike. I hope that considering when pigs is pigs and when they isn't helps us figure out when human differences rightly matter and when they don't.

TWENTY-TWO

TALKING LIFE, TALKING HOGS

THIS BOOK has been an attempt to track moral and practical connections among disparate things. The initial bond is one that I basically forced between pigs and people. My main excuse is an eccentric career that kept me commuting between sows and students of the liberal arts. But from that initial link came many others, simply waiting to be noticed. I discovered ways that American tchotchke-ites and pork politicians, pathogens and swine scientists, PETA people, environmentalists, viruses and veterinarians, integrators and old-timey hog farmers can all speak to the sort of culture we make. Rather than marshaling them in support of a single vision of a "better" culture, I have tried to highlight more subtle, contestable morals that their tales afford.[1]

One of the morals that comes through most clearly for me is that "the vision thing"—that ambitious, categorical, diagnose-and-fix spirit of eradicator, integrator, and progressive—has problems of its own. From an alien vantage—a hog, a microbe, an interloper—it is a strikingly American vanity. Although, as the saying goes, you do not do surgery in a sewer, there is still a good deal of dignity in dirt. Pigs and sundry pig people remind me how often Americans forget that there are and ought to be limits to human designs. Reformers, entrepreneurs, and cultural critics might be wise to recall how much of life worth living is one conflicted thing after another.

Often, I hear, practices that are supposed to be contradictory turn out to be complementary and vice versa. And such ironics, as amusing as they may be to observe, can be simply painful to experience. But often, as well, try as we might, there is not a damned thing we can do about it. Insofar as we can do something constructive, particular circumstances suggest a clearer path than the category to which they supposedly belong. Between any one being and another, health and disease, money and body or spirit, individual and collective interest, continuity and change, adaptation and loss, the lines of separation are so flimsy that they corral few but eye-catching cases.

This is not, I emphasize, to counsel resignation of the social Darwinist, Forrest Gump, or Alfred E. Neuman ("What me worry?") variety. Many circumstances—injustice, poverty, cruelty, catastrophe—demand immediate redress, even if we suspect that today's remedy might just turn out to be tomorrow's pathology. We do not need foolproof "impact projections" from the home office or clairvoyants before responding to need. We just do the best we can. At least that seems to me what happens as well as what is supposed to happen, given the first two of the Ten Commandments. "Doing right" is an act of faith, of conviction more than calculation. What is borderline blasphemous is yoking that act to cocksure, "ultimate" right. Demagogues and idolaters are the ones who are moved by such certitude or immobilized by the lack of it. But the sanctimony in this warning is itself a sign of the syndrome.

What I like about the colloquial alternative—"shit happens"—is the earthy, black humor that carries its message. It at once faces distress unblinkingly, summons solidarity, and leaves options open, including the opportunity to laugh. Since, it reminds us, good and bad are mixed and follow each other in no particular order (at least none that mortals can know for sure), we should not be too impressed by one or demoralized by the other. With a wry grin, it expresses faith and camaraderie but also evokes enough of the butt end of life to admonish conceit. Text lately featured on T-shirts and coffee mugs sounds this blue note ecumenically:

> TAOIST: Shit happens.
> MUSLIM: If shit happens, it is the will of Allah.
> BUDDHIST: If shit happens, it isn't really shit.
> HINDU: This shit happened before.
> MORMON: This shit is going to happen again.
> JEW: Why does this shit always happen to us?
> CATHOLIC: If shit happens, you must deserve it.
> PROTESTANT: Shit won't happen if I work harder.
> FUNDAMENTALIST: Shit happens because the Bible says so.
> AGNOSTIC: Shit may have happened; but then again, maybe not.
> ATHEIST: Since I don't believe in that shit, it won't happen to me.
> HEDONIST: There is nothing like a good shit happening.

Such an anti-didactic message befits souvenirs of the TGE trail.

Of course, I also aim to have traversed more substantial terrain, one that at least affords an overview of recent developments in agriculture and dispositions toward disease (social and spiritual as well as bodily) that will help readers decide how they ought to engage them. The disparate connections, then, are designed to broaden and deepen understanding of what *is,* in fact, "the best we can do," not only in matters that are obviously porcine—livestock management, food safety, animal rights, environmental protection, disease control, and attendant movements for reform—but also their human analogues.

The key one for me, I suppose, is the way we understand our own mortality, what it means to have a body, where it begins and ends, how it is best sustained in relation to other bodies, to imagination, circumstance, and the passage of time. Here the midlife, academic, and Jewish aspects of my profile probably show, but I do not think the vantage is so limited. In the course of researching this book, I was pleased to have met so many people who projected their swine experience in complementary directions. I wish I could follow them all. Alas, this beginning will have to suffice, not only because a publisher's deadline beckons but also because my special connection to the subject has ended.

In the fall of 1996, when I returned from a year of academic duty in Asia, Roger gently but firmly let me know that they really "could not use" me on the farm anymore. Of course, they never really did need me. Although I was diligent and had a decent sense of livestock and machinery, I know that my skills never surpassed those of your average teenager born into the business. Since I worked so few days per month, I always needed more briefing about what was happening than anyone would comfortably supply. Furthermore, the year while I was gone had been a tough one all around. Even the meager wages I received could no longer pass for justified.

My swinological services were especially dispensable. It had become obvious, also during my absence, that Phil's feeder pigs and Roger's fat hogs were not making enough money for the hassle, and they probably would not unless everyone agreed to "modernize" yet again and expand enough to compete with the big guys. Instead, the Stutsmans and Berglunds sold off their hogs and focused on doing better with corn, beans, and cattle. That was the end of it, just one of those shifts in agriculture that, whether chalked up to quirky happenstance or grinding "larger forces," entails painful adjustment.

I will long remember interrupting the last phases of work on this book to help Phil take the last of his sows to the IBP packing plant in Columbus Junction. We loaded them on the gooseneck trailer, zigged south and zagged east. Campy c-and-w on the radio covered the mood and the squeals just about right. The occasion was sad but hardly devastating. It was just a bit more struggle than usual to summon a mood that would ease it all down the road.

"Kind of the end of an era," Phil sighed, quarter-to-half in jest.

"Yeah, I guess so," I said. "But I bet you won't miss the stink."

Then someone—I am not sure which one of us—recalled a story that I retell. It originally came from a guy I met in the buffet line at one of those Customer Appreciation Days that ag suppliers host. He told me that he, too, was a hog farmer, though only semi-voluntarily. His father had died recently and left him in charge of the home place, including four groups of sows. He had always been grateful for the cash they generated, but now that Dad was gone, he could freely admit he hated the things. It is a familiar confession even in the Swine Belt. Everyone knows that hogs can try a person's patience. That is the main reason some folks are "just not cut out to be hog men," and apparently he never was. So he had begun selling off the herd, a few head at a time, according to a scheme that he must have been contriving for years.

"Every time I get rid of a group, I'm pulling out their fences. I'm making sure there's no turning back—just in case I'm tempted," he said.

I laughed.

"Yeah, I got it all figured out—how many hogs I'll have one month, two months, three months down the road. I figure, in January I'll have exactly two left."

"Yeah?" I asked, rising to the straight-man bait.

"Yeah. Two. I'm going to eat one and beat the other to death with a stick."

The humor in the story, of course, hinges on its half-truth. (Another hundred days of feed bills and frustration—for a hundred pounds of food? maybe yes; for a beating? obviously no.) Even the untrue half rightly reminds a person that ordinary life—"in the good old days" as now—can drive you to distraction. Just as the story might help one farmer adjust to the loss of his father, it helped Phil and me smooth some rumble in the ride. The last of the herd, the litters off the sows that Phil sold to IBP, would still be back at his home place when we returned. It was not quite over yet.

A couple of months later, though, I got a call from the meat locker in Kalona asking me if I wanted my hog dressed "as usual" (hams and bacon cured, some brats, the rest left fresh in one-pound packs). It was a surprise gift from Phil, guaranteeing that pork would be more than a figurative bridge from one era to the next.

Through such mixtures of humor and bathos, change acquires meaning. It mutates and cumulates across the countryside and beyond. While, for example, American swinedom has been going big-guy big time, the Dutch have been struggling to cut back. There is just too much pig shit and popular disaffection for the sort of production that fuels a massive export trade. The following explanation came to me, not from some tofu-touting eco-nut but a veterinarian employed by the Ministry of Agriculture to support hog farming in the Netherlands:

> We know in our country we have too much pigs. We are slaughtering about 24 million pigs a year—not that different from the U.S.A., but we only have 15 million people. So, we are only consuming a small percentage of these pigs; the rest is exported to other countries. And the country is too small for such a lot of animals. It's a bad thing for the environment of our country. We are importing a lot of feed from all parts of the world; so a lot of minerals are coming into the soil.
>
> Now the government is softly telling the farmers that we have to decline from 24 million to 12 million, cut in half over the next ten years. And I think it should be a good thing for everybody. Our economy doesn't really depend on the export of meat. It is hard to sustain exports all over the world. We try to be sensitive to these things. Besides, the price will probably increase, so the farmers will get more money for the same kilograms of meat. And it should be good for the consumer, because a lot of consumers are eating too much meat.
>
> The largest supermarket in our country is buying meat in France, because there are a lot of small farms producing pork and poultry there. And it is selling for twice the price. But a lot of people are buying it because of the way animals are treated and the taste. People say it tastes better. People are asking for less meat, but better meat.
>
> See, in our country there is a cooperation between taste and animal welfare. It's part of our culture. Part of our tradition is having respect for animals and respect for nature. We Dutch are very sensible about these things. The developments that you see coming from the U.S. are that animals, that nature is just interesting as an economic source for people.[2]

Such talk from an American public servant would, I think, count as heresy. The U.S. government is supposed to make food abundant and affordable first: nutritious, tasty, and friendly to the environment only second. It has been tremendously difficult for Americans to accept government "meddling" with markets to restrain rather than promote consumption, even when the product is as harmful as tobacco. But many people throughout northwestern Europe are willing to spend more for food or just eat smaller portions, if doing so befits valued relations to the environment and production as well as consumption. While Americans stoically sacrifice just about everything at the altar of "economic reality," these people try to rig an economy more to their liking. For better or worse, they, too, prize small farms, though I think with fewer delusions that "small" or "traditional" must be better or that family farmers are chosen people. Both conservation and change are high-stakes bets that neither nature nor markets are obliged to cover. But in this case with a little encouragement, these Europeans hope, small farms might afford livestock and their caretakers a less industrial, less gleaming and efficient but more serendipitously nurturant circumstance. It is literally as well as figuratively a difference in taste.[3]

Of course, it is not only a matter of taste. An infrastructure capable of comparably monitoring—much less cutting by half—whole economic sectors just does not exist in the United States. In countries that are less sparsely populated, socially and economically polarized, and evangelically free-marketeer, people use the state to challenge and moderate as well as fulfill their dreams.

I was presented with a particularly vivid example in Copenhagen, when I asked the Agricultural Ministry if someone would give me a crash course on Danish swinedom. I was met by a sizable group including the senior officers of the national Veterinary Service, the Animal Welfare Society, and the Slaughterhouses Union. It is hard imagine American counterparts—animal liberationists and butchers—so amiably gathering in Wichita, much less Washington, D.C. But for these Danes it was a commonplace. Since they often meet to iron out differences on farm policy and procedures, they were pleased to regroup merely for the benefit of a stranger.[4] Furthermore, since the European Union subsidizes exports and since all Danish packing plants are producer-owned cooperatives, farmers can engineer a distinctly large share of the "imperatives" to which they respond. Some challenges they limit, and others they leave in place. In short, the dog-eat-dog cruelties that pass for realism in America are, to a much greater extent, materially irrelevant in Denmark. The difference is institutional as well as dispositional. A whole lot more than preaching and tinkering would be required to remodel American agriculture after an alien prototype, even if it were surely better.[5]

Although too anchored in place and time to transport and "apply," these differences might inspire Americans to imagine alternatives of their own, particularly futures that are more than projections from this month's bottom line. They can encourage us to ask: Are there genuine alternatives to "go" and "whoa," a fruitful terrain between the extremes of control and resignation that I find so prevalent on the porcine trail?

Of course, there are many, and many of them evolved on American soil itself. They survive conspicuously, for example, among naturalists, Native Americans, and feminists. But in scouring the globe for modern, syncretic alternatives, I was particularly impressed by examples from the Czech Republic. Time itself there is marked in a way that yokes culture to agriculture. Whatever emanates from Prague, each month in the Czech calendar echoes rural rhythm. June, for example, is called "red" (*Cerven*) after cherries that ripen then, and in anticipation of ripe wheat August is called "sickle" (*Srpen*). February, typically the month of peak peasant debt, is "heavy burden" (*Unor*), but it is followed by fecundity in March, "heavy with child" (*Březen*).

The month that particularly interests me is December. It is called "*Prosinec,*" which means "beg" or "ask," a time to implore the heavens for help in surviving winter weather. But even more than cold and prayer, December is associated with a festival, *zabijačka*—literally, "the slaughter of the pig." Although it might sound gruesome, it is as upbeat as any other seasonal celebration. While May brings visions of birth, blossoms, decorated eggs, and whatnot, December joyously mixes love and death. People huddle together around *zabijačka,* sacrifice the life of a dear animal, and gratefully renew their own. Even the inescapable cold outside becomes God's walk-in cooler. Intermittently for weeks in advance, families gather in ancestral villages, check how the fat hog is doing, then spill its blood and feast on its flesh. Many Christmas cards, like those featuring the beloved imagery of Josef Lada, depict in equal measure the gaiety, gratitude, and gore of a hog kill.[6]

A powerful illustration can be found in "Cutting It Short," the popular novella penned by Bohumil Hrabal:

> I said to my piggies,
> 'Tatty bye, little old piggie-wiggies;
> you're going to make ever such lovely hams!'
> The piggies had no particular desire for such glory, I knew,
> but all of us have
> the one death coming to us,
> and nature is merciful . . .
> Everything alive that has to die in a moment,
> everything is gripped by horror,
> as if the fuses go
> for both man and beast,
> and then you feel nothing
> and nothing hurts.
> That timorousness lowers the wicks in the lamps,
> till life just dimly flickers
> and is unaware of anything in its dread. . . .
> It was Mr. Myclík too who taught me
> to get one extra spare cauldron . . .
> only for boiling sausages and blood puddings

> and brawn and offal and heating fat,
> because whatever cooks in the pan
> leaves something of itself behind,
> and a pig-slaughtering, lady,
> it's the same as a priest serving mass,
> because, after all,
> both are a matter of flesh and blood.[7]

Clearly this ritual entails much more than "pork production." Faith in the interdependent pains and pleasures of life and in the finality of death find common expression in meat.

Through e-mail and essays by Czech students in an English class at Palacky University in Olomouc, I found evidence that such an interpretation remains alive and well among young, urban, well-educated citizens. All of the students, except a few vegetarians (and not excepting some of them), associated *zabijačka* with warmth in winter, family, tradition, intimacy, joy, food and drink aplenty. In response to their American teacher's queasiness with its bloody side, one of them argued:

> For me, the pig slaughter was an unique opportunity to see death in real. They killed the pig I liked a lot, and it was a great shock. And I do not mean any offense—but the Americans should have a chance to see it like this. It makes you think of life and death in real. And then it is a great metaphor: They feed you a lot, they give you buckets of stale beer, and they cut your throat. It is funny.

In response to a particularly graphic image on a Christmas card—a grinning farm family gathered around a blood-dripping carcass—one student wrote:

> Killing the pig, you can see in the card, is not a slaughter. It is just typical for the life in the village to raise a pig and to take care about him for the whole year and then in the winter to kill him. It's not a violent act. . . .
>
> I think it's much better to raise a pig, to kill him, and to live on the meat from this pig than to go to the stores, buy meat, and have in your mind the image of slaughter houses.
>
> So why don't you like the card?[8]

The answer, I think, is hard to articulate because it is not quite a matter of choice. In fact, it is hard to suspend American "common sense" long enough to understand why anyone would ask. No one simply "decides" what is or is not nauseating. Such imagery just hits people in the gut. It feels irresistible like gravity,

RADOSTNÉ VÁNOCE

Czech card (1996) wishes "Merry Christmas" under an image (1954) of *zabijačka* by Josef Lada. Reproduced by permission of Mgr. Josef Lada.

a force made in heaven and lashed deep within. That knot in the stomach is a taut ball of cultural threads, ways of dealing with food, animals, pain, pleasure, disease, the bounds of humanity and mortality that are "just there," taken for granted. Czechs can admire the way they entwine and then graphically unwind in *zabijačka;* Americans are more likely repulsed by the whole business.[9]

Clearly, for example, Americans find the taste of their tenderloin a lot sweeter if they avoid imagining the process by which it arrived on their plate. And keeping it out of mind, under present circumstances, is pretty easy. Outside of the industry itself, relatively few Americans have ever visited a working farm, even fewer actually worked on one, and just short of none has ever stepped foot in a packing plant. It is easy to grow up absolutely clueless about where food comes from, at least before it got into tidy packs at the store.

Naturally enough, then, people use analogues they do know—such as their pets or the cuddly critters that they see in cartoons or zoos—to form an impression of the fate of commercial livestock. Imagine creatures pretty much like Tabby or Spot (or yourself) assembled by the thousands, bound in pens, fattened, prodded, slaughtered, and dismembered. Their butts are sliced and wrapped in supermarket cellophane. The horror of it all is just about overwhelming. Those who contemplate the process are apt to swear off meat entirely. The first to go is the kind that bears the most blatant reminder of its origin, the "red" meat. And if you conjure that image yet more graphically at length, you are apt to suppose that "nature" itself calls upon humans to stop such violence once and for all, to impose the ecological equivalent of a "New World Order."

In characterizing this syndrome so hyperbolically, I do not mean to stand for or against meat. I chose contrasting examples—the Dutch vs. Czech (roughly, Protestant vs. Catholic) instances—precisely to waffle on that issue. One is remorsefully and the other joyously carnivorous. To me, though, that difference is less important than the proportioned sense of mortality that they both bring to a bigger subject of which meat is a part. Albeit to different ends, they both accept responsibility to diverse forms of life and death. They acknowledge that every life depends on the sacrifice of others (whatever the color of their flesh) and that genuine conflicts as well as dependencies exist among us. After a long history of abusing those relations, you would think that humans would have learned that lesson. No individual or species should act as if it knows what is ultimately best for the whole. This uncertainty might be honored in many ways—say, through these examples of Dutch restraint or Czech celebration—but something is amiss in a society when it just makes most people gag.[10]

Among the Americans I have met who face that reality most forthrightly are the animal scientists who study the behavior and psychology of commercial livestock, the Jane Goodalls and Aldo Leopolds of agriculture.[11] Those who specialize in swine are scattered across North America, Western Europe, and Australia, but many of them have crossed paths at a handful of universities (chiefly, Illinois and Purdue in the United States) and collaborated in research and consulting projects. Their impetus is adamantly practical: How do we craft affordable environments in which animals will thrive? How wide should the alleyway in a hog house be? Are sows less likely to act frightened and balk if it is more than twenty-six inches wide? Do they prefer a certain color or texture to the walls? How much does that preference matter? What wattage bulb should glow by the door? What temperature should it be at each hour of the day? Do the sows do better when a herdsmen adjusts the thermostat before bed or if they adjust it themselves? How do we move those animals efficiently? How can a packing plant arrange the actual killing to be as efficient and pain-, panic-, and PSE-free as possible?

As practical as these questions are, they also require looking life and death in the face. And that face looks back at you. Metaphysical and ethical convictions must be clear, especially when you confront a public that prefers ignorance or just gags at the thought of your world. It can make you feel passionate toward your work, at times hurt, ashamed, or angry. Stan Curtis, one of the premier farm animal ethologists in the United States, speculates:

> I believe that there's leakage in attitude from the medical profession into the veterinary profession. I think that for the last fifty years, the attitude has been based on two things.
>
> The era of antibiotics and vaccines has given the medical profession a real feeling of omnipotence. They actually think that they are akin to God, and they raise the expectation in the hearts and minds of their patients. That's one thing.
>
> Number 2—the undercurrent that makes the patients vulnerable to that kind of malarkey—is a fear of death in our culture that has come up in about the same period. We want to deny it. We want nothing to do with it.
>
> Most people have never laid eyes on a dead body until the undertaker has his go at it. I mean, death has been removed from our experience. That's not only death of humans; that's death of animals, too. Butchering, that kind of thing—that all goes on behind closed doors, just like death in a hospital.
>
> When do we see death other than in roadkill? I have a twenty-six-year-old daughter who, as often as not, cries when she sees a roadkill! I mean, Jesus!
>
> There's no rationale there at all. I think this makes the populace as a whole vulnerable to these crazy ideas that some physicians have had—that, by God, they can do anything—or to a veterinary dean who says, "Give me $25 million, and I'll make every animal in the state of California healthy forever."[12]

One of Curtis' most famous students, Temple Grandin, has often voiced similar frustration with popular ignorance of agriculture and denial of death, which she finds distinctly severe in the United States.[13] She turns that challenge back on the livestock industry:

> I find that when I explain slaughter practices, people are relieved. They had imagined things to be worse than they are in reality. . . . Some managers want to deny the reality of what a slaughter plant does. One manager told me he did not want to have the stockyards visible from the road because he wanted his plant to look like a "food factory." This attitude ultimately will create a public-relations disaster. Thirty years ago, everybody saw the inside of a slaughter plant at least once. Today kids think milk comes from a chemical factory, and meat is made

> in the back of the market. As I have stated many times before, we need to clean up our "house" and show it off to the public.[14]

With her many years of professional training and research, packing plant and farm experience around the world, and dedication, Grandin's services are much in demand. Her ability to empathize with individuals of other species has been honed to the point that her feedlot and loading-chute hunches (say, on the redesign of chutes) are almost as dependable as the carefully controlled experiments on which they are indirectly based. Colleagues admit some jealousy that her clients, such as well-heeled packing companies, are not funding even more expensive and reliable research, but they also admit, damn it, she is usually right.

It is perhaps telling that Grandin's greatest fame outside of agriculture came through her leading role in a *New Yorker* article and then a best-selling book by Oliver Sacks.[15] He latches onto her expression—"An Anthropologist on Mars"—and makes it the center of her life. As sensible as that strategy might be for a diagnostician or even for Grandin herself, there are also a couple of curious side effects. First is the way the figure of speech estranges the setting. A casual reader might think that Colorado State University, where Grandin holds a regular faculty position, or the feedlots and packing plants, where she does research, properly belong on another planet, that they require a savant's sensibility. Second is Sacks's emphasis on one of Grandin's distinctions, her autism, as key to her professional success. Although he obviously respects her talents, I fear he overly exoticizes them, rendering her the ethological equivalent of a freak. Experience with her and her colleagues around the world encourages me to think that, although singularly gifted, her expertise in the interaction of humans and livestock is endemic to her profession.

Here, for example, Stan Curtis, who was one of her graduate mentors, describes ways that even farmers may miss some of the wonder of the world of swine:

> If you think about a pig's day, the vast majority of the time the pig is just being a pig. The presence of people is a big event in the life of a pig, particularly if that pig is in a more or less boring environment, which a lot of pigs are. So, the pig behaves in a very special way when human beings are present. As soon as they leave—you can see this on video recordings—they begin behaving like pigs.
>
> If you look at the activity pattern of a pig when no humans are present, they spend about 95 percent of their time lying down. The reason pigs move are to forage for food or to seek water or a mate. If they have all those things provided to them or they aren't in heat or have been castrated and don't have any motivation, they don't move much at all.

> Now, when humans are there, they are stimulated because they like to interact with that strange other being that, judging from experience, is not going to threaten their survival. So they're up and about; they're talking; they're up and begging for food. Basically, trying to get your attention: "Hey, I'm over here! Bring the food over here, if that's what you have in that bucket."
>
> And interestingly enough, they don't need that much food.
>
> Why is it that they're like that?
>
> Well, they're like that because for 99.9-plus percent of the evolutionary history of these wonderful creatures, they didn't have food brought to them on a platter. So they had to rummage and forage for it. And so their appetite is there for them to get them off their butts and out looking for food so they can even survive. Their drive to feed is a strong, strong drive.

Like Grandin, Curtis takes pleasure and gains wisdom in alternating between human and (insofar as possible) porcine perspectives, ever remembering intractable difference as well as connection between the two. The interactions between species are extremely complex, and hence the extraordinary power of humans over them is hard to assess. But the challenge hardly requires interplanetary travel. Inspiring lessons are to be gained right here on earth:

> I'll tell you, we had an experiment going on. When they were students, Temple Grandin and Ian Taylor did them. We gave pigs access to toys, objects to play with—bowling balls, rubber tires, pipes—"manipulanda," the experimental psychologists call them. Anyway, Temple found out that they don't like to play with the toy once it has muck on it. And so then we said, "Well, we have to suspend them, like mobiles over their pens." Temple first identified the fact that they have to be suspended, and then she did preference studies. They have a progression as they grow. To begin, pigs like cloth; then they like rubber; and then the older pigs like chain for the texture of their toy. We studied this ad nauseam.
>
> Well, we had pigs that were in a major experiment, confirming careful observations with observations in a big house. We randomly assigned these toy treatments in half the pens, and half of the pens had no toys. What we found was the common problem in hog houses: Some pigs will turn their feeder into a toilet. They back up to the cup in the line feeder and defecate.
>
> Well, my student Ian Taylor was helping Temple do the work, and he made the casual observation (which is the first step in scientific process, of course), "Hey, the pigs without toys defecate in their feeders; the pigs with toys don't."

And so we designed an experiment to look at that specifically. The conclusion we came to was this: The pig, being a fairly intelligent animal, knew that if it defecated in its cup, then at least once a day (and sometimes twice) a human was going to go in there and clean those turds out of that feeder. And what did it do, those fifteen or twenty seconds each time that person went in there? It nuzzled and pulled and turned him into a toy. And so they were putting two and two together. They realized, "Hey, I'll do that again and see if that works . . . Hey, by golly, it did! I crapped into the feeder, and they came in the next morning." And they did it everyday.

And so—this interaction with humans is the question, and I don't think most farmers realize that they're actually being used by their pigs sometimes. As Stephen Budiansky says, we're flattering human beings way too much to think that "we domesticated animals." He says, "Hell, it's just the opposite."[16] I think farmers don't realize when they're hog toys.

As farmers we know about those behaviors that matter—How do we get the animals from one place to another? What kind of footing do we need to have it so that the boar can copulate and not fall down?—those kinds of things. But as far as what's going on cognitively with that animal? I don't think farmers think much about it. . . .

The truth of the matter is that we have a lot more work to do on this front, but as far as we know domestic animals are caught at the level of cognitive development of approximately a two- or two-and-a-half-year-old child. That does not mean that they don't recognize something that they've seen before when they are re-presented with that something. But it does mean that they can't fantasize. Out of sight, out of mind. There are whole series of experiments that would serve as evidence for that conclusion.

In particular, animals do not seem to daydream, like cartoonists draw, a house cat with those ideas in little bubbles. This is anthropomorphism, you know, casting human qualities and capabilities onto a nonhuman fellow creatures. . . . We think we know—when it comes to feeling if it's cold outside or if the wind is hurting our face or if the sun is too bright, or something like that. "Hell, we're all experts in that!" But that's where we err. Because we aren't pigs.[17]

To their credit, animals scientists use their recognition of difference and interdependence as a reason to assume more, rather than less, responsibility for the quality of the interaction, right down to the moment of death. Temple

Grandin, in particular, has devoted a lot of her effort to the design of approaches to killing floors and to schedules that encourage sensitivity among the people who do the killing. She is a strong advocate not only of sound animal science and efficiency but also of spiritual ties among species. In publications, for example, she documents admirable similarities among the traditions of aboriginal, Jewish, and Islamic peoples. She advises modern meat packers and consumers to learn from their rituals of respect. Since she has dedicated her own professional life to farm animals, she is understandably self-conscious about her own role in ending their lives. She explains:

> The author's own ritual is to face the plant and bow her head down when first approaching it. She has also written "Stairway to Heaven" [a prayerful poem by a friend] or "Valhalla" [the name of the great hall where Odin, the Norse deity, receives the souls of battle heroes] on some of the drawings for new systems. The braces and supports on one slaughtering system were designed utilizing the Greek Golden Mean and a mathematical sequence which determines the behavior of many things in nature. Humans do not really know what happens after death. A ritual act of submission before one kills an animal acknowledges the unknown that haunts all people.[18]

When I asked Temple Grandin for more about these ideas, she sent to me a stack of technical reports and a bit of fiction, a radio monologue by Garrison Keillor that was excerpted in *Meat and Poultry* magazine:

> This is the time of year when people slaughtered hogs on the farm, back when people still did that. . . . It was quite a thing for a kid to see. To see living flesh, the living insides of another creature. I'd expect to be disgusted by it but I wasn't; I was fascinated. I got as close as I could. . . .
>
> Rollie [Hochstetter] was the last in Lake Wobegon to slaughter on the farm. One year he had an accident: The knife slipped, and an animal that was wounded got loose and ran across the yard before it fell. Rollie never kept hogs after that. He didn't feel he was worthy of it.
>
> That's all gone. Children growing up in Lake Wobegon will never have a chance to see adults going about the ritual of slaughter that goes back centuries. . . .
>
> It was a powerful experience, when life and death hung in the balance. Now it's all gone, like so much of that life I saw when I was a kid—a life in which people made do, in which people made their own, in which people lived off the land . . . lived between the ground and God. . . . It wasn't a simple life. It was simple for me because I was a child and my happiness was looked after by other people. But it was not simple for the others. Ever.[19]

Grandin unpacked the moral she saw in this story, one that came not from other planets, towns, or times but here and now.

> Garrison Keillor's "Hog Slaughter" is a vivid recounting not of death, but of respect for life, and for that I believe it's an important story for our industry. . . .
>
> It is a sobering experience to be a caring person, yet to design a device to kill large numbers of animals. When I complete a project I am left with a feeling of great satisfaction, but I usually cry all the way to the airport.
>
> While city people never see meat animals die, meat plant workers who stun, shackle and stick animals on high-speed slaughter lines get an overdose of death. Most of them become detached and do their jobs as if they are stapling boxes. A few will become sadistic, but others, such as rabbis at kosher plants, perform slaughter as a sacred ritual. . . . It is up to the managers to preserve the ritualistic character of the slaughter, to hold on to that respect. . . . The animal should be able to walk in with dignity.
>
> When I was at Arizona State University, my roommate [Gloria Tester], who happened to be blind, summed up how I feel. After she touched cattle walking up a ramp [on its way to slaughter] she wrote:
>
> > The Stairway to Heaven—
> > is dedicated to those people
> > who desire to learn the meaning of life,
> > and not to fear death.
> > You, through respect for these animals,
> > can come to respect your fellow man as well.
>
> Touch, listen, and remember.[20]

It is hard to improve on such advice. It does, though, leave an awful lot to the imagination: What exactly is it that we ought to touch? What should we hear and remember? And how do we best show respect?

Lessons of the TGE trail seem less programmatic than spiritual. Surely we do not all need to rush out and cuddle the nearest swine (or virus, or scientist, or farmer). Sometimes, they tell us, things are best left alone. And sometimes not. If pigs, pathogens, and people actually do have anything in common, it is experience with such alternatives, pursued to extremes. Since backers are absurdly tenacious one moment and careless the next, bets are bumped long after they should have been called. There are just too many, too dynamic and conflicted circumstances to rely on any singular scheme—efficiency, intimacy, populism, realism, "go," "whoa," or whatever. Judging from porcine experience, we might show each other the greatest respect by facing each challenge with the messy mix of faith and fear that birth, death, and dinner replenish. I am not sure if this message from the hog house is better characterized as Romantic or Talmudic, but it is one of the sweeter things that just might stick to your boots.

NOTES

CHAPTER 1

1. Variants of the popular song, "Old MacDonald Had a Farm," date from at least the early eighteenth century. The first published American version (featuring farmer "MacDougal") dates from 1917. Roger Lax and Frederick Smith, comps., *The Great Song Thesaurus* (New York: Oxford University Press, 1989), p. 339. "The real McCoy" is an idiom with pedigree too obscure even for the *Oxford English Dictionary*. "Uncle Sam," the beanpole embodiment of the nation, is himself a descendant of the meatpacking trade. Elinor Lander Horwitz, *The Bird, The Banner, and Uncle Sam: Images of America in Folk and Popular Art* (Philadelphia: J. B. Lippincott, 1976), pp. 92-106.
2. *Places in the Heart* (Tri-Star, 1984); *Red River* (United Artists, 1948); *Rio Bravo* (Warner Brothers, 1959); *Green Acres* (CBS, 170 episodes, 1965-1971); *The Beverly Hillbillies* (CBS, 216 episodes, 1962-1971).
3. Alan Nadel, *Containment Culture: American Narratives, Postmodernism, and the Atomic Age* (Durham, NC: Duke University Press, 1995); Miroslav Holub, *Shedding Life: Disease, Politics, and Other Human Conditions* (Minneapolis: Milkweed Editions, 1997); Susan Sontag, *Illness as Metaphor, and AIDS and Its Metaphors* (New York: Doubleday, 1990).

CHAPTER 2

1. Robert Newton Peck, *A Day No Pigs Would Die* (New York: Alfred A. Knopf, 1973). Peck credits the book to the influence of "my father, an illiterate farmer and pig-slaughterer whose earthy wisdom continues to contribute to my understanding of the natural order and the old Shaker beliefs deeply rooted in the land and its harvest." *Contemporary Authors*, New Revision Series (Detroit: Gale Research Co., 1983), vol. 31, p. 342.
2. Ellis Parker Butler, *Pigs is Pigs* (New York: McClure, Phillips and Co., 1906).
3. E. B. White, *Charlotte's Web* (New York: Harper and Row, 1952). In its first thirty years, the book attracted rave reviews, twenty translations, and sales totaling more than 6 million volumes. In 1976, teachers, librarians, publishers, and children's authors queried by *Publishers Weekly* rated *Charlotte's Web* the single best children's book to have been written in two centuries. In a 1990 poll, 300,000 children, just like their peers ten and twenty years earlier, rated the book their favorite. Scott Elledge, *E. B. White: A Biography* (New York: Norton, 1984), p. 299; Beverly Gherman, *E. B. White: Some Writer!* (New York: Atheneum, 1992), p. 92; Lucy Rollin, "The Reproduction of Mothering in *Charlotte's Web*," *Children's Literature* 18 (1990), pp. 50-52. White is also well known for his contribution to William Strunk, Jr., *The Elements of Style*, 3rd ed. (New York: Macmillan, 1979).
4. John W. Griffith, *Charlotte's Web: A Pig's Salvation* (New York: Twayne Publishers, 1993), p. 8.
5. Gherman, *E. B. White*, pp. 84-93; Griffith, *Charlotte's Web*, pp. x-xi; Peter F. Neumeyer, ed., *The Annotated Charlotte's Web by E. B. White, Pictures by Garth Williams*, (New York: HarperCollins Publishers, 1994), pp. 2, 218, 237, 239.
6. E. B. White, "Death of a Pig," *The Atlantic Monthly* (January 1948), pp. 31-32.

7. Barbara Noske, *Humans and Other Animals: Beyond the Boundaries of Anthropology* (London: Pluto Press, 1989), pp. 10-11; Neumeyer, ed., *The Annotated Charlotte's Web*, p. 224. White once cautioned, "When you read it, just relax. Any attempt to find allegorical meanings is bound to end disastrously, for no meanings are in there. I ought to know." *Letters of E. B. White*, ed. Dorothy Lobrano Guth (New York: Harper and Row, 1976), p. 373. But given White's meticulous style, it is hard to believe that the symbolism is pure projection on the part of the reader. See for example, Neumeyer, ed., *The Annotated Charlotte's Web*, pp. 2, 9-15.
8. William Anderson, "Garth Williams After Eighty," *The Horn Book Magazine* 69:2 (March 1993), pp. 181-186; Zena Sutherland and Mary Hill Arbuthnot, *Children and Books* (New York: HarperCollins, 1991), p. 145. Note that the voice of Miss Piggy also belongs to an Englishman, Frank Oz, born Frank Oznowicz in Hereford in 1944.
9. Richard P. Horwitz, ed., *Exporting America: Essays on American Studies Abroad* (New York: Garland Publishing, 1993).
10. The address, originally entitled "Shit Happens," was delivered at the annual meeting of the Mid-America American Studies Association in Minneapolis, April 17, 1993. An expanded version was published as "Shit Happens: An American Studies Engagement," *American Studies*, 34:2 (Fall 1993), pp. 69-79.
11. Steven Watts, "The Idiocy of American Studies: Poststructuralism, Language, and Politics in the Age of Self-Fulfillment," *American Quarterly* 43:4 (December 1991), pp. 625-660. "Forum," *American Quarterly* 44:3 (September 1992), includes: Barry Shank, "A Reply to Seven Watts's 'Idiocy,'" pp. 439-448; Nancy Isenberg, "The Personal Is Political: Gender, Feminism, and the Politics of Discourse Theory," pp. 449-458; and Steven Watts, "Reply to Critics," pp. 459-462.
12. Richard Horwitz, "Teaching About Method," *American Studies*, 31:1 (Spring 1990), pp. 101-116.
13. Hans Reichenbach, *Experience and Prediction: An Analysis of the Foundations and the Structure of Knowledge* (Chicago: University of Chicago Press, 1938), pp. 6-7; Frederick Suppe, ed., *The Structure of Scientific Theories*, 2nd ed. (Urbana: University of Illinois Press, 1977); and Paul Hoyningen-Huene, "Context of Discovery and Context of Justification," *Studies in History and Philosophy of Science* 18:4 (December 1987), pp. 501-15.
14. Gene Wise, "'Paradigm Dramas' in American Studies: A Cultural and Institutional History of the Movement," *American Quarterly*, 31:3 (Bibliography 1979), pp. 293-337.
15. Michel Negroponte, *Silver Valley* (1983); Ross McElwee, *Sherman's March* (1985); Jeffrey K. Ruoff, interview with Michel Negroponte, June 15, 1988, Soho, New York City.

CHAPTER 3

1. The most likely cause of death among pigs before weaning is being laid on by a fat hog, most likely its mother. According to hog farmers, that is how 40 percent of the pre-weaning deaths occurred in the United States in 1990, and nearly 50 percent in 1995. USDA, APHIS, Veterinary Services, "Part III: Changes in the U.S. Pork Industry, 1990-1995," *Swine '95: Grower/Finisher* (Fort Collins, CO: Centers for Epidemiology and Animal Health, NAHMS, 1997), p. 12.
2. Even with great increases in farm efficiency, everyday life has its share of losses. Over the most recent five-year period, 1990-1995, an average of 6 to 8 percent of every litter was stillborn, and 9 to 12 percent of the remainder died before weaning. The good news is that the odds of survival are about nine out of ten; the bad news is that, since litters average about ten pigs and whole groups farrow simultaneously, a pile of corpses is also nearly inevitable. USDA, "Part III: Changes in the U.S. Pork Industry," pp. 10-13.
3. "If all the soil conservation acronyms, scientific jargon, no-tills, zone-tills, ridge-tills and mulch-tills have you dazed and confused, don't feel bad. People who study these things for a living aren't in much better shape. 'Part of the problem is that we tend to talk about pure systems,' says Ohio State Extension ag engineer Randall Reeder. 'But in real life, farmers are blending ideas from all sorts of systems to find what works best for them.'" "Straight Talk on Tillage Terms," *Farm Industry News,* Special Issue on Mastering Mulch (November 1993), p. 8.

4. According to the National Animal Health Monitoring System, in the twelve months of 1995 as in 1990, about 6 percent of all hog operations in the United States readily admitted bouts with TGE. USDA, "Part III: Changes in the U.S. Pork Industry," p. 18.

CHAPTER 4

1. "Hog engineers talk about swine 'dunging patterns,' by which of course they mean where in their travels pigs relieve themselves. It turns out that pigs like to be clean—in some senses—and unlike cows are aware enough of their bodily functions to stay clear of both their feed and their sleeping places. It also turns out that pigs like *a view* while they eliminate. Unlike chickens and cows, they are not only self-aware beings, they also have well-developed aesthetic sensibilities." Mark Kramer, *Three Farms: Making Milk, Meat, and Money from the American Soil* (Cambridge, MA: Harvard University Press, 1987), p. 137.
2. Arthur O. Lovejoy blazed this trail for humanists in *The Great Chain of Being: A Study of the History of an Idea* (New York: Harper, 1960), as Donald Griffin did for scientists in such works as *The Question of Animal Awareness: Evolutionary Continuity of Mental Experience* (New York: Rockefeller University Press, 1976), *Animal Thinking* (Cambridge, MA: Harvard University Press, 1984), and *Animal Minds* (Chicago: University of Chicago Press, 1992).
3. Zhu Bajie is usually associated with the classic novel *Xiyouji* [Journey to the West] by Wu Cheng'en, who lived from approximately 1500 to 1582. Zhu has a human body but a pig's head with a long snout, narrow eyes, and big ears.
4. See, for example: Felix Pitre with Christy Hale, ill., *Juan Bobo and the Pig: A Puerto Rican Folktale* (New York: Lodestar Books, 1993).
5. Probably the best anthropological source on swine in New Guinea is Roy A. Rappaport, *Pigs for the Ancestors: A Ritual in the Ecology of a New Guinea People*, new ed. (New Haven, CT: Yale University Press, 1984). See also: Peter D. Dwyer, *The Pigs That Ate the Garden: A Human Ecology from Papua New Guinea* (Ann Arbor: University of Michigan Press, 1990); Michael Fullingim, "Of Pigs, Men, and Life: A Glimpse at Wiru Society," in *Nucleation in Papua New Guinea Cultures,* ed. Marvin K. Mayers and Daniel D. Rath (Dallas, TX: International Museum of Cultures, 1988), pp. 23-35; Jane Goodale, *To Sing with Pigs Is Human: The Concept of Person in Papua New Guinea* (Seattle: University of Washington Press, 1995); Lorraine Dusak Sexton, "Pigs, Pearlshells, and 'Women's Work': Collective Response to Change in Highland Papua New Guinea," in *Rethinking Women's Roles: Perspectives from the Pacific,* ed. Denise O'Brien and Sharon W. Tiffany (Berkeley: University of California Press, 1984), pp. 120-152; David C. Wakefield, "Dog-Pigs and Other People," in *Studies in Componential Analysis,* ed. Karl J. Franklin (Ukarumpa via Lae, Papua New Guinea: Summer Institute of Linguistics, 1989), pp. 65-71.

   Certainly no library of swinalia could be without: Kent Britt, "The Joy of Pigs," *National Geographic,* 154:3 (September 1978), pp. 398-415; Stanley E. Curtis, "Pigs and People," in *Swine Nutrition*, ed. Elwyn R. Miller, Duane E. Ullrey, and Austin J. Lewis (Boston: Butterworth-Heinemann, 1991), pp. 3-34; Don Muhm, "Pork Potpourri - Pigs and Stuff," in *Iowa Pork and People: A History of Iowa's Pork Producers* (Clive, IA: Iowa Pork Foundation, 1995), pp., 207-250; Marilyn Nissenson, *The Ubiquitous Pig* (New York: Abrams, 1992); Michael Ryba, *The Pig in Art* (London: Orbis Publishing, 1983); or Frederick C. Sillar and Ruth Mary Meyler, comp., *The Symbolic Pig: An Anthology of Pigs in Literature and Art* (Edinburg: Oliver and Boyd, 1961).

   Among the premier pig images in American art are "Pretty as a Pig" by Andy Warhol and "Portrait of the Pig" by Jamie Wyeth. Lincoln Kirstein, "James Wyeth," in *An American Vision: Three Generation of Wyeth Art,* comp. Brandywine Conservancy (Boston: Little Brown, 1987), p 160. The most powerful images may be those by Sue Coe in "Porkopolis—Animals and Industry" (1988-1989). Sue Coe, *Paintings and Drawings* (Metuchen, NJ: Scarecrow Press, 1985) and *Dead Meat* (New York: Four Walls Eight Windows, 1995).

   Exemplary of more specialized works from sundry corners of the academy are: Timothy Cook, "Upton Sinclair's *The Jungle* and Orwell's *Animal Farm:* A Relationship Explored," *Modern Fiction Studies* 30:4 (Winter 1984), pp. 696-703; Erika Doss, *Spirit Poles and Flying Pigs: The Politics of Public Art in Contemporary America* (Washington, DC: Smithsonian Press,

1995); Stephen Duplantier, "Family Pigs in Central Louisiana: The Boucherie des Habitants in Avoyelles Parish," *Louisiana Folklore Miscellany* 5:3 (1983), pp. 9-18; Emilio Faccioli, Comune di Reggio Emilia, Assessorato alia cultura, ed. and comp., *L'Eccellenza e il trionfo del porco: immagini, uso e consumo del maiale dal XIII secolo ai nostri giorni,* new ed. (Milan: Mazzotta, 1982); Marvin Harris, "Pig Lovers and Pig Haters," in *Cows, Pigs, Wars and Witches: The Riddles of Culture* (New York: Random House, 1975), pp. 35-57; Theodore C. Humphrey and Lin T. Humphrey, *"We Gather Together": Food and Festival in American Life* (Ann Arbor, MI: UMI Research Press, 1988); Irving Massey, *The Gaping Pig: Literature and Metamorphosis* (Berkeley: University of California, 1976); John R. Perry, "Blackmailing Amazons and Dutch Pigs: A Consideration of Epic and Folktale Motifs in Persian Historiography," *Iranian Studies; Journal of the Society for Iranian Studies* 19:2 (1986), pp. 155-165; David Rothel, *Great Show Business Animals* (San Diego, CA: A. S. Barnes, 1980); Jacques Verroust, Michel Pastoureau, Raymond Buren, eds., *Le cochon: histoire, symbolique et cuisine du porc* (Paris: Sang de la terre, 1987); Richard A. Watson, "A Pig's Tail," *Latin American Literary Review* 15:29 (January-June 1987), pp. 89-92; Allan White, "Pigs and Pierrots: The Politics of Transgression in Modern Fiction," *Raritan: A Quarterly Review* 2:2 (Fall 1982), pp. 51-70.

6. Milo Kearney, *The Role of Swine Symbolism in Medieval Culture: Blanc Sanglier* (Lewiston, NY: E. Mellen Press, 1991), p. 6. See also: Faccioli, *L'Eccelenza e il trionfo del porco*; and Julian Wiseman, *A History of the British Pig* (London: Duckworth, 1986).
7. Nissenson, *The Ubiquitous Pig,* pp. 119, 116. Note, though, that in modern German the idiom "Schwein haben" (literally, "to have swine") means "to be lucky." Wilfried Dickhoff, "For an Architecture of Vor-Sch(w)ein," in *A House for Pigs and People/Ein Haus für Schweine und Menschen,* ed. Carsten Höller and Rosemarie Trockel (Köln: Verlag der Buchhandlun Walter König, 1997), p. 44.
8. Apparently, the meaning of the term "marrano" has drifted quite a bit since the Inquisition, roughly, from New Christian (a variety of *converso*), to closeted Jew, to infidel. In some places (e.g., Honduras) it may have gone full circle to where it simply is another word for "pig."
9. Joel Greenberg, "Israeli Jew is Convicted of Insulting Islam," *The New York Times* (December 31, 1997), p. A7. There is a long history and large literature on the ways that some people—especially people in the West—have maligned others by accusing them of resembling or mating with animals. See: L. Perry Curtis, Jr., *Apes and Angels: The Irishman in Victorian Caricature* (Washington, D.C.: Smithsonian Institution, 1971); John W. Dower, *War Without Mercy: Race and Power in the Pacific War* (New York: Pantheon, 1986), pp. 77-93, 147-156, 341-342, 242; Richard Drinnon, *Facing West: The Metaphysics of Indian-Hating and Empire-Building* (New York: New American Library, 1980); Thomas F. Gossett, *Race: The History of an Idea in America* (Dallas, TX: Southern Methodist University Press, 1963); Stephen Jay Gould, *The Mismeasure of Man* (New York: Norton, 1981); Margaret T. Hogden, *Early Anthropology in the Sixteenth and Seventeenth Centuries* (Philadelphia: University of Pennsylvania Press, 1964); and V. G. Kiernan, *The Lords of Human Kind: Black Man, Yellow Man and White Man in an Age of Empire* (Boston: Little, Brown, 1969).

   For a discussion of this animal-human association in relations between African and European Americans, where the dynamic has been particularly cruel, see: Gould, *The Mismeasure of Man,* pp. 71, 125-126; Hogden, *Early Anthropology*, pp. 417-418; and especially Winthrop D. Jordan, *White Over Black: American Attitudes Toward the Negro, 1550-1812* (Chapel Hill: University of North Carolina Press, 1968), pp. 28-32, 199-233, 305-310.
10. Kearney, *The Role of Swine Symbolism*, p. 322.
11. The dialogue in *Babe* includes a few, passing references to Babe as a boy (and none as a girl), but the name would fit either, and the imagery itself includes many full-frame views of a female rear end. The conflict among these codes may account for the fact that I have yet to meet anyone (including farm kids) who can confidently recall whether the pig was one or the other. "The movie keeps Babe's gender deliberately vague, [but] because of all the ground-level, from behind shooting . . . only females were used (obvious reasons)." Kenneth Chanko, "This Pig Might Fly," *Entertainment Weekly* 288 (August 18, 1995), p.16. Some of the gender-bending also can be discerned in the promotional material on the film, on Babe's voice (Christine Cavanaugh), and on the trainer (Karl Lewis Miller) featured on the World Wide Web during the film's run: http://movieweb.com/movie/babe/babe.txt. The obvious exception in this regard (the evasion of sexu-

ality in pig pop culture) is Miss Piggy, whose gender and sex are even more adamantly defined than her species. But she may have gained popularity precisely because of the way she toys with sexual identity. Her absurdly hyperfeminine, flirtatious demeanor flouts mass-media conventions as well as the contemporary regard for "male chauvinist pigs" among the liberals who produced the show and who were the parents of the target audience. Furthermore, as with Babe, the character of Miss Piggy is produced through a gender bend. Frank Oz supplies her voice. See also Don Muhm, "No Boar is *Playboar*," *Iowa Pork and People*, pp. 242-243.

12. Edited from telephone conversation with Bob Tannehill, Library Director, Chemical Abstracts, Columbus, Ohio, May 31, 1994.
13. John Sedgwick, "Brotherhood of the Pig," *GQ: Gentlemen's Quarterly* 58 (November 1988), p. 304.
14. John Brooks, *Once in Golconda: A True Drama of Wall Street 1920-1938* (New York: Harper and Row, 1952), pp. 61-63, 271-272.
15. J. H. Breakell and Co., "Piglet Pin," print advertisement, *New Yorker* (July 11, 1994), p. 41.
16. Telephone interview, Breakell and Co, April 17, 1995.
17. The decision of the U.S. Court of the Second District of New York was delivered on September 25, 1995 by Justice Kimba Wood, *Hormel Foods Corp. v. Jim Henson Productions Inc.* (no. 95 CIV. 5473 KMW), *U.S. Patent Quarterly*, 2nd Series, Vol. 36, pp. 1812-1824. An appeal was argued before the Second Circuit Court of Appeals on October 31, 1995, and denied on January 4, 1996 (no. 1010, Docket 95-7977), *Federal Reporter*, 3rd Series, Vol. 73, pp. 497-508. In distinguishing this case from the *Universal City Studios* precedent (involving "Donkey Kong" dilution of the value of "King Kong") the court opined: "In this case, the names 'Spa'am' and SPAM are quite similar, but the 'characters' are a wild boar puppet and a can of luncheon meat—significantly farther apart than two gorillas." *Hormel Foods Corp. v. Jim Henson Productions Inc.*, 36 USPQ2d, p. 1822.

CHAPTER 5

1. Harkin's joke is apparently a variant on one Truman told at an Iowa plowmen's competition in 1948: "No man should be allowed to be president who does not understand hogs, or hasn't been around a manure pile." Don Muhm, *Iowa Pork and People: A History of Iowa's Pork Producers* (Clive, IA: Iowa Pork Foundation, 1995), p. 243. Photo caption, *Daily Iowan* (April 25, 1995), p. 1. The *Little House on the Prairie* books and TV series are among the strongest and most popular evocations of "family farming," self-reliant, intimate, independent. Their image is doubly misleading, not only because they fictionalize the memoirs of Laura Ingalls Wilder, but also because her daughter, Rose Wilder Lane, revised the memoirs to harmonize with her profound disaffection for New Deal programs. Linda Kerber, "Women and Individualism in American History," *The Massachusetts Review* (Winter 1989), pp. 604-605; Stephanie Coontz, *The Way We Never Were: American Families and the Nostalgia Trap* (New York: Basic Books, 1992), pp. 168-176.
2. For an example of a more dispassionate, point-by-point analysis, see: USDA, APHIS, Veterinary Services, "Part III: Changes in the U.S. Pork Industry, 1990-1995," *Swine '95: Grower/Finisher* (Fort Collins, CO: Centers for Epidemiology and Animal Health, NAHMS, 1997).
3. "Task Force Dragged into Fierce Creston Debate About Hog Lots," *Iowa City Press-Citizen* (September 10, 1994), p. 5B.
4. See, for example, Patty Cantrell, "Is the Family Farm an Endangered Species?" *Ms.* (March/April 1997), pp. 33-37.
5. The words "populist" and "progressive," of course, have unique and quite specific referents in American history, especially when capitalized (as in the Populist or People's Party, that in 1896 endorsed the presidential candidacy of William Jennings Bryan, and the Progressive Party, associated with the presidential aspirations of Theodore Roosevelt and then Robert La Follette, 1912-1924). I here use the terms much more loosely and, hence, pretty interchangeably. I generally prefer "populist" to connote a rural, blue-collar, backward-looking or demagogic quality and "progressive" for a more urban, bourgeois and hip, forward-looking or utopian one. But I use both to signal opposition to corporate domination, presumably in defense of humbler folk.

6. Michael G. Kammen, *Mystic Chords of Memory: The Transformation of Tradition in American Culture* (New York: Alfred A. Knopf, 1991).
7. For a nice example of "realism" in the hog trade magazines, see the special issue of *National Hog Farmer* (May 15, 1994), which was heavily dedicated to the boom outside the Swine Belt and to heroic adaptations in the Midwest. For a counterexample in the populist mode, see the press releases from the Humane Farming Association or Lenor Yarger, "Iowa's Hog Hell," *Icon* (January 25, 1996), pp. 4-5, based almost entirely on the testimony of Sharon and Ken Petrone who had long been involved in organizing to stop vertical integrators in Iowa.
8. John Greenwald, "Hogging the Table," *Time* (March 18, 1996), p. 76.
9. Michael Satchell, "Hog Heaven—and Hell," *U.S. News and World Report* (January 22, 1996), p. 55.
10. According to its sponsor, the Church Land Project headquartered in Des Moines, 130 people registered for "Community, Church and Large-scale Hog Production: Theology and Resolution of Hog Production Conflicts in Rural Communities." Among the main speakers were: the Reverend Jerry Avise-Rouse (Mt. Ayr Larger United Methodist Church, Mt. Ayr, Iowa), the Reverend Gil Dawes and Barb Grabner (PrairieFire Rural Action), Bernard Evans (St. John's University), William Heffernan, (Department of Rural Sociology, University of Missouri), Carmen Lampe (First Baptist Church, Mt. Ayr, Iowa), Barb Mathias (Iowa Council for International Understanding), Barbara Ross (Diocese of Jefferson City, Missouri), Mark Schultz (Land Stewardship Project, St. Paul, Minnesota), and Denise Turner (Christian Church, Trenton, Missouri). *Church Farmland News* 4:3 (April 1995), p. 2; "Churches and Hogs," *Des Moines Sunday Register* (November 27, 1994), p. 3J; Citizens Task Force on Livestock Concentration, *A Citizens' Report: Recommendations for the 1995 Iowa Legislature on Concentrated Livestock Production* (October 12, 1994); "Profit Is Only Motive for Some Forms of Ag," *The Bishop's Bulletin* (January 1995), p. 7.

    On bigotry and its role in U.S. populism in general, see: John Higham, *Strangers in the Land: Patterns of American Nativism, 1860-1925* (New Brunswick: Rutgers University Press, 1988); V. O. Key, *Southern Politics in State and Nation,* 2nd ed. (Knoxville: University of Tennessee Press, 1984); Catherine McNicol Stock, *Rural Radicals: Righteous Rage in the American Grain* (Ithaca, NY: Cornell University Press, 1996); Walter T. K. Nugent, *The Tolerant Populists: Kansas Populism and Nativism* (Chicago: University of Chicago Press, 1963); Jeffrey Ostler, "The Rhetoric of Conspiracy and the Formation of Kansas Populism," *Agriculture History* 69 (Winter 1995), pp. 1-27; Jeffrey Ostler, "Why the Populists Party Was Strong in Kansas and Nebraska but Weak in Iowa," *The Western History Quarterly* 23 (November 1992), pp. 451-474; and C. Vann Woodward, *Tom Watson: Agrarian Rebel* (London: Oxford University Press, 1938). Historians and political scientists have long debated the relative importance of nativism among rural progressives before the Great Depression. Woodward and Nugent, for example, nicely parse the extremes. Whether fundamental in the grassroots or superficial in the posture of a few leaders, white supremacy, xenophobia, anti-Catholicism, and anti-Semitism were undeniably evident in otherwise progressive movements through the late nineteenth and early twentieth centuries in the United States. Armed compounds in the Utah or Texas outback—uniformly celebrating the "little guy" and his adamantly Northern European and Protestant lineage—show those connections remain strong today. Chip Berlet and Matthew N. Lyons, *Too Close for Comfort: Right-wing Populism, Scapegoating, and Fascist Potentials in U.S. Political Traditions* (Boston: South End, 1996).

    Xenophobia was clearly a resource in the hysteria surrounding Indiana Packers Company, which built a 300,000-square-foot plant in Delphi, Indiana. Among the key complaints was that it was "foreign owned," a joint venture of Ferruzzi of Italy and Mitsubishi of Japan, even though at the time the NPPC was working furiously to remove European Community and Japanese barriers to U.S. ventures. National Pork Producers Council, *1993-94 Issues Handbook: "Communicating the Views of America's Pork Producers"* (Des Moines, IA: National Pork Producers Council, 1994), p. 35.
11. Assessing the impact of Circle Four Farms (a four-year-old, 600,000-hog joint venture of North Carolina pork magnates) on the 1,000 residents of Milford, Utah, Mayor Mary Wiseman concluded, "They've been a godsend;" Milford farmer Joey Leko countered: "It's like the devil came to Milford." "Giant Hog Farm Divides Tiny Community," *Iowa City Press-Citizen* (November 10,

1997), p. 11B. Gary L. Benjamin, "Industrialization in Hog Production: Implications for Midwest Agriculture," *Economic Perspectives* 21:1 (January/February 1997), pp. 5-7.

CHAPTER 6

1. Popular fears seem inversely related to the actual risk. Despite the variety of circumstances of billions of meals served from 1973 to 1977, the Centers for Disease Control and Prevention detected a total of only 119 "outbreaks" of human disease in the United States where the vehicle of transmission was pork or ham; subsequently that number has fallen by more than 75 percent. Although the number of daily meals has greatly increased, over a more recent five-year period (1988 to 1992) there were only 29 confirmed outbreaks in the United States. USDA, APHIS, Veterinary Services, "Part III: Changes in the U.S. Pork Industry, 1990-1995," *Swine '95: Grower/Finisher* (Fort Collins, CO: Centers for Epidemiology and Animal Health, NAHMS, 1997), p. 34.
2. National Pork Producers Council, *1993-94 Issues Handbook: "Communicating the Views of America's Pork Producers"* (Des Moines, IA: National Pork Producers Council, 1994), pp. 26-27 and 72-73.
3. Since the Food Quality Protection Act, which reformed the Delaney Clause, was enacted only in August 1996, its effects are uncertain at this time.
4. According to the National Swine Survey, in 1995 about 90 percent of all hog farms in the United States included antibiotics in the regular rations of commercial hogs "for disease prevention or growth promotant purposes." Less than 50 percent inject antibiotics intravenously, but the practice became significantly more common from 1990 to 1995. USDA, APHIS, Veterinary Services, *Antibiotic Usage in Premarket Swine* (Fort Collins, CO: Centers for Epidemiology and Animal Health, NAHMS, 1997), p. 1; USDA, "Part III: Changes in the U.S. Pork Industry," p. 20.
5. Marlys Miller, "FDA is Paving the Way for Drug Compliance," *Pork '93* (June 1993), pp. 42-44; Joe Vansickle, "Voluntary Tetracycline Withdrawal Urged," *National Hog Farmer* (September 15, 1996), p. 15; Marlys Miller, "HACCP, Coming to a Packer Near You," *Pork '97* (March 1997), pp. 44-46; John Gadd, "Life Without Growth Promotants," *National Hog Farmer* (March 15, 1998), pp. 52-53.
6. Much of this information comes from Marcie Anthone, director of Research and Strategic Planning for Bozell Worldwide, the advertising agency engaged by the National Pork Producers Council.
7. Karen McMahon, "Global Markets Hold Promise," *National Hog Farmer* (January 15, 1995), p. 51. The consumer profile comes from Sue Levy of Bozell Worldwide in Chicago, who conducted a national survey, part of the 1994 "Pork Chain Quality Audit" for the National Pork Producers Council. Marlys Miller, "Identifying Links in the Quality Chain," *Pork '94* (June 1994), p. 39; Dale Miller, "Farm-to-Fork Quality Audit," *National Hog Farmer* (May 15, 1994), p. 13.
8. National Pork Producers Council, *1993-94 Issues Handbook,* pp. 72-73; "Integrators: Heart Association Certifies Smithfield's Lean Pork," *Pork '96* (December 1996), p. 48.
9. David Gerrard, "The Other White Meat: Be Careful What You Wish For," *Pork '97* (November 1997), p. 34. Bankers agree: "Primarily responsible for the changes under way in the U.S. pork industry are today's discriminating consumers." Alan Barkema and Michael L. Cook, "The Changing U.S. Pork Industry: A Dilemma for Public Policy," *Federal Reserve Bank of Kansas City Economic Review* (Second Quarter, 1993), p. 49. John Lawrence and Rodney Goodwin, "Claiming Real Values of Terminal Sire Lines," *National Hog Farmer* (June 1, 1995), p. 28.
10. "Super Bowl Blitz," *Des Moines Sunday Register* (January 15, 1995), p. 3J. Some producers deeply resented the use of a couple of million dollars of their check-off money for 150 seconds of commercial advertising during the Super Bowl. "I'd vote to take next year's Super Bowl budget and split it between pork quality/stress gene research and finding some solution to PRRS [a deadly reproductive and respiratory syndrome in pigs]." Dale Miller, "Opinion Page," *National Hog Farmer* (March 15, 1996), p. 15. "Super Bowl Kick-off," *National Hog Farmer* (November 15, 1996), p. 6; Jane Messenger, "Has Pork Found Its Glass Slipper" and "Pork: Now the Meat of Many Choices," *Pork '97* (March 1997), pp. 26-34.

CHAPTER 7

1. In standard ag ridicule, PETA is decoded "People Eating Tender Animals." CORE (the Congress on Racial Equality) was established in 1942 to promote racial harmony during and following World War II. In 1960 Martin Luther King, Jr., who three years earlier helped organize black clergy in the Southern Christian Leadership Conference, inspired the organization of allied students in the Student Nonviolent Coordinating Committee (SNCC) in Raleigh, North Carolina.

2. The pie target, Dainna (Jellings) Smith of West Union, Iowa, recalls the event as her "most miserable as Iowa Pork Queen" and "A Gooey Present from PETA" in Don Muhm, *Iowa Pork and People: A History of Iowa's Pork Producers* (Clive, IA: Iowa Pork Foundation, 1995), pp. 103-104; "News Update," *National Hog Farmer* (October 15, 1993), pp. 8-9.

   Since the Animal Liberation Front is avowedly outlaw, information on the organization is difficult to confirm. See "the ALF bible"—Peter Singer, *Animal Liberation: A New Ethics for Our Treatment of Animals* (New York: New York Review, 1975); U.S. Department of the Treasury, Bureau of Alcohol, Tobacco and Firearms, *The Animal Liberation Front in the 90's* (Washington, D.C.: Tactical Intelligence Branch, 1992); David Henshaw, *Animal Warfare* (London: Fonatana, 1989); Michael P. T. Leahy, *Against Liberation: Putting Animals In Perspective,* rev. ed. (New York: Routledge, 1994); Roderick F. Nash, *The Rights of Nature: A History of Environmental Ethics* (Madison: University of Wisconsin Press, 1989); David T. Hardy, *America's New Extremists: What You Need to Know About the Animal Rights Movement* (Washington, D.C.: Washington Legal Foundation, 1990); Susan Sperling, *Animal Liberators: Research and Morality* (Berkeley: University of California Press, 1988).

3. Lenore Yarger, "Iowa's Hog Hell," *Icon* (January 25, 1996), p. 4.

4. Note that the rights attributed here to animals greatly resemble those (a) earlier attributed to slaves in opposing the institution of slavery or (b) denied slaves in defending it and more recently (c) attributed to human fetuses in opposing abortion. Furthermore, note that among those reasoning from natural rights in the same period are America's "judicial conservatives" (e.g., Robert Bork and Clarence Thomas) whom "liberal" senators grilled in confirmation hearings.

5. Jeremy Rifkin, *Beyond Beef: The Rise and Fall of Cattle Culture* (New York: Dutton, 1992). Note also a historic association between animal and women's rights movements not only in the United States but also in the United Kingdom. Sperling, *Animal Liberators.*

6. Barbara Grabner, "Willie Nelson, FARM AID Lend a Hand," *Corporate Hog Update* (June 1995), p. 16; "2,000 Protest Corporate Hog Farms," *The Cedar Rapids Gazette* (April 2, 1995), p. 1; Marlys Miller, "What's Really Going on Here," and "Township Rallies Against Premium Standard Farms," *Pork '95* (May 1995), pp. 4, 71-72; "Hog-Farm Protesters Continue March Into Iowa," *Iowa City Press-Citizen* (April 20, 1995), p. 8A.

7. "Video Debunks Animal Agriculture Myths," *Pork '94* (January 1994), p. 17; "Espousing Animals' Roles in Our Lives," *Pork '94* (August 1994), p. 5.

8. P. R. English and S. A. Edwards, "Animal Welfare," in *Diseases of Swine*, 7th ed., ed. Allen D. Leman et al. (Ames: Iowa State University, 1992), p. 907, emphasis added. In defining "welfare," English and Edwards (p. 902) cite the "five freedoms" that were key principles in British farm animal welfare codes and that many animal scientists advocate as a global standard:

   1. Freedom from malnutrition; i.e., the diet should be sufficient in both quantity and quality to promote normal health and vigor.
   2. Freedom from thermal and physical discomfort, which means that the environment (e.g., housing) should be neither excessively hot nor cold, nor should it impair normal rest or activity.
   3. Freedom from injury or disease; i.e., the husbandry system should minimize the risk of injury or disease and any cases that do occur should be recognized and treated without delay.
   4. Freedom to express most normal patterns of behavior.
   5. Freedom from fear and stress.

J. F. Webster, "Meat and Right: Farming as if the Animals Mattered," in *New Perspectives in Pig Production*, ed. A. M. Petchey (Aberdeen: North of Scotland College of Agriculture, 1987), pp. 5-10.

9. Marlys Miller, "They're Counting on Your Complacency," *Pork '94* (July 1994), p. 5.

10. Iowa loyalist Connie Meyer wrote to her trade magazine's "Reader Mailbox" to compliment a previously published letter from Greg Gunthrop, an Indiana producer: "I really liked his statement that farms in his area that are still in existence were paid for by pigs and dairy cows that were kept outside. This should make commodity groups sit up and take notice that we do not need to spend mega bucks to research odor. We just need to let family farmers raise pigs the way pigs need to be raised. If commodity groups had given the mega bucks spent on odor research to ten new, young family farmers to get started in the pork industry, maybe hog farmers would still be considered good neighbors and friends in Iowa." "Kudos for Outdoor Rearing," *Pork '97* (November 1997), p. 6.

    Paul Pitcher, a professor in the Swine Health and Production Management Department of the Center for Animal Health and Productivity and the University of Pennsylvania School of Veterinary Medicine, estimates that the majority of his students are vegetarians. Paul Pitcher, e-mail letter to Deb Hyck posted on Swine-L, October 12, 1995. See also the response from Mike Varley, Department of Animal Physiology and Nutrition, University of Leeds, Leeds, England, October 9, 1995. Faculty and staff at veterinary colleges that I visited in Indiana, Iowa, Ohio, and Pennsylvania also uniformly confirmed these generalizations.

11. The Humane Farming Association, *Bringing Home the Bacon: A Look Inside the Pork Industry* (San Francisco: The Humane Farming Association, 1991), p. 3.

12. "Sows Chew Wherever They Are," *Pork '95* (May 1995), p. 16; P. H. Hemsworth, "Behavioral Problems," in *Diseases of Swine*, pp. 653-659, especially p. 657.

13. See for example, "Study Shows Pigs Respond to Attention," *Pork '97* (April 1997), pp. 11-12.

14. "Put Your Hog Handling Skills to the Test," *Pork '97* (January 1997), pp. 60-61; Ian A. Taylor et al., "Design of Feeders for Swine: Kinematics, Behavior, and Individuality," *Proceedings of the Sixth International Congress on Animal Hygiene* (Skara, Sweden: ICAH, 1988), pp. 390-398; Ian J. H. Duncan, "The Science of Animal Well-being," keynote address to the combined meeting of the American Society of Animal Science and the International Society for Applied Ethology, Pittsburgh, PA, August 8-11, 1992, *Animal Welfare Information Center Newsletter* 4:1 (January-March 1993), pp. 1, 4-7; John Gadd, "European Update," *National Hog Farmer* (January 15, 1994), pp. 54-57; "Reduce Pig Stress," *National Hog Farmer* (October 15, 1995), p. 51; "Ask Your Veterinarian: New Health Diagnostics," *Pork '94* (August 1994), p. 11; "Environment Key to Feed Intake," *Pork '95* (April '95), pp. 44-46; Lora Duxbury-Berg, "Indoor Versus Outdoor Intensive Pork Production," *National Hog Farmer* (January 15, 1995), pp. 58-60; "Eliminating Vices Pays at Market," *Pork '95* (February 1995), p. 15.

15. Marlys Miller, "A Shopper's Guide to Farrowing House Flooring," *Pork '93* (December 1993), p. 21; Ron Brunoehler, "Open-air Housing for Happier Hogs," *Farm Industry News* (May/June 1994), p. 10; "A Simple Treat Keeps Boars Happy," *Pork '93* (June 1993), p. 28; Paul Hemsworth, "Handle with Care: Mishandled Hogs Perform Poorly," *National Hog Farmer* (November 15, 1993), p. 58.

16. Interviews with Jack L. Albright, Alan M. Beck, Tim Emerick, and Julie Morrow-Tesch, Department of Animal Sciences and Center for Applied Ethology and Human-Animal Interaction, School of Veterinary Medicine, Purdue University, and Brent Ladd, Rural Center for Food Animal Well-being, West Lafayette, Indiana, May 30-31, 1994. Morrow-Tesch moved to Purdue from the USDA Meat Animal Research Center in Clay Center, Nebraska. The most useful review of the relevant literature in meat science is J. E. Cannon et al., "Pork Quality Audit: A Review of the Factors Influencing Pork Quality," *Journal of Muscle Foods* 6 (1995), pp. 369-402. See also: Marlys Miller, "Does the Stress Gene Make You Money?" *Pork '96* (September 1996), pp. 54-55; David Gerrard, "The Other White Meat: Be Careful What You Wish For," *Pork '97* (November 1997), pp. 34-36.

17. Many market hogs carrying the "stress gene" (about 36 percent) are apt to grade PSE, but the 20 percent figure is for those who are *not* carriers. The causes of PSE are extraordinarily well understood mainly because PSE in pigs is also the model of choice for understanding a human disorder, malignant hyperthermia (MH), which can cause neurological, liver and kidney disease,

even death in humans. Interview with Lauren Christian, Iowa Pork Industry Center, Ames, Iowa, September 27, 1996; C. Michael Knudson et al., "Distinct Immunopeptide Maps of the Sacroplasmic Reticulum Ca^{2+} Release Channel in Malignant Hyperthermia," *The Journal of Biological Chemistry* 265:5 (February 15, 1990), pp. 2421-2424; C. F. Louis et al., "Malignant Hyperthermia and Porcine Stress Syndrome: A Tale of Two Species," *Pig News and Information* 11:3 (1990), pp. 341-344; Junichi Fujii et al., "Identification of a Mutation in Porcine Ryanodine Receptor Associated with Malignant Hyperthermia," *Science* 253 (July 26, 1991), pp. 448-451; Miller, "Does the Stress Gene Make You Money?"

18. Cannon et al., "Pork Quality Audit," p. 375; Lora Duxbury-Bert, "Smart Handling Reduces PSE," *National Hog Farmer* (May 15, 1996), p. 41.
19. "Research to Prove Happy Pigs Are Productive Pigs," *National Hog Farmer* (February 15, 1995), p. 36; "Carcass Quality: Nervous Pigs Produce More PSE," *Pork '93* (June 1993), pp. 17-18.
20. Cannon et al., "Pork Quality Audit," p. 384. An act of the U.S. House of Representatives (H.R. 5680) and Senate (S. 296) bars trade in and demands swift euthanasia of "nonambulatory livestock." See National Pork Producers Council, *1993-94 Issues Handbook: "Communicating the Views of America's Pork Producers"* (Des Moines, IA: National Pork Producers Council, 1994), p. 67.

CHAPTER 8

1. Elizabeth Bloom, "North-Central Iowa at Center of Debate," *Waterloo Courier* (April 17, 1994), p. B4-5.
2. "The Urgent Quest for Odorless Pigs," *Chicago Tribune* (June 27, 1994), p. 12. The microbiologist here quoted was Roderick Mackie from the University of Illinois. Steven P. Rosenfeld, "Conference Sniffs Out Hog Issues," *The Cedar Rapids Gazette* (June 15, 1994), p. 5C.
3. The unsuccessful campaign in Oklahoma was in response to expansions in the hog industry led by Tyson Foods, PIC, and Seaboard Industries. "Oklahomans Rally at State Capitol," *Pork '95* (May 1995), p. 75. Pat Stith and Joby Warrick, "Hog Battles Brewing," *The [Raleigh] News and Observer* (February 26, 1995), p. 9A.

    On February 25, 1997 when the commissioners of Craven County, NC agreed to halt the construction or expansion of large livestock operations, "environmental and health risks" were their justification. The state-wide moratorium, enacted in the Clean Water Responsibility and Environmentally Sound Policy of 1997, affected only construction defined as "new" (initiated March 1, 1997 to 1999) and was negotiated at some cost to less purely environmental concerns. "North Carolina County Imposes Moratorium," *Pork '97* (April 1997), p. 58; "N.C. Governor Asks for Two-year Moratorium," *Pork '97* (May 1997), pp. 81-82; Marlys Miller, "Will the Window Slam Shut?" *Pork '97* (May 1997), p. 5; "North Carolina Approves Hog Ban," *National Hog Farmer* (September 15, 1997), p. 31.
4. John Copeland and Janie Simms Hipp edit the summaries published by the National Center for Agricultural Law Research and Information in Fayetteville, Arkansas.
5. Rodney Barker, *And the Waters Turned to Blood: The Ultimate Biological Threat* (New York: Simon and Schuster, 1997); William J. Broad, "Scientist at Work, JoAnn M. Burkholder: In a Sealed Lab, A Warrior Against Pollution," *The New York Times* (March 25, 1997), p. B9; "Manure Eyed as State Shuts Down River," *Pork '97* (November 1997), p. 102.
6. "FDA Backs Off on Lower Selenium Levels," *Pork '94* (October 1994), pp. 52-54; "The Selenium Issue Is Still Up in the Air" and "Dietary Selenium Not Harmful to Environment," *National Hog Farmer* (August 15, 1994), pp. 6, 52.
7. John Gadd reports that the ten liters of average daily hog waste require 200 grams of dissolved oxygen while the 240 liters of human waste require only 70 grams. He figures that a 2,000-sow hog farm produces the biological oxygen demand (BOD) of 100,000 people. John Gadd, "European Update," *National Hog Farmer* (November 15, 1993), p. 38; and "European Update," *National Hog Farmer* (September 15, 1994), p. 48.

8. "Dye Reduces N, P in Manure," *National Hog Farmer* (December 15, 1994), p. 20; Gadd, "European Update," p. 39; Alan L. Sutton, "Less Protein Cuts Odor Compounds," *National Hog Farmer* (March 15, 1998), pp. 29-30.
9. The "Iowa Farm and Rural Life" poll was conducted in the spring of 1989. On work conditions in hog confinement buildings, see: Livestock Industry Task Force Committee, College of Agriculture, Iowa State University, *Job Creation in Animal Agriculture in Iowa: A Preliminary Report* (Ames: Iowa State University, 1989), pp. 27-30; Dale Miller, "Straight-Talkin' Vet Gets My Vote," *National Hog Farmer* (February 15, 1997), p. 10; Joe Vansickle, "Confinement Health Risks," *National Hog Farmer* (November 15, 1997), p. 31.
10. Lora Duxbury-Berg, "Producer Recalls Tragic Tale of Manure Pit Deaths," *National Hog Farmer* (May 15, 1994), pp. 24-30; John George, "Exorcising the Monster From Your Pit," *Pork '96* (November 1996), p. 56.
11. Jane Gaydos, Letter to the Editor, *The [Raleigh] News and Observer,* Part 2, "Early Response," in "Boss Hogs 2: The Sequel" on the World Wide Web, http://www.nando.net.sproject/hogs/hfhome.htm, 1995.
12. "Award in Odor Case" and "Blow to Big Hog Lots?" *Des Moines Sunday Register* (September 4, 1994), p. 3J; "Iowa Ag Zone Protection Fails to Halt Suit," *Pork '94* (October 1994), pp. 54-55; Jeff De Young, "It's Not the Money," *The Cedar Rapids Gazette* (October 16, 1994), p. 9D; "Hog Pit Odor Drives Away Iowa Family," *The Cedar Rapids Gazette* (November 20, 1994), p. 28A.
13. Marlys Miller, "There Is Something in the Air," *Pork '94* (August 1994), p. 5. Mike King, "The New Smell of Money," *Pork '94* (September 1994), p. 22.
14. The survey, conducted by the University of Missouri-St. Louis Public Policy Research Centers, queried public opinion through random telephone calls to more than a 1,000 people in Missouri. Lora Duxbury-Berg, "Close Neighbors Rank Pork Production Most Favorable," *National Hog Farmer* (May 15, 1996), p. 11; "Swine Odor Conference Stirs Public Opinion," *Pork '94* (August 1994), p. 44.
15. Dirck Steimel, "Economy vs. Environment?" *Des Moines Sunday Register* (October 10, 1993), p. 1-2J. Joe Vanisckle, "Expansion Dream Survives Roadblocks," *National Hog Farmer* (June 15, 1994), pp. 12-18.
16. Although Land O' Lakes dominates the national branded butter market, the company still bills itself "a network of community cooperatives and family farmers." Organized as the Minnesota Cooperative Creameries Association in 1921, it changed its name to Land O' Lakes in 1926 on the suggestion of two members, a husband and wife who tended their own herd.
17. "Hog Lot Owners Win Battle Over Project," *The Cedar Rapids Gazette* (October 29, 1995), p. 9B, emphasis added.
18. Dale Miller, Opinion Page, *National Hog Farmer* (June 15, 1995), pp. 8, 10.
19. See, for example: *Pork '94* (February 1994), p. 27; and "News Update," *National Hog Farmer* (September 15, 1994), p. 8.
20. Iowa pork producer Allan Mallie, for example, follows a courteous regimen: "'A week before I apply [manure to fields] I send post cards to everyone within a one-mile radius and explain this will be the time and please excuse it, that we will do our best to make it short and sweet.' He also asks neighbors to notify him if they are planning an engagement at their house during that period, and he tries to work with them. 'I want a healthy community and health relationships,' he says." Marlene Lucas, "No Viruses, Just Hogs," *The Cedar Rapids Gazette* (February 2, 1998), p. 2B.
21. "A Menu Featuring Less Nitrogen, Phosphorous," *Pork '95* (April '95), pp. 34-38.
22. "World Demand for Pork Booms," *National Hog Farmer* (September 15, 1994), p. 50; "Improved Feed Efficiency Benefits Environment," *National Hog Farmer* (March 15, 1996), p. 13; Joe Roybal, "Organic Nonsense," *Beef* (October 1996), p. 4. For an overview, see Dennis T. Avery, *Saving the Planet with Pesticides and Plastic: The Environmental Triumph of High-Yield Farming* (Indianapolis: Hudson Institute, 1995).
23. "Iowa Nuisance Bill Favors Producers," *National Hog Farmer* (June 15, 1995), p. 17.
24. Michael Satchell, "Hog Heaven—and Hell," *U.S. News and World Report* (January 22, 1996), p. 58; John Greenwald, "Hogging the Table," *Time* (March 18, 1996), p. 76.
25. Dale Miller, "Opinion Page," *National Hog Farmer* (August 15, 1995), p. 16.

26. "Huge Spill of Hog Waste Fuels an Old Debate in North Carolina," *The New York Times* (June 25, 1995), p. 13; "Whooooweeee! 25 Million-gallon Spill of Wastes Revs Up Debate Over Big Hog Lots in North Carolina," *The Des Moines Register* (June 27, 1995), p. 1A.
27. Ronald Smothers, "Waste Spill Brings Legislative Action: North Carolina Still Loves Hog Farms," *The New York Times* (June 30, 1995), p. 8A.
28. "Whooooweeee!" *The Des Moines Register;* Karen McMahon, "Massive Lagoon Spill Pollutes River," *National Hog Farmer* (July 15, 1995), p. 31; "Lagoon Spills Create Regulatory Concern," *National Hog Farmer* (August 15, 1995), p. 17; "$110,000 Fine for Manure Spill," *National Hog Farmer* (September 15, 1995), p. 20; "N.C. Regulators Move Against 'Big Spill' Owners," *Pork '95* (September 1995), pp. 63-65; "N.C. Fines Oceanview $110,000 for Spill," "Fines Aren't Only Oceanview Problem," and "Major N.C. Problem: Lagoons Too Full," *Pork '95* (October 1995), pp. 44-46; "Oceanview Farms Gets Lagoon-Spill Fine Cut," *Pork '96* (June 1996), p. 101.
29. Ronald Smothers, "Waste Spill Brings Legislative Action."
30. Barbara Grabner, "Manure Spills Shake Hog Industry," *Corporate Hog Update* (June 1995), p. 3; Marlys Miller, "Not a Banner Day," *Pork '95* (August, 1995), p. 5; Jay P. Wagner, "Could Big Spill Occur Here?" *The Des Moines Register* (June 27, 1995), p. 5A; "Boss Hogs 2: The Sequel."
31. "N.C. Lagoon Inspectors' Report Due This Month," *Pork '95* (September 1995), p. 65; "Boss Hogs 2: The Sequel."
32. "Manure Spill Shows Factory Farm Danger, Opponents Say," *Iowa City Press-Citizen* (July 22, 1995), p. 8A.
33. Joe Vansickle, "Manure Spills Plague Premium Standard Farms," *National Hog Farmer* (October 15, 1995), p. 9.
34. Members of the Campaign for Family Farms demanded that the NPPC stop supporting large-scale hog operations, which seems reasonable enough. The protestors cannot be faulted simply because few were actually hog farmers. But they can be faulted for ignorance of federal statutory restrictions on such organizations (e.g., the NPPC could not, as demanded, contribute to the Campaign for Family Farms) and for ignoring the fact that most of the 85,000 voting members of the NPPC were family farmers. "NPPC Leader Cited" and "Group Protests NPPC," *National Hog Farmer* (September 15, 1995), p. 6. See also Jerry Perkins, "A Cool Head in the Midst of Heated Battle: Farm Leader Keppy Takes Reasonable Approach to Family-farm Woes," *Des Moines Sunday Register* (February 12, 1995), p. 2V; Joe Vansickle, "NPPC Memo Stirs Debate," *National Hog Farmer* (March 15, 1997), p. 39; "Pork Group Reimburses Checkoff Funds," *Iowa City Press-Citizen* (April 5, 1997), p. 7A.
35. Karen McMahon, "Averting a Manure Spill," *National Hog Farmer* (September 15, 1995), p. 17. Karen McMahon, "North Carolina Tightens Manure Regulations, *National Hog Farmer* (March 15, 1996), p. 28. "Iowa Blocks Expansion" and "New North Carolina Hog Farm Halted," *Pork '96* (July 1996), p. 42.
36. "Inspections Find Most in Compliance," *National Hog Farmer* (March 15, 1996), p. 28.
37. After 1994 the NPPC more aggressively promoted its response to mounting public pressure for government intervention on behalf of environmental protection.
38. John D. Copeland of the National Center for Agricultural Law in Fayetteville, Arkansas, provides a summary of the case and producers' responses to it in "Dire Consequences," *National Hog Farmer* (September 15, 1995), pp. 12-15.

CHAPTER 9

1. Stephanie Coontz, *The Way We Never Were: American Families and the Nostalgia Trap* (New York: Basic Books, 1992), pp. 73-74.
2. For a carefully reasoned and alarming assessment, see the report of a vice president of the Federal Reserve of Chicago, Gary L. Benjamin, "Industrialization in Hog Production: Implications for Midwest Agriculture," *Economic Perspectives* 21:1 (January/February 1997), pp. 2-13; and James V. Rhodes, "The Industrialization of Hog Production," *Review of Agricultural Economics* 17:2 (May 1995), pp. 107-118.

3. "Few Surprises in Pig Report," *National Hog Farmer* (April 15, 1994), p. 32; Michael Satchell, "Hog Heaven—and Hell," *U.S. News and World Report* (January 22, 1996), pp. 55-58; Bill Fleming, "Market Share Myths Broken," *National Hog Farmer* (March 15, 1994), p. 57; Ronald Smothers, "Slopping the Hogs, the Assembly-line Way," *The New York Times* (January 30, 1995), p. A8; Marty Strange, "Control Corporate Farming: Iowa Needs to Protect Itself," *Des Moines Register* (May 5, 1994), p. 1C; Rita Koselka, "$Oink, $Oink," *Forbes* (February 3, 1992), p. 56.
4. Jim Barnett, "Raising a Stink," *Raleigh News and Observer* (July 18, 1993), p. 10A; Satchell, "Hog Heaven—and Hell," p. 57; "Pork Industry Rapidly Restructuring," *Corporate Hog Update* (June 1995), p. 9.
5. "Few Surprises in Pig Report;" "Big Farms' Hog Market Share Grows," *Des Moines Sunday Register* (January 1, 1995), p. 3J; "Iowa Steers Clear of Elite Pork Image," *Iowa City Press-Citizen* (January 18, 1995), p. 10A; "Pork Conference," *Des Moines Sunday Register* (January 1, 1995), p. 3J; Satchell, "Hog Heaven—and Hell," p. 58.
6. Compact D/SEC (Disclosure, Inc.), "Complete Company Records, Smithfield Foods Inc.," *Compact Disclosure Annual Report 1995* (June 1996); Al V. Krebs, "The North Carolina Connection," *Corporate Hog Update* (June 1995), pp. 10-11; Koselka, "$Oink, $Oink," pp. 54, 56; Gregory E. David, "Bionic Pigs?" *Financial World* (November 9, 1994), pp. 32-33. In fiscal 1997 Smithfields acquired 8.7 million of the 10.5 million hogs killed in its eastern operation from just four of the six largest U.S. pork producers. Those four—Murphy, Carroll's, Prestage, and Maxwell/Goldsboro—staff the board of Smithfield and dedicate their entire production to Smithfield on long-term contract. "Smithfield Reports Record Sales," *Pork '97* (October 1997), p. 64. See also the results of the "Pork Powerhouse" surveys reported by Betsy Freese in *Successful Farming* 92:10 (October 1994), pp. 20-24, 93:10 (October 1995), pp. 20-22; and 94:10 (October 1996), pp. 27-32.
7. Jim Barnett, "Raising a Stink," *The [Raleigh] News and Observer* (July 18, 1993), p. 10A. Karen McMahon, "Westward Bound!" *National Hog Farmer* (May 15, 1994), p. 36; "Huge, Fully Integrated Utah Project Launched," *Pork '94* (January 1994), p. 104; Satchell, "Hog Heaven—and Hell," p. 59; "Circle 4 Gilts Farrow; More to Come," *Pork '95* (August 1995), p. 58; Karen McMahon, "A New Horizon in the Hog Industry," *National Hog Farmer* (May 15, 1996), pp. 12-14.
8. Bill Fleming, "What It Takes to Get a Loan," *National Hog Farmer* (October 15, 1993), p. 44.
9. "Want to Expand? Your Feed Company May Help," *National Hog Farmer* (October 15, 1993), pp. 24-26.
10. Karen McMahon, "Co-Op Seeks Growth to 200,000 Sows," *National Hog Farmer* (March 15, 1996), pp. 23-24.
11. The production of terminal sire lines can be considered a predictable outcome of specialized breeding like that which led to the distinction of wool and mutton sheep or dairy and beef cattle nearly two centuries ago. In the case of hogs, "Generally, colored breeds (Duroc and Hampshire) exhibit superior growth and carcass characteristics while the white breeds (Large Whites and Landrace) excel in the maternal traits." Hence seed and market stock tend to favor different genetics. J. E. Cannon et al., "Pork Quality Audit: A Review of the Factors Influencing Pork Quality," *Journal of Muscle Foods* 6 (1995), p. 378.
12. "PIC Buys NPD," *Pork '95* (June 1995), pp. 76-77; "PIC in Korea," *Pork '94* (April 1994), p. 73.
13. "Murphy to Crank Up New Feed Mill," *Pork '93* (December 1993), p. 57; "News Update," *National Hog Farmer* (December 15, 1994), p. 10.
14. Marvin L. Hayenga, "Pork Slaughter Industry; New Plants, New Players," *National Hog Farmer* (May 15, 1994), pp. 84-86.
15. "Construction Speeds on Utah Operation," *Pork '95* (June 1995), p. 82; "Michigan Packer Enters 10-year Agreement," *Pork '95* (January 1995), p. 88; Karen McMahon, "War Chest Builds Contract Units," *National Hog Farmer* (November 15, 1997), pp. 42-43.
16. "Tyson Targets China for Expansion," *Pork '95* (January 1995), p. 90; Dirck Steimel, "Meeting the Challenge of a Glut of Meat," *Des Moines Sunday Register* (November 27, 1994), p. 2J; Smothers, "Slopping the Hogs, the Assembly-line Way."
17. Initiative 300 "grandfathered" operations like National Farms; so they could continue but not expand. Alan Barkema and Michael L. Cook, "The Changing U.S. Pork Industry: A Dilemma for Public Policy," *Federal Reserve Bank of Kansas City Economic Review* (Second Quarter, 1993), p. 60; "State Laws Can Pen You In," *Pork '94* (July 1994), pp. 16-17.

18. In 1994 the fifty largest pork production companies contracted with growers to finish about 80 percent of the hogs they marketed. Two-thirds of those finishers were "independent"—that is, they were not owned by a feed company or packer. "Growth of Contracts Fuels 'Mega' Expansion," *National Hog Farmer* (September 15, 1994), p. 44.
19. Although the trade magazines regularly feature play-by-play on the laws and regulations affecting agriculture, a convenient, progressive accounting of those affecting hog production in particular appears in legislative, regulatory, and judicial summaries compiled by Barbara Grabner as editor of PrairieFire Rural Action in *Corporate Hog Update* (September-October 1994) and (June 1995). See also: "States Consider Hog Issues," *National Hog Farmer* (June 15, 1995), p. 19.
20. As a result, megaproducers are apt to be accused of "violation of democratic principles and disruption of community values." Kendal Thu of the Institute of Agricultural Medicine and Occupational Health at the University of Iowa, quoted in Joann Alumbaugh, "The Tiny Township and the Pork Powerhouse," *Hog Producer,* Program for the 1995 World Pork Expo (June 9-11, 1995), p. 25.
21. Alumbaugh, "The Tiny Township and the Pork Powerhouse."
22. Collins, who was president of the Iowa Pork Producers Association from 1961 to 1965, was actually sanguine about the integrator threat. Forty years ago, as far as he was concerned, family farms "already are integrated." By that he meant that farmers had for decades bred and raised their own animals, grown their own feed, provided their own labor and management. And they never were very "free" in arranging loans or marketing their goods. I am relying here on citations from Collins' lead article, "Why We Won't Lose Our Hog Business," for the monthly insert "The Iowa Farm and Home Register" published by the *Des Moines Register* back in the 1950s. Don Muhm, *Iowa Pork and People: A History of Iowa's Pork Producers* (Clive, IA: Iowa Pork Federation, 1995) pp. 34, 74. It was also more than thirty years ago that the British author Ruth Harrison coined much of the vocabulary now making the hog war rounds. See for example: *Animal Machines: The New Factory Farming Industry* (London: V. Stuart, 1964).

CHAPTER 10

1. Of 250 "producers, packers, veterinarians, university specialists, and business and financial executives" polled, two-thirds agreed that the pork industry would soon resemble poultry—be more capital and less labor intensive, with larger herds, fewer owners, more wage-workers, and lower profit margin per head. Mike King, "Survey Predicts Big Changes by Year 2000," *Pork '93* (June 1993), pp. 54, 56; Tony M. Forshey, "Veterinarians on Call," *National Hog Farmer* (July 15, 1995), p. 32.
2. Gary L. Benjamin, "Industrialization in Hog Production: Implications for Midwest Agriculture," *Economic Perspectives* 21:1 (January/February 1997), p. 4.
3. Mike King, "Real-Time Ultrasound Reaches More Producers," *Pork '93* (June 1993), pp. 46-50; Livestock Industry Task Force Committee, College of Agriculture, Iowa State University, *Job Creation in Animal Agriculture in Iowa: A Preliminary Report* (Ames: Iowa State University, 1989), p. 73; "Carcass Composition in a Flash," *Pork '96* (November 1996), p. 11.
4. See research by Gary L. Cromwell, University of Kentucky, "Pigs Don't Select Balanced Diet," *National Hog Farmer* (December 15, 1994), p. 32; Alan Bell, "It Pays to Discriminate Between Barrows and Gilts," *Pork '94* (June 1994), pp. 46-50; Gregory E. David, "Bionic Pigs?" *Financial World* (November 8, 1994), pp. 32-33; and "Genetic Super Pig Still Not a Reality," *Pork '94* (April 1994), p. 86. A laboratory in Iowa is even developing a way to vaccinate pigs against fat. "Vaccinating Pigs Against Fat Possible," *National Hog Farmer* (February 15, 1995), p. 53.
5. For a practical introduction to the technology, emphasizing its rewards when compared to "natural service," see: "The Art of AI," *National Hog Farmer*, Blueprints for Top Mangers 24 (April 15, 1997), pp. 6-57, 66, and the advertisements throughout the issue. See also: "AI Explosion is No 'Surprise,'" *Pork '94* (November 1994), pp. 24-25.
6. Peggy Hawkins, a veterinarian with Pfizer Animal Health, cautioned, "There are risks. . . . There may be a disease or two that we haven't identified. Or a virus could change and we'd face a new version to deal with." Marlys Miller, "Does AI Come with a Clean Bill of Health?" *Pork '95*

(September 1995), p. 36. See also "AI Health Concerns," *Pork '95* (May 1995), p. 12; "Why You Should Consider AI," *Pork '95* (January 1995), pp. 15-17.

7. "Check Up on Your Biosecurity Plan," *Pork '97* (April 1997), pp. 16-18; Mike King, "Plan Now to Avoid Building Disasters," *Pork '94* (August 1994), p. 23; Marlene Lucas, "No Viruses, Just Hogs," *The Cedar Rapids Gazette* (February 2, 1998), p. 2B.
8. According to veterinarian Kirk Clark, "Whatever disease pigs get, they mainly get from the sow after birth. So it makes sense to wean pigs and separate them from the mothers as early as possible." "Study: Early Weaning Is Best Prevention," *Pork '96* (May 1996), p. 18; Interview with Kirk Clark, Purdue University, West Lafayette, IN, May 31, 1994.
9. Marlys Miller, "Early Weaning: A Way to Expand," *Pork '94* (November 1994), pp. 32-35; Lee Whittington, "Is SEW Worth the Extra Cost," *National Hog Farmer* (June 15, 1996), p. 38; David Bishop, "SEW from the Sow's Perspective," *Pork '96* (May 1996), p. 28; Joe Vansickle, "Segregated Early Weaning: Pitfalls Tarnish the 'Perfect System,'" *National Hog Farmer* (May 15, 1996), pp. 48-50.
10. Karen McMahon, "Segregated Early Weaning Trial Produces Efficient Pigs," *National Hog Farmer* (February 15, 1995), p. 47; Howard Hill, "Segregated Early Weaning Unfolds: An Historical Perspective of SEW Details Its Growth," *National Hog Farmer* (Fall 1994), pp. 16-18; Whittington, "Is SEW Worth the Extra Cost"; Keith Thornton, "Are You Ready for Multiple Sites?" *Pork '95* (February 1995), p. 39; Joe Vansickle, "Weaner Pig Market Takes Off," *National Hog Farmer* (September 15, 1995), p. 8; Gary Parker, "Switching to SEW," *National Hog Farmer* (February 15, 1995), p. 46; Tom Stein, "How Do Your Numbers Compare," *Pork '93* (June 1993), pp. 36-40; Tom Stein, "How Much Does Weaning Age Cost You?" *Pork '94* (November 1994), pp. 60-62; "Uniform In, But Not Uniform Out," *Pork '97* (November 1997), pp. 20-22.
11. "Technologies Valued at $11 Per Pig," *Pork '95* (October 1995), p. 12; Alan Barkema and Michael L. Cook, "The Changing U.S. Pork Industry: A Dilemma for Public Policy," *Federal Reserve Bank of Kansas City Economic Review* (Second Quarter, 1993), p. 57.
12. Many surveys show that larger producers are using more advanced technologies. Comparing large (5,000+ head sold per year) and small (500 to 1999) producers, 1994-1995: Large units were more likely to wean early (40 percent of large operations wean by 20 days vs. only 2.2 percent of small); to use all-in/all-out (82 vs. 56 percent); to sell by carcass-merit where there is a premium for quality (66 vs. 39 percent); to qualify for premiums at the market (78 vs. 22 percent claim to have "consistently" received at least $2 in bonus per head); and to split-sex feed, use computerized record keeping, and artificially inseminate. Marlys Miller, "Are the 'Big' Guys Beating You?" *Pork '94* (September 1994), p. 5; USDA, APHIS, Veterinary Services, "Sources of Pigs Entering the Grower/Finisher Phase on U.S. Pork Operations," and "Feed Management by U.S. Pork Producers," *Info Sheet* (June 1996); USDA, APHIS, Veterinary Services, "Part II: Reference of 1995 U.S. Grower/Finisher Health and Management Practices," *Swine '95: Grower/Finisher* (June 1996), pp. 21, 14, 16; Marty Strange, "Control Corporate Farming: Iowa Needs to Protect Itself," *Des Moines Register* (May 5, 1994), p. 1C; Benjamin, "Industrialization in Hog Production," pp. 3-4, 9.
13. Strange was, among other things at the time, program director for the Center for Rural Affairs in Walthill, Nebraska. Strange, "Control Corporate Farming"; Chris Hurt et al., "Industry Evolution: By 2000, Packers Expected to Coordinate Pork Production," *Feedstuffs*, Special Issue (August 24, 1992), p. 1; Livestock Industry Task Force Committee, *Job Creation in Animal Agriculture in Iowa: A Preliminary Report*, p. 78.

CHAPTER 11

1. Alan Barkema and Michael L. Cook, "The Changing U.S. Pork Industry: A Dilemma for Public Policy," *Federal Reserve Bank of Kansas City Economic Review* (Second Quarter, 1993), p. 51; Marlys Miller, "What the Numbers Don't Tell You," *Pork '95* (October 1995), p. 5; Marlys Miller, "Those Who Work Smarter Win," *Pork '96* (May 1996), p. 5; Lora Duxbury-Berg, "Poised to Survive," *National Hog Farmer* (June 15, 1996), pp. 28-34; Alan Bell, "Who Will Raise the Hogs?" *Pork '97* (May 1997), pp. 26-30.

2. By many measures (certainly in comparison to poultry), the hog industry is still small. In 1994, 95.7 million market hogs came from 149,000 farms, more than 80 percent marketing under 1,000 head per year (17.4 percent of the total). Only 66 operations marketed more than 50,000, but their contribution was approximately the same share (16.8 percent) of the total. The number of producers marketing more than 5,000 is likely to increase slightly, and the number marketing over 50,000 to increase dramatically (e.g., to double their market share by the millennium), but that still would leave more than 17,000 producers who raise fewer than 3,000 head per year. Bill Fleming, "Market Share Myths Broken," *National Hog Farmer* (March 15, 1994), p. 57; Dirck Steimel, "Small Hog Farmers Face a 'Revolution,'" *Des Moines Sunday Register* (May 22, 1994), p. 1J; U.S. Department of Commerce, Bureau of the Census, *Statistical Atlas of the United States* (Washington, D.C.: Government Printing Office, 1925), pp. 308-309; U.S. Department of Commerce, Bureau of the Census, *Agricultural Atlas of the United States* (Washington, D.C.: Government Printing Office, 1992) Census of Agriculture, vol. 2, part 1, pp. 128-134.
3. Karen McMahon, "The Pork Powers: Iowa No. 1 and No. 2 North Carolina," *National Hog Farmer* (May 15, 1995), pp. 12-14.
4. Steve Cornett, "The Pork Explosion," *Beef* (May 1995), p. 33. As predictably as trade magazines find good news in every shakeout, investigative journalists and academics find the bad. See, for example: Peggy F. Barlett, *American Dreams, Rural Realities: Family Farms in Crisis* (Chapel Hill: University of North Carolina Press, 1993); Mark Friedberger, *Shake-out: Iowa Farm Families in the 1980s* (Lexington: The University Press of Kentucky, 1989); Howard Kohn, *The Last Farmer: An American Memoir* (New York: Harper and Row, 1988).
5. Marlys Miller, "Independents Have a Place in the Future," *Pork '95* (October 1995), pp. 18-21; Milton C. Hallberg, Charles Abdalla, and Paul B. Thomson, *Performance in Animal Agriculture: A Framework for Multidisciplinary Analysis,* Center for Biotechnology Policy and Ethics Discussion Paper 96-8 (College Station: Texas A & M, 1996); Milton C. Hallberg, "Assessing the Performance of Animal Agriculture," *CSTPE Newsletter* 6:2 (September/October 1996), p. 1.
6. Karen McMahon, "Westward Bound!" *National Hog Farmer* (May 15, 1994), p. 35; Barkema and Cook, "The Changing U.S. Pork Industry," p. 58.
7. Of course, there is a certain irony in counting as "saved" the number of pigs that make it to slaughter, but the ones that are not so "saved" (most likely stillborn, savaged or crushed by a sow) seem lamentable casualties. USDA, APHIS, Veterinary Services, "Part III: Changes in the U.S. Pork Industry, 1990-1995," *Swine '95: Grower/Finisher* (Fort Collins, CO: Centers for Epidemiology and Animal Health, NAHMS, 1997), pp. 10-16; Kerry Keffaber, "Plan Ahead for a Clean Bill of Health," *Pork '94* (March 94), p. 62; "The Sky Isn't Falling," *Pork '95* (December 1994), p. 5; Dale Miller, "Opinion," *National Hog Farmer* (February 15, 1994), p. 10; "Sows Producing More Pigs, More Pork Per Year," *Pork '97* (April 1997), p. 13.
8. Note, however, that Mueller factors out the market price received (which, by his own estimate, accounts for 27 percent of the variation) as if it were independent of size. Small farmers constantly complain that packer bonus plans greatly favor bigger operators. Given, too, his focus on Illinois relatively early in hog wars, he may miss the size advantage of truly gargantuan farms. By some accounts, for example, size becomes a significant factor only at a threshold (e.g., 50,000 head marketed per year) considerably higher than he would have found in Illinois at the time. Bill Fleming, "Size Does Not Determine Profitability," *National Hog Farmer* (November 15, 1993), p. 50; "As Hog Prices Fall, Costs Grow More Critical," *Pork '94* (August 1994), p. 18.
9. Alan Bell, "It Pays to Discriminate Between Barrows and Gilts," *Pork '94* (June 1994), p. 48; Bill Slakey, "Despite Large Operations' Advantages, Small Farms Sometimes Just as Efficient," *Waterloo Courier* (April 17, 1994), p. B4; Howard Hill, "Segregated Early Weaning Unfolds: An Historical Perspective of SEW Details Its Growth," *National Hog Farmer* (Fall, 1994), p. 10; Steimel, "Small Hog Farmers Face a 'Revolution,'" p. 2J.
10. Joe Vansickle, "Major Upheaval Forecast in Next Decade," *National Hog Farmer* (May 15, 1995), p. 23; USDA, APHIS, Veterinary Services, "Shedding of Salmonella by Finisher Hogs in the U.S." *APHIS Info Sheet* (January 1997), p. 1.
11. In 1995 Tyson traded its pork slaughter plant in Marshall, Missouri, to Cargill in exchange for its broiler operations in Georgia and Florida. "Tyson, Cargill Swap Hog, Poultry Operations," *Pork '95* (September 1995), p. 67; Miller, "What the Numbers Don't Tell You"; "Cargill/Tyson Exchange," *National Hog Farmer* (August 15, 1995), p. 15.

12. Doug Barten, "Iowa Has a Future, But Things Change," *Pork '95* (April 1995), p. 8; Marlys Miller, "Don't Let Independence Block Progress," *Pork 94* (March 94), p. 5; "New Ideas for Independent Competition," *Pork 94* (March 94), pp. 30-33; Lora Duxbury-Berg, "Producer Loop Assures Market Access," *National Hog Farmer* (April 15, 1995), pp. 15-20; Karen McMahon, "Bell Farms Builds Independent Alliances, " *National Hog Farmer* (Feburary 15, 1997), pp. 18-21; National Pork Producers Council, *Leveraged Marketing: A Competitive Strategy for the '90s* (Des Moines: National Pork Producers Council, 1993).
13. In 1996 U.S. pork exports for the first time exceeded $1 billion in value. Joe Vansickle, "Second in Pork Exports," *National Hog Farmer* (January 15, 1997), p. 6; "U.S. Pork Exports Rise More Than Imports," *Pork '95* (April 1995), p. 68; Karen McMahon, "Global Markets Hold Promise," *National Hog Farmer* (January 15, 1995), p. 51; "Pork Exports Take Off," *National Hog Farmer* (October 15, 1995), p. 8; Miller, "Opinion."
14. National Pork Producers Council, *Employer/Employee Relations in the Pork Industry* (Des Moines, IA: NPPC, 1991), pp. 7, 12-13, 15, 18, 21, 23, 30-31.
15. Al V. Krebs, "Corporate Profile—Seaboard Expands Into Pork," *Corporate Hog Update* (September-October 1994), p. 6.
16. Dale Miller, "Opinion Page," *National Hog Farmer* (May 15, 1994), p. 8; McMahon, "Westward Bound!" pp. 34-35; Karen McMahon, "Oklahoma Building Frenzy Doubles Sow Numbers," *National Hog Farmer* (May 15, 1994), pp. 36-37; Marlene Lucas, "New Dent in Pork Crown," and Dave Gosch, "Columbus Junction Residents Try to Put Positive Spin on Layoffs," *The Cedar Rapids Gazette* (March 2, 1997), pp. 1A, 10A; Joe Vansickle, "Hog Shortage Spurs Plant Closing," *National Hog Farmer* (September 15, 1997), p. 19.
17. Ronald Smothers, "Slopping the Hogs, the Assembly-line Way," *The New York Times* (January 30, 1995), p. A8; McMahon, "Westward Bound!" p. 36.

CHAPTER 12

1. For a more favorable interpretation of alternative names, see Marlys Miller, "Are You a Hog Farmer or a Food Producer?" *Pork '97* (March 1997), p. 5.
2. "Tyson, Cargill Swap Hog, Poultry Operations," *Pork '95* (September 1995), p. 67; Michael Fritz, "Contented Pigs? Tyson Foods to Market Leaner Pork," *Forbes,* 144 (August 7, 1989), pp. 118-119; Joe Vansickle, "Pork Takes Aim at Top Meat Spot," *National Hog Farmer* (May 15, 1995), p. 552; Rita Koselka, "$Oink, $Oink," *Forbes* (February 3, 1992), p. 54. Gary L. Benjamin, "Industrialization in Hog Production: Implications for Midwest Agriculture," *Economic Perspectives* 21:1 (January/February 1997), pp. 2-13.

   Feeders of all sorts have long complained of deceptive contract procedures. In response the NPPC has repeatedly asked the Secretary of Agriculture to assure that the Grain Inspection, Packers and Stockyards Administration reports actual pricing, rather than allowing what many producers believe are secret, monopolistic deals. In 1994 and 1995 there were particularly intense rumors that select contract feeders were getting prices and premiums that others were denied, that there was, in effect, regulatory tolerance of trade restraint. "Time to Get Organized," *Pork '94* (December 1994), p. 6; Stephen Mohling, "More Pork Forum Insight," *Pork '94* (July 1994), p. 8; Karen McMahon, "Iowa Producers Ask For Packer Price Reports," *National Hog Farmer* (March 15, 1996), p. 9; "Producer Group Takes Issue with NPPC" and "NPPC Asks USDA to Probe Market Access," *Pork '95* (October 1995), pp. 42-43; Marty Strange, "Control Corporate Farming: Iowa Needs to Protect Itself," *Des Moines Register* (May 5, 1994), p. 1C; Bill Fleming, "Packer Contracts Grow as Producers Seek Security," *National Hog Farmer* (May 15, 1995), pp. 24-30; Marlys Miller, "Independents Have a Place in the Future," *Pork '95* (October 1995), p. 20.
3. "North Carolina Branded Pork to Japan," *Pork '94* (July 1994), pp. 49-50; "Should You Wear a Shirt and Tie?" *Pork '94* (August 1994), p. 14; Alan Bell, "'Mountain of Meat' Clouds Profit Outlook," and "Hog Cycle May Be Changing," *Pork '94* (September 1994), pp. 40, 38.
4. The Swine Enterprise Summary is produced annually at Iowa State University. Marlys Miller, "The Sky Isn't Falling," *Pork '94* (December 1994), p. 5.
5. South Dakota Agricultural Statistics Service, *1994 Hogs Facts* (Sioux Falls, SD: SDASS, 1995).

6. These figures were compiled by John Lawrence, a hog farmer before turning livestock economist for the Iowa State University Extension Service. Don Muhm, "Pork Producers Lost Money 11 of 12 Months Last Year," *Des Moines Register* (February 12, 1995), p. 10V; John D. Lawrence, "What's Ahead for Iowa's Pork Industry?" in Don Muhm's *Iowa Pork and People: A History of Iowa's Pork Producers* (Clive, IA: Iowa Pork Foundation, 1995), pp. 252-264.
7. "Competition Squeezing Small Pork Producers," *Iowa City Press-Citizen* (November 12, 1993), p. 6D; Chris Hurt et al., "Industry Evolution: By 2000, Packers Expected to Coordinate Pork Production," *Feedstuffs,* Special Issue (August 24, 1992), pp. 1, 18-19.
8. April 1993 to 1994, hog prices dropped $2.74/cwt, but the margin for retailers and packers increased $6.80 (approximately $3.70 for retailers and $3.10 for packers). This trend continued in 1994, when in just six months farmers received 13.8 cents less per pound, while packers received 2.9 cents more and retailers 7.2 cents more. Bill Fleming, "Wider Packer, Retailer Margins Cut Hog Profits," *National Hog Farmer* (June 15, 1994), p. 28; "Low Hog Prices Leave 'Margin' for Error," *Pork '94* (December 1994), p. 20; "News Update," *National Hog Farmer* (November 15, 1994), p. 9; "Packer Margins Drop From Fall '94 Highs," *Pork '95* (September 1995), pp. 67-68; Jay P. Wagner, "Exodus of Pork Producers Foreseen," *Des Moines Sunday Register* (November 27, 1994), pp. 1-2J.
9. "Can Cash Prices Rebound into 1995?" *Pork '94* (November 1994), p. 28; Joe Vansickle, "Pork Takes Aim at Top Meat Spot," *National Hog Farmer* (May 15, 1995), p. 552. While small operators and investors play penny-ante, short-term markets, large operators play the long haul. CEOs can use temporary downturns to grab stock for both the company and themselves. By judiciously investing just $1.8 million between 1976 and 1992, for example, Smithfield's Joseph Luter III acquired stock worth $47 million. Koselka, "$Oink, $Oink," p. 56
10. "Can Cash Prices Rebound into 1995?"; Joe Vansickle, "Economic Facts You Need to Know Before You Expand," *National Hog Farmer* (August 15, 1994), pp. 22-24.
11. Vansickle, "Economic Facts You Need to Know"; "Can Cash Prices Rebound into 1995?"
12. These figures were assembled by Glenn Grimes and Ron Plain, University Missouri agricultural economists. Bill Fleming and Karen McMahon, "Brace Yourself! Hog Crunch Ahead," *National Hog Farmer* (April 15, 1994), p. 22.
13. "Texas Company Mulls 27,000-Sow operation," *Pork '95* (September 1995), p. 70.
14. "Expansion Delayed," *National Hog Farmer* (June 15, 1995), p. 18; Dirck Steimel, "These Little Piggies Went to Market—Or Did They?" *Des Moines Sunday Register* (September 26, 1993), p. 1J; "The Opportunity to Compete," *Pork '96* (June 1996), p. 5; "Tyson Getting Out of Pork, Beef Processing," *Pork '96* (June 1996), p. 101; "What's Behind Strong Spring Prices?" *Pork '96* (May 1996), p. 75; "Sow Slaughter Picks Up," *National Hog Farmer* (May 15, 1996), p. 6.
15. Ronald Smothers, "Slopping the Hogs, the Assembly-line Way," *The New York Times* (January 30, 1995), p. A8.
16. "Corporate Producers Expand in Isolated Fringe Areas," *National Hog Farmer* (May 15, 1994), pp. 44-49; Karen McMahon, "Oklahoma Building Frenzy Doubles Sow Numbers," *National Hog Farmer* (May 15, 1994), pp. 36-39; John Gadd, "European Update," *National Hog Farmer* (September 15, 1994), p. 48.
17. "Congressmen Honored," *National Hog Farmer* (October 15, 1995), p. 8; "Congress Gives GATT Sweeping Approval," *Pork '95* (January 1995), pp. 71-73; "U.S. Calls Korean Trade Barrier Unfair," *Pork '95* (January 1995), pp. 86-87; "PSF Plant Gains OK to Export to Europe," *Pork '95* (August 1995), pp. 54-55; "Foreign Trade Network Forms," *Pork '93* (December 1993), pp. 51-52; Strange, "Control Corporate Farming"; "U.S. Wants Access to Chinese Market," *Pork '96* (December 1995), pp. 50-51. The original organizers of APEX were Doskocil/Wilson Foods, Farmland Industries, FDL, Hormel, IBP, Seaboard Farms and W&G Marketing.
18. According to long-standing USDA regulations, the "fresh" chicken and turkey sold in markets could be nearly 10 percent water added in processing, costing consumers about $1 billion per year, but "fresh" beef and pork could not contain any added water at all. On July 23, 1997, U.S. District Court Judge Ronald Longstaff ruled that the relevant USDA meat processing and labeling regulations were "arbitrary and capricious," in violation of the U.S. Administrative Procedure Act. "News Update," *National Hog Farmer* (July 15, 1994), p. 8.

19. For example, 1993 House Resolution 4, funding USDA surplus commodity programs, supplied $74 million to purchase 65 million pounds of pork. "Who's Welfare is at Stake?" *Pork '95* (May 1995), pp. 69-71; "USDA Adds More Pork to School Lunch," *Pork '95* (August, 1995), p. 57.
20. "Pork Funds Survive Panel's Knife," *Pork '94* (July 1994), pp. 40-41; "Swine Odor Conference Stirs Public Opinion," *Pork '94* (August 1994), p. 44; "Murphy Family Farms Names New President," *Pork '95* (April 1995), p. 66.
21. "Senate Restores Full Market Promotion Program," *Pork '95* (June 1995), p. 89; "News Update," *National Hog Farmer* (November 15, 1994), p. 9; "News Update," *National Hog Farmer* (September 15, 1994), p. 8.
22. Roy C. Ferguson, President, Ferguson Group Ltd., Letter to the Editor, *Pork '93* (June 1993), p. 8.
23. "South Dakota Offers Grant to Morrell Plant," *Pork '95* (January 1995), p. 82.
24. Karen McMahon, "Oklahoma Building Frenzy Doubles Sow Numbers," *National Hog Farmer* (May 15, 1994), p. 38; Dirck Steimel, "Guymon Getting a Big Plant,'" *Des Moines Sunday Register* (May 22, 1994), p. 2J.
25. "Seaboard Bonds Rile Critics in Kansas," *Pork '96* (March 1996), p. 72. A common, populist strategy in resisting state support for integrators is to advocate returning regulatory authority to the county level. As it turns out, though, those local levels may be even more vulnerable to the big guys. In the first year of local options in Kansas, for example, eighteen counties opened themselves to integrators, no doubt in part inspired/coerced by the boom in Missouri. Lora Duxbury-Berg and Joe Vansickle, "Corporate Farming Debate Heats Up," *National Hog Farmer* (November 15, 1994), pp. 22-32.
26. J. E. Cannon et al., "Pork Quality Audit: A Review of the Factors Influencing Pork Quality," *Journal of Muscle Foods* 6 (1995), p. 384.
27. Note that a "typical packer" is one that kills about 5,000 head per day. Marlys Miller, "Where Do You Stand on PSS?" *Pork '95* (February 1995), p. 5. Mike King, "PSS Gene Divides the Pork Industry," *Pork '93* (October 1993), pp. 38-40; John Lawrence and Rodney Goodwin, "Claiming Real Values of Terminal Sire Lines," *National Hog Farmer* (June 1, 1995), p. 27.
28. Lauren Christian, "Clarifying the Impact of the Stress Gene," *National Hog Farmer* (June 1, 1995), p. 47.
29. Marlys Miller, "Does the Stress Gene Make You Money?" *Pork '96* (September 1996), pp. 53-58; Mike King, "PSS Gene Test To Get U.S. Patent," *Pork '94* (June 1994), pp. 72-73; Mike King, "PSS Testing Evolves," *Pork '95* (February 1995), pp. 46-47; Christian, "Clarifying the Impact of the Stress Gene," pp. 46-47; Dale Miller, "Farm-to-Fork Quality Audit," *National Hog Farmer* (May 15, 1994), p. 21; D. G. McLaren and Andrew Coates (of PIC), "PIC Not 'Pushing' Use of Stress Gene," *National Hog Farmer* (January 15, 1994), p. 8; "PIC Guarantees Stress Gene Status," *Pork '95* (May 1995), p. 74; "PIC Will Eliminate Halothane Gene," *Pork '97* (December 1997), p. 60.
30. Dave Barry, "Dave Barry's Vacation in Iowa," *The Des Moines Register* (August 20, 1995), p. 1B. See also: Lenore Yarger, "Iowa's Hog Hell," *Icon* (January 25, 1996), p. 4.
31. Center for Rural Affairs, *Spotlight on Pork* (Columbia: University of Missouri Center for Rural Affairs, 1994); Lora Duxbury-Berg, "Why the Task Force Wants a Change," *National Hog Farmer* (November 15, 1994), p. 32.
32. In December 1997 the U.S. Environmental Protection Agency sued Smithfield Foods for $20 million, alleging nearly 7,000 violations of the Clean Water Act in just two Smithfield plants, 1991 to 1996. On August 8, 1997, Smithfield was assessed a $12.6 million fine, "the largest civil penalty for water pollution in U.S. history." "Smithfield Penalized for Water Pollution," *Pork '97* (September 1997), p. 89. Pat Stith and Joby Warrick, "Who's in Charge?" *The [Raleigh] News and Observer* (February 26, 1995), p 8A.
33. Stith and Warrick, "Who's in Charge?" p. 9A; Al V. Krebs, "The North Carolina Connection," *Corporate Hog Update* (June 1995), p. 11.
34. Michael Satchell, "Hog Heaven—and Hell," *U.S. News and World Report* (January 22, 1996), p. 59; Stith and Warrick, "Who's in Charge?" p 1A.
35. Pat Stith and Joby Warrick, "Big Pork Helps Its Friends," *The [Raleigh] News and Observer* (February 26, 1995), p. 9A.

CHAPTER 13

1. "Pork World Embraces Networking," *Des Moines Register* (March 3, 1993), p. 8S; Ted Williams, "Assembly Line Swine," *Audubon* (March-April 1998), pp. 26-33.
2. "Spelling Out PSF," *National Hog Farmer* (September 15, 1994), p. 20; Laura Jereski and Randall Smith, "Morgan Stanley's Pig-farm Dream Wallows In Losses, Spatters Clients," *Wall Street Journal* (May 22, 1996), p. C15; Gary L. Benjamin, "Industrialization in Hog Production: Implications for Midwest Agriculture," *Economic Perspectives* 21:1 (January/February 1997), p. 8.
3. Karen McMahon, "A $1 Billion Boost," *National Hog Farmer* (August 15, 1994), p. 45.
4. "Premium Standard Packs Huge Impact," *Pork '94* (October 1994), p. 58; Steve Cornett, "The Pork Explosion," *Beef* (May 1995), p. 36; "Premium Standard Enters Japan's Retail Arena," *Pork '95* (September 1995), p. 67.
5. "Premium Standard Packs Huge Impact"; McMahon, "A $1 Billion Boost."
6. Of course, PSF is not alone in stimulating local development. For example, in 1973 when Circle Four Farms broke ground in Milford, Utah, the city was declining along with mines and the railroad. The town had not issued a single building permit in a dozen years. Four years later Circle Four had a payroll of over $6 million and 300 local employees, making it the county's largest employer and the de facto underwriter of more than 60 new homes and other buildings under construction in the fall of 1997. "Giant Hog Farm Divides Tiny Community," *Iowa City Press-Citizen* (November 10, 1997), p. 11B.
7. Joe Vansickle, "Keying on the Environment," *National Hog Farmer* (September 15, 1994), pp. 16-20.
8. "Premium Standard Farms Opens Processing Plant," *Pork '94* (December 1994), p. 39.
9. "Lincoln Township vs. Premium Standard Farms: A Timeline," *In Motion Magazine*, http://www.inmotionmagazine.com/hogtl.html (1997), pp. 1-3. In the fall of 1996 on the advice of the Land Stewardship Project (LSP), Erickson Diversified Corporation announced that it would no longer carry PSF pork in its chain of eighteen grocery stores (Econofoods, More 4, Erickson's, and Food Bonanza) in Minnesota and Wisconsin. According to LSP director Mark Shultz, PSF was selected largely because of its good-corporate-citizen posture. Discrediting PSF would implicitly discredit other integrators who made no so such pretense. I called him because I was curious why the Land Stewardship Project would help Erickson (itself part of an economically concentrated food industry) gain such low-cost, positive PR. Furthermore, Erickson opted to replace its pork by buying from Hormel (hardly a tarnish-free institution) rather than, say, hiring local, unionized in-store meat cutters. Nevertheless, Shultz crowed: "We applaud Erickson for making a choice not to support the industrial system that's ruining rural communities." Charlie Arnot, Director of Public Relations for PSF objected, "Unfortunately, Erickson made this decision after hearing only one side of the story." "Econofoods Drops Pork Products Made by Premium Standard," *Iowa City Press-Citizen* (October 23, 1996), p. 7A; Telephone Conversations with Mark Shultz, October 23 and 24, 1996; Arnot did not return my calls; Paul Sturtz and Scott Dye, "Why Boycott Premium Standard Farms," *In Motion Magazine*.

CHAPTER 14

1. For a classic anthropological reference to this phenomenon, see Robert Redfield, *The Little Community and Peasant Society and Culture* (Chicago: The University of Chicago Press, 1956), pp. 134-135.
2. In December 1996 I phoned Paul to ask him to review my portrayal of our interaction. He referred me to his former student and research colleague, Kendall Thu, whom he said he trusted to respond in their common interest. After Kendall reviewed several drafts of this chapter—each draft incorporating more revisions that he advised—a number of points of disagreement remained. Those points are identified in the notes.

3. After reviewing drafts of this chapter (January-February 1997), Kendall repeatedly insisted that they were never predisposed to endorse a side. Although I hope readers will take that insistence into account, I also hope that the evidence that I cite renders it unconvincing.
4. Kendall has objected to the implication that they did not welcome my advice. In fact, I do not think they did, though I am pleased to help Kendall publicize his intention.
5. Kendall reminds me that the title of the newspaper article was ascribed by someone else (hence, not "theirs") and that they were "issuing a warning about the downside of large-scale hog operations. No more, no less." Throughout our communications about the honesty and fairness of my interpretation of their work (January-February 1997), Kendall has (with remarkably inconsistency, from my point of view) insisted that their stance in celebration of one side ("independent" family farms) and in opposition to another (large-scale operators) is neither true nor fair. They do not buy what I call "the two-side story" and resent being so "pigeonholed." Any appearance that they stand on "a side," Kendall insists, is more my doing than theirs.

   I hope that the evidence I provide convinces readers, as it convinces me, that this insistence is, at best, wishful thinking. When your research is initiated, supported, and endorsed by one set of people (who have no qualms about representing themselves as a side), when that same research is panned by foes (who, you allege, aim to suppress your research), and when journalists who follow the legal and political wrangling on a daily basis agree on the alignment, I am persuaded that the work itself is fairly characterized as "on a side." Kendall and Paul's claims to nonpartisanship (like family farmers' claims to "independence" or integrators' to "community vitality") seem to me pretensions of rhetorical convenience. They are embraced when soliciting support and dropped when boasting of conviction.
6. Kendall says that the newspaper quotation (in particular, the reporter's use of the word "laughable") is inaccurate. Personal correspondence, February 13, 1997.
7. Kendall M. Thu and E. Paul Durrenberger, "Large-scale Hog Farming vs. Quality of Life," *Des Moines Register* (March 8, 1994), p. 3B; "Large Pig Farms a Threat to Jobs, Researchers Say," *Iowa City Press-Citizen* (February 22, 1994), p. 8A.
8. Compare the following positions, both taken in 1994: (1) "We have no vested interest or political predisposition as either defenders of family farms or proponents of industrial agriculture. . . . Our being outsiders to the process we wish to understand is not simply a statement of political neutrality but a methodological necessity because people involved in the process are less able to isolate and articulate the tacit knowledge that informs their thought than those outside it." And (2) "The assumptions of neoclassical economics . . . is the ideology of robber barons masquerading as an academic discipline. These assumptions are promoted to hide and justify political manipulation. . . . we call on anthropologists to utilize their fieldwork skills to expose the nature of state socio-political systems." E. Paul Durrenberger and Kendall M. Thu, "Agricultural Industrializing in Iowa: The Role of Science in the Formulation and Promotion of Policies for the Proliferation of Intensive Hog Production Technology," proposal to the Studies in Science, Technology and Society Program of the National Science Foundation, June 1, 1994, pp. 3, 10, 35; Kendall M. Thu and E. Paul Durrenberger, "Human Dimensions of Public Policy in U.S. Industrial Agriculture: A Role for Anthropology," paper presented at the 93rd Annual American Anthropological Association Meeting, November 30-December 4, 1994, Atlanta, Georgia, pp. 8, 18.
9. Kendall says that they "have never sought headlines." Inflammatory forays in the popular press (such as that original op-ed piece that they penned) plus a steady stream of public appearances and self- or quasi-vanity publications—consistently framed as straightening out official misinformation—lead me to disagree.
10. Kendall contests my grantsmanship interpretation, in particular, my inference that Kelley was hoping that Kendall would help earn his keep.
11. Kelley J. Donham and Kendall M. Thu, "Relationships of Agricultural and Economic Policy to the Health of Farm Families, Livestock, and the Environment," *Journal of the American Veterinary Medical Association* 202:7 (April 1, 1993), pp. 1084-1091.
12. Thanks are due Kendall Thu for correcting my recollection of this part of his Norwegian experience. In his view, his relationship to folk experience is one of "appropriately situating" rather than "discounting" (my word). From my point of view, his capacity to situate their experience without much on-the-ground detail and without their capacity to situate his amounts to "discounting." Personal correspondence, January 22, 1997.

13. Kendall's interpretation is that Paul meant no personal offense, just criticism of my work. Since Paul had never before been privy to my work on this subject or asked about it, I trust readers could understand why I took it personally. Personal correspondence, February 13, 1997.
14. The lawyer, Deborah Van Dyken, kindly confirmed this overall picture in a phone conversation, January 15, 1997. She said that she was the one who initiated the contact and accompanied Paul and Kendall on most of the trip. After the Tar Heel packing plant opponents "threw a bash for them," other organizers "showed them around" a few key sites for three or four days. Thanks, again, to Kendall for reminding me that "several integrators . . . could not accommodate us" in offering an "opportunity to present their side of the story." Kendall says they subsequently "extended a request through the NPPC to have conversations with the heads of the firms. So far, no takers."
15. Consider the following opening sentences from an article published in the spring of 1993: "The shape of the U.S. pork industry is changing dramatically, as pork production shifts into the hands of fewer, larger farmers with closer ties to processors and consumers. The changing shape of the pork industry, the nation's second largest meat industry, points to the loss of thousands of small hog farms in the United States." And note three things: (1) This article was published nearly a year before Durrenberger and Thu's exposé; (2) it was penned, not by some academic crashing the barricades, but by an officer of a U.S. Federal Reserve Bank; and (3) it was well-known among both "the industry" and its opponents. It was cited in the newsletter, *Corporate Hog Update* (June 1995), which is published by PrairieFire Rural Action long after it was handed to me by an officer of the National Pork Producers Council. He grabbed it off the top of a stack that had been photocopied for anyone who was interested. So it is hard to believe that the job-threatening dimensions of integration were new or some elite-kept secret. Alan Barkema and Michael L. Cook, "The Changing U.S. Pork Industry: A Dilemma for Public Policy," *Federal Reserve Bank of Kansas City Economic Review* (Second Quarter, 1993), p. 49. Similar warnings attracted popular press decades ago. See, for example, Bernard Collins, "Why We Won't Lose Our Hog Business," *The Iowa Farm and Home Register* (1959), cited in Don Muhm, "'Little Guy' Can Compete in Hog Farming," *Des Moines Register* (April 30, 1994), p. 7A; and Ruth Harrison, *Animal Machines: The New Factory Farming Industry* (London: V. Stuart, 1964).
16. Kendall objected to my interpretation of their agenda. Personal correspondence, February 13, 1997. See the following by E. Paul Durrenberger and Kendall M. Thu: "Our Changing Swine Industry and Signals of Discontent," *Iowa Groundwater Quarterly* 5:4 (December 1994), p. 6; "Pigs and the Anthropology of Trouble," *The Iowa Anthropology Newsletter* (Fall 1994), pp. 6, 10, 11; "Industrial Agricultural Development: An Anthropological Review of Iowa's Swine Industry," paper presented at the 20th Annual Conference of the National Association of Rural Mental Health, Des Moines, Iowa, July 1-4, 1994.
17. See, for example: Amos Turk, James W. Johnston, Jr., and David G. Moulton, eds., *Human Responses to Environmental Odors* (New York: Academic Press, 1974); Amos Turk, James W. Johnston, Jr., and David G. Moulton, eds., *Methods in Olfactory Research* (New York: Academic Press, 1975); Howard R. Moskowitz, Andrew Dravnieks, William S. Cain, and Amos Turk, "Standardized Procedure for Expressing Odor Intensity," *Chemical Senses and Flavor* 1:2 (August 1974), pp. 235-237; and Amos Turk and Jonathan Turk, *Environmental Science,* 4th ed. (Philadelphia: Saunders College Publishing, 1988).
18. Kendall claims that he later did not apply for NPPC funds (targeted for an issues focus that he helped put together) "for reasons of maintaining independence."
19. I refer here mainly to Thu and Durrenberger, "Industrial Agricultural Development." This work has been republished in various forms but with approximately the same method and conclusion.

    In addition to conceding some statistical fine points, Kendall has asked me to emphasize that (1) "the question was whether there was any indication that volume of hogs via any structure vs. number of hog farmers was related to better conditions in rural areas" rather than simply whether big guys spread the wealth; (2) they, too, have and in the article itself volunteer misgivings about this particular analysis; (3) they argue for better ethnography to counter potentially misleading statistics; and (4) their findings should carry the additional force of "forty years of research in this area backing what we're alluding to in this article." Personal correspondence, February 13, 1997.

20. The technical matters I have most in mind are the relevance of the data to the research question (basically, should Iowa go the way of North Carolina?) and potential distortions introduced in coding the data.
21. Kendall objects that "we did *not* say one caused the other." Personal correspondence, February 13, 1997.
22. Thu and Durrenberger, "Industrial Agricultural Development," pp. 10, 11, 12, 13.
23. Paraphrased from phone conversations with Charlie Arnot, communications director for PSF, the week of April 5, 1994. Note that Kendall claims that they never misrepresented, in fact, "never wrote a bad word about PSF." Personal correspondence, February 13, 1997.
24. Paraphrased from phone conversation with Kelley Donham, April 11, 1994.
25. The reorganization plan was filed in U.S. Bankruptcy Court in the District of Delaware (docket number 96-1033) on July 2, confirmed on September 6, and closed on December 27, 1996. Given the stature of its investors, especially majority owner Morgan Stanley, PSF's reorganization was headline financial news. PSF was gobbled up by a yet bigger operation. In January 1998 Continental Grain agreed to buy 51 percent of PSF, including all of its Missouri and Texas operations. Hence, in challenging PSF populists may actually have helped to increase concentration and integration in agriculture. "Continental Grain Purchase," *National Hog Farmer* (February 15, 1998), p. 6; "Continental Buys Majority Share of PSF," *Pork 98* (February 1998), pp. 72-73.
26. Karen McMahon well covered the main plot of the bankruptcy proceedings for *National Hog Farmer.* See her articles, such as "Premium Standard Farms' Bankruptcy: 'Newco' Picks up the Pieces," *National Hog Farmer* (August 15, 1996), pp. 10-19, as well as "PSF Reorganizes Under Bankruptcy Law," *Pork '96* (August 1996), p. 50.
27. Kendall M. Thu, "Piggeries and Politics: Rural Development and Iowa's Multibillion Dollar Swine Industry," *Culture and Agriculture* 53 (1995/1996), p. 23.

    In reviewing a draft of this chapter, Kendall vehemently objected to my conclusion that they believe or ever said that science is superior to common sense. Since I read their work as saying it frequently, I am not sure how to respond except to share with readers both their objection and my sources. The most forgiving explanation that I can come up with is that they see nothing condescending in claiming that their scientific procedures are the key to contextualizing and correcting popular beliefs.

    Prior subjects of Durrenberger's research have raised similar objections, claiming that folk beliefs are more reasonable than his methods allow and that, in effect, his numbers were rigged. See, for example, J. Stephen Thomas et al., "Independence and Collective Action: Reconsidering Union Activity Among Commercial Fisherman in Mississippi," *Human Organization* 51:2 (Summer 1992), pp. 151-154; and Thoroddur Bjarnason and Thorolfur Thorlindsson, "In Defense of a Folk Model: The 'Skipper Effect' in the Icelandic Cod Fishery," *American Anthropologist* 95:2 (1993), pp. 371-394, especially p. 391, where they conclude: "We agree with Pálsson and Durrenberger that one should not take whatever people say about their lives at face value. People are, however, experts on their own conditions and should therefore be taken at least as seriously as the scholars observing them."

    See also: Durrenberger and Thu, "Agricultural Industrializing in Iowa"; and Thu, ed., *Understanding the Impacts of Large-scale Swine Production: Proceedings from an Interdisciplinary Scientific Workshop, June 29-30, 1995, Des Moines, Iowa* (Iowa City: University of Iowa, 1996). This last publication is the one that Kendall claims best represents their work at the time (a couple of years after our meeting at Joe's place), and I must agree that it is by far the most carefully reasoned. Along with researchers from the land-grant colleges that Paul and Kendall elsewhere smear, the contributors unite in a single, droning moral: The key to resolving hog wars is more money for scientific research.
28. Scientists have tended to favor the analysis of brute "impacts," easily measured, immediate results. Attention to long-term, indirect, and syncretic effects are humbling. See, for example: Edward Tenner, *Things Bite Back: Technology and the Revenge of Unintended Consequences* (New York: Alfred A. Knopf, 1996).
29. John W. Dower, *War Without Mercy: Race and Power in the Pacific War* (New York: Pantheon, 1986), p. 138 and see pp. 118-146. See also: Hussein Fahim, "Foreign and Indigenous Anthropology: The Perspectives of an Egyptian Anthropologist," *Human Organization* 36:1 (1977), pp. 80-86; Mafhoud Bennoune, "What Does It Mean to Be a Third World Anthropologist?" *Dialectical*

Anthropology 9 (1985), pp. 357-364; Caroline Brittell, *When They Read What We Write: The Politics of Ethnography* (Westport, CT: Bergin and Garvey, 1993); Alexander Leighton, *Human Relations in a Changing World: Observations on the Use of the Social Sciences* (New York: Dutton, 1949), p. 128; Clyde Kluckhohn, *Mirror for Man: The Relation of Anthropology to Modern Life* (New York: McGraw-Hill, 1949), pp. 171-178, 200.

30. Durrenberger and Thu, "The Expansion of Large Scale Hog Farming In Iowa: The Applicability of Goldschmidt's Findings Fifty Years Later," *Human Organization* 55:4 (November 4, 1996), pp. 409-415; Walter Goldschmidt, *As You Sow: Three Studies in the Social Consequences of Agribusiness,* with a foreword by Senator Gaylor Nelson (Montclair, NJ: Allanheld, Osmun and Co., 1978), especially pp. 455-487. Kendall insists that I misread their connection to Goldschmidt. The interest, he says, is that even if Goldschmidt succeeded in publishing his work, he failed to alter "basic material conditions, e.g. control over production and distribution of basic resources, namely food, that shapes society." Personal correspondence, February 13, 1997.
31. Muriel Ann Weir, "*Pamphlet No. 5* and the Freedom to Publish at Iowa State College," M.S. thesis, Iowa State University, 1976.
32. Oswald H. Brownlee, *Wartime Farm and Food Policy, Pamphlet No. 5: Putting Dairying on a War Footing* (Ames: The Iowa State College Press, 1943). The pamphlet sold for twenty cents.
33. For examples of the dimensions of the conflict, see such news coverage as "Oleo Argued by Senators: Connally Tells Scorn of 'Greasy Butter,'" *Des Moines Register* (May 21, 1943), p. 14; or "The Margarine vs. Butter Debate" and "Warning to Dairy Interests," *Des Moines Register* (May 27, 1943), p. 4.
34. Weir, "*Pamphlet No. 5,* pp. 197-200. Harry Eugene Allison, "Competition Between Butter and Margarine," M.S. thesis, Pennsylvania State College, 1949.
35. John Gadd, "European Update," *National Hog Farmer* (September 15, 1993), p. 58.
36. "Poised to Survive," *National Hog Farmer* (June 15, 1996), pp. 28-34; and Marlys Miller, "Those Who Work Smarter Win," *Pork '96* (May 1996), p. 5.
37. Karen McMahon, "Premium Standard Farms' Bankruptcy," pp. 14-15. A prominent banker, Robert Niehaus, brought pre-Exodus Egypt to mind when he described the string of plagues that befell PSF as "a little bit like the Bible." Laura Jereski and Randall Smith, "Morgan Stanley's Pig-farm Dream Wallows in Losses, Spatters Clients," *Wall Street Journal* (May 22, 1996), p. C15.

CHAPTER 15

1. Jane Adams, "Resistance to 'Modernity': Southern Illinois Farm Women and the Cult of Domesticity," *American Ethnologist* 20:1 (February 1993), pp. 89-113.
2. According to the National Safety Council, U.S. agriculture accounts for about 1,500 deaths (300 of them children) and 140,000 disabling injuries each year. That rate—about 50 deaths and 500 disabling injuries per 100,000—is five times the national occupational average. *Accidental Facts* (Chicago: National Safety Council, 1989), pp. 1-32; Kelley J. Donham and Kendall M. Thu, "Relationships of Agricultural and Economic Policy to the Health of Farm Families, Livestock, and the Environment," *Journal of the American Veterinary Medical Association* 202:7 (April 1, 1993), pp. 1084-1091; Patricia J. Woznick, "Domains of Subjective Well-being in Farm Men and Women," *Journal of Family and Economic Issues* 14:3 (Summer 1993), pp. 97-114.
3. Most grain farmers spend early spring commuting to the Soil Conservation Service (SCS), poring over plats and aerial photos, trying to figure out what can be done both to plant effectively and to "stay in the program"—in other words, to conform to conservation conditions tied to 1980s-1990s farm subsidy, insurance, tax, and loan programs.
4. Even in Iowa—by a host of measures one of the nation's most agricultural states—relatively few people live on farms, fewer than one in ten. Of the 1.5 million Iowans who live outside metropolitan areas, only about 67,000 derive most of their income directly from agriculture. "Rural Policy: More than Farm Policy," *Des Moines Sunday Register* (April 23, 1995), p. 1C.
5. Halina M. Zaleski, Ph.D., Extension Swine Specialist, University of Hawaii, "Re: Practicing Without a License," Swine-L post, December 7, 1995, and personal communication, December 12, 1995.

6. The following is an abbreviated and edited version of e-mail correspondence, December 14-17, 1992.

CHAPTER 16

1. The following condensation was edited from interviews, phone conversations, and e-mail exchanges with Robin Gross in 1993. Most of the material was excerpted from an interview in Des Moines, Iowa, January 6, 1993.
2. Bruce Brown, *Lone Tree: A True Story of Murder in America's Heartland* (New York: Crown, 1989).

CHAPTER 17

1. Annette Kolodny, *The Land Before Her: Fantasy and Experience of the American Frontiers, 1630-1860* (Chapel Hill: University of North Carolina Press, 1984); Annette Kolodny, *The Lay of the Land: Metaphor as Experience in American Life and Letters* (Chapel Hill: University of North Carolina Press, 1984); Richard W. B. Lewis, *The American Adam: Innocence, Tragedy and Tradition in the Nineteenth Century* (Chicago: University of Chicago Press, 1955); Leo Marx, *The Machine in the Garden: Technology and the Pastoral Ideal in America* (New York: Oxford University Press, 1964); Henry Nash Smith, *Virgin Land: The American West as Symbol and Myth* (Cambridge, MA: Harvard University Press, 1950); Paul B. Thompson, *Sacred Cows and Hot Potatoes: Agrarian Myths and Agricultural Policy* (Boulder, CO: Westview Press, 1992); Paul B. Tompson, "Agrarianism and the American Philosophical Tradition," *Agriculture and Human Values* 7 (Winter 1990), pp. 3-9; Imhoff Vogeler, *The Myth of the Family Farm: Agribusiness Dominance of U.S. Agriculture* (Boulder, CO: Westview Press, 1981).
2. USDA Census data are usefully interpreted and reported in periodicals of trade associations and sundry government bureaus. Most useful are the occasional reports of the National Swine Survey, especially "Part III: Changes in the U.S. Pork Industry, 1990-1995," *Swine '95: Grower/Finisher* (Fort Collins, CO: Centers for Epidemiology and Animal Health, NAHMS, 1997).
3. In 1991, 60.6 percent of the swine farm employees had some formal education beyond high school, and more than a quarter had four-year degrees. More than a quarter of the people holding a "farm manager" or "assistant manager" position and more than 5 percent of all employees had earned a graduate degree. National Pork Producers Council, *Employer/Employee Relations* (Des Moines, IA: NPPC, 1991), pp. 22-24.
4. Murray R. Benedict, *Farm Policies of the United States, 1750-1950: A Study of Their Origins and Development* (New York: Twentieth Century Fund, 1953); Wendell Berry, *The Unsettling of America: Culture and Agriculture* (San Francisco: Sierra Club Books, 1977); Willard Cochrane, *The Development of American Agriculture: A Historical Analysis* (Minneapolis: University of Minnesota Press, 1979); Douglas R. Hurt, *American Agriculture: A Brief History* (Ames: Iowa State University Press, 1994); Kramer, *Three Farms: Making Milk, Meat, and Money from the American Soil* (Cambridge, MA: Harvard University Press, 1987); Peter H. Lindert, "Historical Patterns of Agricultural Policy," in *Agriculture and the State: Growth, Employment, and Poverty in Developing Countries,* ed. C. Peter Timmer (Ithaca, NY: Cornell University Press, 1991), pp. 29-83. Hans J. Michelmann, Jack C. Stabler, and Gary G. Storey, eds., *The Political Economy of Agricultural Trade and Policy: Toward a New Order for Europe and North America* (Boulder, CO: Westveiw Press, 1990); John Opie, *The Law of the Land: Two Hundred Years of American Farmland Policy* (Lincoln: University of Nebraska Press, 1987); Sonya Salamon, *Prairie Patrimony: Family, Farming, and Community in the Midwest* (Chapel Hill: The University of North Carolina Press, 1992); Thompson, *Sacred Cows and Hot Potatoes;* Vogeler, *The Myth of the Family Farm.*

5. Gender-based wage discrimination in the hog industry might be "explained" by noting that a higher percentage of women than men work in farrowing houses than finishing units. The average salary for working the same number of hours in the former is lower than for the latter. In effect, the older and fatter the hogs that you tend, the higher your pay—one of those correlations that "just so happens" to penalize women for their alleged "talent" in tending to births and newborns. NPPC, *Employer/Employee Relations,* pp. 12-16. See also: Jane Adams, "Resistance to 'Modernity': Southern Illinois Farm Women and the Cult of Domesticity," *American Ethnologist* 20:1 (February 1993), pp. 89-113; Deborah Fink, *Agrarian Women: Wives and Mothers in Rural Nebraska* (Chapel Hill: University of North Carolina Press, 1992); Deborah Fink, *Open Country Iowa: Rural Women, Tradition and Change* (Albany: State University of New York Press, 1986); Katherine Jellison, *Entitled to Power: Farm Women and Technology, 1913-1963* (Chapel Hill: University of North Carolina Press, 1993); Christine C. Kleinegger, "Out of the Barns and Into the Kitchens: Farm Women's Domestic Labor, World War I to World War II," Ph.D. dissertation, State University of New York at Binghamtom, 1986; Salamon, *Prairie Patrimony.*
6. For a brief introduction to the history of this Euro-American vision see: Jack P. Greene, *The Intellectual Construction of America: Exceptionalism and Identity From 1492 to 1800* (Chapel Hill: University of North Carolina Press, 1993).
7. John Demos, *Entertaining Satan: Witchcraft and the Culture of Early New England* (New York: Oxford University Press, 1982); Carol F. Karlsen, *The Devil in the Shape of a Woman: Witchcraft in Colonial New England* (New York: Norton, 1987).

CHAPTER 18

1. Emily Martin, *Flexible Bodies: Tracking Immunity in America from the Days of Polio to the Age of AIDS* (Boston: Beacon Press, 1994).
2. USDA, APHIS, Veterinary Services, *Antibiotic Usage in Premarket Swine* (Fort Collins, CO: Centers for Epidemiology and Animal Health, NAHMS, 1997), p. 2; USDA, APHIS, Veterinary Services, "Part III: Changes in the U.S. Pork Industry, 1990-1995," *Swine '95: Grower/Finisher* (Fort Collins, CO: Centers for Epidemiology and Animal Health, NAHMS, 1997), p. 28.
3. Interview with Jeff Zimmerman, College of Veterinary Medicine, Iowa State University, Ames, IA, June 3, 1993.
4. William R. Clark, *The Experimental Foundations of Modern Immunology*, 3rd ed. (New York: Wiley, 1986); William D. Foster, *A History of Medical Bacteriology and Immunology* (London: Heinemann Medical, 1970); Ronald Hare, *Pomp and Pestilence: Infectious Disease, Its Origins and Conquest* (New York: Philosophical Library, 1955); Martin, *Flexible Bodies*; Scott H. Podolsky and Alfred I. Tauber, *The Generation of Diversity: Clonal Selection Theory and the Rise of Molecular Immunology* (Cambridge: Harvard University Press, 1997); Arthur M. Silverstein, *A History of Immunology* (San Diego: Academic Press, 1989); Alfred I. Tauber, *Organism and the Origins of Self* (Boston: Kluwer Academic, 1991); David Wilson, *The Science of Self: A Report of the New Immunology* (London: Longman, 1972).
5. Edited from an interview with John E. Butler and Imre Kacskovics, Department of Microbiology, University of Iowa Medical School, Iowa City, Iowa, February 18, 1994.
6. Although Jeff said his training was "to a large degree, field-based," he also characterized that approach as expensive and frustrating because "you end up with these results that you can't explain. . . . To answer the kind of questions we're trying to answer about transmission or pathological effects, so on and so forth, you can't do those in the field. All you can say is 'This is what happened. But who knows why the hell it happened?'" Edited from Zimmerman interview.
7. Linda J. Saif and R. D. Wesley, "Transmissible Gastroenteritis," in *Diseases of Swine*, 7th ed., ed. Allen D. Leman et al. (Ames: Iowa State University Press, 1992), p. 362 and see pp. 362-386; Prem S. Paul, "Research Update on Transmissible Gastroenteritis of Swine," in *Historical Overview and Research Update on Enteric Diseases of Swine*, Regional Research Committee-Project NC-62 (Manhattan: Agricultural Experiment Station, Kansas State University, 1990), pp. 17-27. For examples of mainstream work that clinicians consult, see: Joel P. Siegel, Laura L. Hungerford, and William F. Hall, "Risk Factors Associated with Transmissible Gastroenteritis in

Swine," *Journal of the American Veterinary Medical Association* 199:11 (December 1, 1991), pp. 1579-1583; Catherine Templeton and Jane Carpenter, "Epidemiology of Transmissible Gastroenteritis in a Large Herd," *The Compendium on Continuing Education for the Practicing Veterinarian*, North American edition 13:9 (September, 1991), pp. 1475-1481; Gay Y. Miller and James B. Kliebenstein, "The Economic Impact of Clinical Transmissible Gastroenteritis for Swine Producers Participating in the Missouri Mail-in-Record Program," *Preventive Veterinary Medicine* 3:5 (November, 1985), pp. 475-488.

8. Clyde A. Kirkbride, *Control of Livestock Diseases* (Springfield, IL: Charles C Thomas, 1986), p. 7; Karen McMahon, "The Salmonella Puzzle," *National Hog Farmer* (September 15, 1997), pp. 16-18.
9. Note also that the most common, pathogenic serotypes in swine account for a diminishing portion of the total. The top three dropped from 82 to 58 percent of clinical cases from 1990 to 1995. In other words, illness is becoming associated with the shedding of yet more diverse serotypes. USDA, APHIS, Veterinary Services, "Shedding of Salmonella by Finisher Hogs in the U.S." *APHIS Info Sheet* (January 1997); USDA, "Changes in the U.S. Pork Industry," p. 33.
10. Producers blamed scours for nearly a quarter of all pre-weaning deaths in 1990 and for 15 percent in 1995. USDA, "Part III: Changes in the U.S. Pork Industry," pp. 12, 13, 15.
11. Dale Miller, "Straight-Talkin' Vet Gets My Vote," *National Hog Farmer* (February 15, 1997), p. 10. Miller is quoting noted swine specialist Paul Armbrecht of Lake City, Iowa.
12. Paul Armbrecht has also advocated this sort of "acclimatization" or "'immune equalization' using vaccination, exposure to cull animals, backfeeding, whatever it takes to get the new animals' (status) equal to the sow herd." Miller, "Straight-Talkin' Vet Gets My Vote."
13. In response to efforts like those of PSF to control disease in massive, hilltop, biosecure buildings, veterinary epidemiologist Jeff Zimmerman expressed a common concern: "The epidemiology is really interesting, because basically, they are dealing with a huge population—I mean whether it's a human population or a pig population or a squirrel population—getting that many individuals in one spot at one time. It is going to be really tough to deal with, I think." Edited from Zimmerman interview.
14. Jad Adams, *AIDS: The HIV Myth* (New York: Saint Martin's Press, 1989); John Lauritsen, *Poison by Prescription: The AZT Story* (New York: Pagan Press, 1989); Jon Rappoport, *AIDS Inc.: Scandal of the Century* (San Bruno, CA: Human Energy Press, 1989); and Randy Shilts, *And the Band Played On: Politics, People, and the AIDS Epidemic* (New York: Penguin Books, 1988).
15. Edited from Butler interview.
16. See, for example, Sol M. Michaelson and Bernard J. Schreiner, "Comparative Biology in the Selection of Experimental Subjects for Cardiac Pathophysiologic Investigations," in *Defining the Laboratory Animal*, Fourth International Symposium on Laboratory Animals, 1969 (Washington, DC: National Academy of Science, 1971), pp. 121-147.

CHAPTER 19

1. Several pathogens were eradicated much earlier, and some have been known to re-emerge. Well after the hog cholera virus was supposedly eliminated, traces of it were found in a sea lion dead on a California beach and then in five species of fish. It is hence somewhat miraculous that commercial swine remain unchallenged. Scientists routinely cite this case to show how unpredictably microbes move in the environment. T. J. L. Alexander and D. L. Harris, "Methods of Disease Control," in *Diseases of Swine,* 7th ed., ed. Allen D. Leman et al. (Ames: Iowa State University Press, 1992), pp. 808-834; Clyde A. Kirkbride, *Control of Livestock Diseases* (Springfield, IL: Charles C Thomas, 1986), pp. 67-82; USDA, Animal and Plant Health Inspection Service, *Hog Cholera and Its Eradication: A Review of U.S. Experience*, APHIS 91-55 (Washington: USDA, 1981).
2. Kirkbride, *Control of Livestock Diseases*, p. 81. Since 1884 when the first U.S. eradication program began, thirteen diseases have been successfully eradicated (about half the number targeted), only one in the last twenty years. USDA, APHIS, "Facts About APHIS: Domestic Animal Health Programs," http://www.aphis.usda.gov/oa/domanima.html (May, 1998), pp. 1-4.

3. Kirkbride, *Control of Livestock Diseases*, p. 77.
4. USDA, *Hog Cholera and Its Eradication*, p. 62.
5. "Paul Reimer" is a pseudonym for a person whom I interviewed in June 1994. As with other informants, I present in the following pages edited excerpts from tape-recorded conversation, but in this case I also have altered some potentially identifying information—names of places and associates, dates, and organizational affiliations—to preserve his anonymity. Hence this material is intended only to evoke the mode of a participant's memories, not the specific actions of any specific person in the past. When I showed him a couple of chapters from this book, Dr. Reimer was deeply concerned that I was mischaracterizing him, the scientific community, and the public with which he had been proudly associated for decades. He insisted that—contrary to his reading (but not mine) of those chapters—prior eradication efforts were undertaken only after careful study and justification, that the international scientific community is bound by mutual respect and affection, and that insofar as possible, the interests of a diverse public were always foremost in mind. I find that memory reasonable and honorable, and I recommend it to readers. In any case, he did not want his name or that of associates to appear in a book that used "inappropriate language" (both inflammatory references to eradication and four-letter words). I disguise his identity out of respect for that wish.
6. Reimer interview.
7. Note that Reimer credits the basic eradication approach to a European tradition begun some fifty years before his involvement:

   > It came from Europe, particularly the British. People kind of looked up to them, to their capabilities and intelligence, and they were adamant about eradication of diseases.... You see, Britain had all of the researchers on problems like contagious pleuropneuomonia, which was a real problem in cattle. And the British eradicated it. So they all became heroes, almost a paternity of researchers on these kinds of diseases. One of them became the first chief of the Bureau of Animal Industry, which became the USDA.... The Danes and the Dutch were also big leaders and were very close to the British. So the British kind of relied on their relationship with the Danes and the Dutch. The Germans were always in and out in this fraternity. You could never pin down the French as to where they were going to be. The Italians were even worse. But you had this international community which started late in the last century, before I got involved with it.

   Reimer interview.
8. Ibid.
9. Ibid.
10. Ibid.
11. See, for example: Joe Vansickle, "The Many Faces of PRRS," *National Hog Farmer* (March 15, 1996), p. 26.
12. Interview with William L. Mengeling, The National Animal Disease Center, Ames, Iowa, June 2, 1993; Joe Vansickle, "U.S. Not Immune to Foreign Diseases," *National Hog Farmer* (February 15, 1998), pp. 10-13.
13. Compared to the United States, the Netherlands has put less emphasis on universal vaccination (the blitzkrieg) and more on sentinels and spot control: "We have a policy here in the Netherlands of non-vaccination, reporting disease very early, and isolating and destroying those herds where there is an outbreak, and very intensive control of neighboring herds where the disease has spread already. . . . The monitoring is thorough, and farmers are compensated for animals that are destroyed. Denmark has a similar system." Interview with A. P. van Nieuwstadt at the Central Veterinary Institute, Department of Virology, the Ministry of Agriculture, Lelystad, Netherlands, August 13, 1993. In the wake of major disease outbreaks in the Netherlands and Taiwan, 1997-1998, such policies have been gravely challenged.
14. Interview with veterinarian Eldon K. Uhlenhopp, College of Veterinary Medicine, Iowa State University, Ames, Iowa, October 20, 1994.
15. Reimer interview.
16. Kirkbride, *Control of Livestock Diseases*, p. 5. See also: Steve Sonka, "Old Disease Continues to Nag Herds," *Pork '95* (September 1995), p. 58.
17. Reimer interview.

18. Ibid. In subsequent correspondence (1997), Reimer offers this clarification of why specific diseases get targeted for eradication:

    Number one, it's got to be a disease that can't be eradicated in our country by the producers and their veterinarians alone. It's got to be one that is highly infectious and contagious. It's one that the experts on the disease say that is technically feasible to eradicate and one that economic experts, familiar with the disease, advocate its eradication from a cost/benefit standpoint. Since such a disease causes all kinds of problems in the production and marketing of their animals and others are fearful of the disease spreading to their herds, there is enough anxiety on the part of the livestock industry involved that they are demanding action by USDA, their state Department of Agriculture, their legislature and Congress to achieve funds for the program. That is what I meant when I said: whether a disease is targeted for eradication depends on how many complaints and concerns are produced from its incidence in our country.

19. My grab sample included about sixty-five informants with professional appointments. I tape-recorded informal interviews with about a third of them. I mainly aimed for a sustained exchange with at least two people at the top and at least one in the trenches of each bureaucratic jurisdiction. Particular individuals were identified by snowballing references (i.e., "Who else should I contact to get the fullest possible range of perspectives"). It is actually a pretty small world. Just about every professional knows every other one and readily recommends contrasting as well as confirming sources.

20. The most common subject of U.S. complaint in this regard is Mexico. Scientists—chiefly Dutch, Danish, and American—are forever developing reliable tests for previously "unknown" (i.e., only recently isolated) pathogens. When Mexican officials (generally, not themselves scientists or clinicians) hear about a "new" disease (e.g., PRRS) in the United States, they may use it to justify halting U.S. exports, even though blood serum archives establish that the disease has been ubiquitous on both sides of the border for 30 to 40 years. In northern Europe (especially the Netherlands) the main subject of complaint is Italy. For example, in the early 1990s after an Italian packer misdiagnosed traces of disease in an imported hog (confusing harmless SVD with foot-and-mouth), the EU embargoed Dutch exports. To lift it, the Dutch agricultural ministry had to agree to an absurdly thorough and expensive testing program—to prove that a disease that was long ago eradicated in the north would not in the future be spread to the south, where, everyone knows, it was a problem. If anything, the threat of contagion was in the opposite direction, but that hardly matters: importers outnumber exporters in the EU. On a related tiff between the European Union and the United States, see: "U.S. to Inspect European Meat Plants," *Pork '97* (May 1997), p. 81; "Trade War Averted," *National Hog Farmer* (May 15, 1997), p. 11; Lucy Walker, "European Report: Agriculture," [KLM] *Holland Herald* (August 1993), pp. 26-28; John Gadd, "Immunity—Producer's Achilles Heel," *National Hog Farmer* (September 15, 1997), p. 24.

21. A variant on this story traveled all the way from Indiana to Hong Kong. Bette Werner, chair of the 1995 Pork Fest, announced the cancellation of the popular kiss-the-pig contest. They were short the requisite porcine kissee. Producers, she said, who might have offered a piglet feared that it would be stressed and exposed to disease. Despite farmers' corroboration of the official explanation, the reporter assumed it was a lie. I bet it wasn't. "A Load of Hogwash," *South China* [Hong Kong] *Morning Post* (September 17, 1995), p. 16.

22. John Butler explains, "It would be very difficult to get money from the U.S. Department of Agriculture to work on TGE, because they would regard the problem as more or less solved. Certainly not totally resolved—they're always looking for a better fix. But it's not something that's threatening to wipe out the pig population in North America." Interview with John E. Butler, Department of Microbiology, University of Iowa Medical School, Iowa City, Iowa, February 18, 1994.

23. Iowa Department of Agriculture and Land Stewardship, "Infectious and Contagious Diseases," *Iowa Administrative Code* (July 27, 1988), 21-64.1 (163).

24. Jerry Kunesh, for example, a Field Service veterinarian with the Department of Veterinary Clinical Science, explained to me: "TGE and pseudorabies are not that different. We've got exactly the same technology for both. For both cases we have developed vaccine, and in both cases we have serological tests which are quite accurate. The difference and the thing that makes pseudorabies

so much more of a concern to the farmer is not the disease itself but the regulation of the disease. If you get pseudorabies, you're automatically quarantined and the situation must be cleaned up." Interview with Jerry Kunesh, College of Veterinary Medicine, Iowa State University, Ames, July 27, 1993.

25. Interview with Walter Felker, Director, Bureau of Animal Industry, Iowa Department of Agriculture and Land Stewardship, Des Moines, October 1, 1993. I am grateful, as well, to Lawrence Birchmier, pseudorabies coordinator in the same department. Birchmier helped me navigate the various codes and rules.

26. See, for example: Dirck Steimel, "Pig Dealer May Have Violated State Laws, *The Des Moines Register* (October 23, 1993), pp. 6-7A.

27. Interview with Mark W. Welter, vice president of research, Ambico, Inc., Dallas Center, Iowa, May 9, 1995. See: Gerald R. Fitzgerald and C. Joseph Welter, "The Effect of an Oral TGE Vaccine on Eliminating Enzootic TGE Virus from a Herd of Swine," *Agri-Practice* 11:1 (January/February 1990), pp. 25-29; Gerald Fitzgerald and C. Joseph Welter, "Oral Vaccination Against TGE Reduces Mortality," *PIGS—Misset* (July/August 1990), p. 27; Gerald R. Fitzgerald, Mark W. Welter, and C. Joseph Welter, "Improving the Efficacy of Oral TGE Vaccination," *Veterinary Medicine* 81 (February 1986), pp. 184-187; Bill Fleming, "The 'New' TGE," *National Hog Farmer* (October 15, 1991), pp. 50-54; R. H. Schlueter, "Oral TGE Vaccination Cuts Virus Shedding," *National Hog Farmer* (February 15, 1995), p. 9; Mark W. Welter, Michelle P. Horstman, C. Joseph Welter, and Lisa M. Welter, "An Overview of Successful TGEV Vaccination Strategies and Discussion on the Interrelationship Between TGEV and PRCV," *Coronaviruses*, ed. H. Laude and J. F. Vautherot (New York: Plenum Press, 1994), pp. 463-468; Al Oppedal, "TGE Eradication is Possible," *Hog Farm Management* (October 1989); David Reeves et al., "Roundtable Discussion: Transmissible Gastroenteritis," *Agri-Practice* 13:7-9 (July/August - October, 1992); Joel P. Siegel, Laura Hungerford, and William F. Hall, "Risk Factors Associated with Transmissible Gastroenteritis in Swine," *Journal of the American Veterinary Medical Association* 199:11 (December 1, 1991), pp. 1579-1583. I am also grateful for the advice of research and sales representatives whom I met at the World Pork Expo, 1993-1995.

28. Prem S. Paul, Xiao-Ling Zhu, and Eric Vaughan, "Current Strategies for the Development of Efficacious Vaccines for Transmissible Gastroenteritis in Swine," *Proceedings of the 92nd Annual Meeting of the United States Animal Health Association,* Little Rock, Arkansas, 1988, pp. 429-442. Kunesh interview, 1993.

Edward Bohl—who was part of the U.S. team who (following a Japanese lead) first isolated TGEV and who authored the chapter on TGE in the editions of *Diseases of Swine* that preceded those by his student, Linda Saif—put it this way:

> To simplify things: I would say that most vaccines provide some protection under some conditions. But, none of them are highly effective, I would say, under all conditions. Now, then, the economics gets involved: "Well, is it worthwhile for a farmer to spend money on a product that isn't too effective?" Contention comes in: "We've done research here that would show cell culture virus given a certain way will provide some immunity." Now, here again: "What do you mean by 'some immunity?' And at what point is 'some immunity' economically feasible to use?" With any vaccine you study, you have to get involved in these types of questions.

Edited from an interview with Edward H. Bohl, Wooster, Ohio, June 1, 1994.

29. See for example: "Holding Down Immunity Spurs Growth, Efficiency," *National Hog Farmer* (May 15, 1994), p. 88; Keith Thornton, "Are You Ready for Multiple Sites?" *Pork '95* (February 1995), pp. 38-49; "Vaccinations Costs Include Growth Lull," *Pork '95* (May 1995), p. 22. Interview with Paul Yeske, Swine Veterinary Center, Saint Peter, Minnesota, October 20, 1994.

Note that I am not here saying that infection is "natural" or desirable in itself—only that counter measures are sometimes worse, a fact which is widely recognized even by scientists favorably predisposed to eradication. See: Scott Dee, "PRRS Remains Formidable Foe," *National Hog Farmer* (July 15, 1995), p. 22; Joseph Rudolphi, "PRRS Virus: One Year Later," *Pork '95* (February 1995), p. 50; Gordon Spronk, "PRRS: A Few Good Answers Producers Can Live With," *National Hog Farmer* (November 15, 1994), p. 42; Joe Vansickle, "Segregated Early Weaning:

Pitfalls Tarnish the 'Perfect' System," *National Hog Farmer* (May 15, 1996), pp. 48-50; Joe Vansickle, "Why SEW Herds Break with Disease," *National Hog Farmer* (May 15, 1996), p. 56.

30. "Closing your eyes, clicking your heels together three times and repeating, 'SEW cures all, SEW cures all' won't work." Julian Edwards, "Why and When Do I Need to Vaccinate?" *Pork '96* (April 1996), p. 42. Interview with Paul Yeske, Swine Veterinary Center, Saint Peter, Minnesota, October 20, 1994.

31. Linda Saif explains:

> In the pathogen-host relationship the usual thinking is that, if a pathogen evolves to live in a more harmonious relationship with the organism, then that will be a more suitable pathogen to persist. In other words, if it kills the host, it's maybe not the ideal pathogen. The more ideal pathogen is one that more or less lives in a more harmonious relationship with the host. And that's probably one of the reasons that maybe PRCV now is persisting in Europe. Unlike TGE, which killed off its host in terms of baby pigs, that virus seems to now more or less persist in the host without leading to the mortality that was evident in TGE. So it seems to maybe have taken the niche of TGE virus.

Interview with Linda Saif, Department of Food/Animal Health, Veterinary Preventive Medicine, Ohio Agriculture Research and Development Center, Wooster, Ohio, June 1, 1994.

32. Edited from van Nieuwstadt interview. Corroborating interviews include those with Kelly Lager, Bill Mengeling, and Ronald Wesley at the National Animal Disease Center in Ames, Iowa, June 2, 1993, with Linda Saif, and with Prem S. Paul, Enteric Virus Research Laboratory, Veterinary Medical Research Institute, College of Veterinary Medicine, Iowa State University, Ames, Iowa, June 4, 1993. See: Joe Vansickle, "Mutation Offers TGE Hope," *National Hog Farmer* (January 15, 1994), p. 70; A.P. van Nieuwstadt and J.M.A. Pol, "Isolation of a TGE Virus-Related Respiratory Coronavirus Causing Fatal Pneumonia in Pigs," *Veterinary Record of the Journal of the British Veterinary Association* 124:2 (January 14, 1989), pp. 43-44; A.P. van Nieuwstadt and J. Boonstra, "Comparison of the Antibody Response to Transmissible Gastroenteritis Virus and Porcine Respiratory Coronavirus, Using Monoclonal Antibodies to Antigenic Sites A and X of the S Glycoprotein," *American Journal of Veterinary Research* 53:2 (February 1992), pp. 184-190; A.P. van Nieuwstadt, T. Zetstra, and J. Boonstra, "Infection With Porcine Respiratory Coronavirus Does Not Fully Protect Pigs Against Intestinal Transmissible Gastroenteritis Virus," *Veterinary Record of the Journal of the British Veterinary Association* 125:3 (July 15, 1989), pp. 58-60.

33. The usual referents in this regard are ethological studies of the behavior of domesticated swine when allowed free range in "pig parks," such as the one that D. G. M. Wood-Gush and A. Stolba inspired near Edinburgh, Scotland. Their work and kindred projects originally were intended to establish a baseline—"natural preferences"—of swine to which commercial agriculture might aspire. More recently, animal scientists are very skeptical of the idea of so fixing "nature," especially when pig parks are very alien to most swine experience for many generations. Interview with Stan Curtis, Pennsylvania State University, State College, June 3, 1994.

34. Interview with John Vahle at the College of Veterinary Medicine, Iowa State University, Ames, Iowa, June 3, 1993.

35. Ibid. My thanks to John Butler for introducing me to this hypothesis and to Joel Weinstock for explaining it to me. Joel V. Weinstock and David Elliott, "Does Crohn's Disease Result from Failure to Acquire Helminthic Infections During Childhood," typescript (Department of Medicine, University of Iowa, May, 1997), pp. 11, 2, 4-5, 9; "Parasitic Diseases of the Liver and Intestine," ed. Joel V. Weisntock, *Clinic in Gastroenterology* 25:3 (1996).

36. Butler interview.

37. Uhlenhopp interview.

38. Edited from Curtis interview. He explains, "A clinician really needs to run by the seat of his pants . . . to take partial information, quickly draw a conclusion, and then get on with it. The scientist who does that doesn't get tenure. It's two different mind sets." Nevertheless, he jokes, "some of my best friends are veterinarians," and he is proud do have been the first non-veterinarian to whom the American Association of Swine Practitioners granted the Dunne Award. "You know, we're not enemies, but I just don't buy some of their nonsense." The Dunne Award is granted in memory

of Howard Dunne, the microbiologist whose achievements include developing the "autosow," the sterile neonate technology to which John Butler refers. See: Stephen Budiansky, *Man and Beast: The Origins of Domesticated Animals and of Nature* (New York: W. Morrow, 1992), and *Nature's Keepers: The New Science of Nature Management* (New York: Free Press, 1995); Stanley E. Curtis, *Environmental Management in Animal Agriculture* (Ames: Iowa State University Press, 1983).

39. Edited from an interview with Jeff Zimmerman, College of Veterinary Medicine, Iowa State University, Ames, Iowa, June 3, 1993. He refers here to *Diseases of Swine,* 7th ed.

40. Saif interview. Linda J. Saif and R. D. Wesley, "Transmissible Gastroenteritis," in *Diseases of Swine*, 7th ed., pp. 362-386. She emphasizes this clarification:

> The HIV scenario I noted was only to give an example of what pathogens do to the host when the immune system is compromised by the infectious agent or if there is no active immunity to a fatal pathogen in a susceptible population. If an animal is immunocompromised and cannot develop an active immune response, then the pathogen will likely produce a fatal outcome. In other words, if the disease is eradicated, then no active immunity is needed against the infectious agent, if it no longer exists in the population. However, if pigs are not actively immune to TGE and TGEV still exists in swine and is introduced into a naive, non-immune herd, then the infection will be severe with high mortality in the highly susceptible non-immune neonatal pigs.

Linda J. Saif, personal correspondence, June 3, 1997.

41. Richard Preston, *The Hot Zone* (New York: Random House, 1994); John Winthrop, *Winthrop Papers* (Boston: Massachusetts Historical Society, 1943), Vol. 2, pp. 120, 141 and Vol. 3, p. 167. Among the few who explicitly (though inconsistently) challenge the latest round of virus demonics are C. J. Peters and Mark Olshaker in *Virus Hunter: Thirty Years of Battling Hot Viruses Around the World* (New York: Doubleday, 1997). Once investigated, this memoir warns, most of the allegedly "emergent" viruses turn out to be "old" or, if "new," much less deadly than unglamorous, old-fashioned cousins. A sympathetic reviewer whines, "More than three times as many people die of malaria *every day* than have been killed by Ebola virus in all of history. Yet it's Ebola that people find 'scary'!" Ed Regis, "Pathogens of Glory," *The New York Times Book Review* (May 18, 1997), p. 18.

CHAPTER 20

1. Interview with Ronald Wesley, The National Animal Disease Center, Ames, Iowa, June 2, 1993.
2. This chapter was inspired by recent feminist "theorizing of the body," as in the works of Judith Butler, *Gender Trouble: Feminism and the Subversion of Identity* (New York: Routledge, 1990) and *Bodies That Matter: On the Discursive Limits of "Sex"* (New York: Routledge, 1993). Although the impetus is persuasive, Butler's prose style seems off-putting and the "body" of which she speaks strangely cold, as if eviscerated. Here, then, I am trying to pump some fluids and flesh back onto the bones, a purpose resembling that of Susan Bordo, for example, in *Unbearable Weight: Feminism, Western Culture, and the Body* (Berkeley: University of California Press, 1993) and *Twilight Zones: The Hidden Life of Cultural Images from Plato to O. J.* (Berkeley: University of California Press, 1997). See: Alison M. Jaggar and Susan R. Bordo, *Gender/Body/Knowledge: Feminist Reconstruction of Being and Knowing* (New Brunswick, NJ: Rutgers University Press, 1989). See also a recent, immensely popular novel that pursues these themes in porcine form: Marie Darrieussecq, *Pig Tales: A Novel of Lust and Transformation* (New York: The New Press, 1997). My title, "the body porcine" comes in morphing Walt Whitman ("the body electric") and Victoria Principal ("the body Principal"). Walt Whitman, "Enfans d'Adam," *Leaves of Grass, Facsimile Edition of the 1860 Text*, ed. Roy Harvey Pearce (Ithaca, NY: Cornell University Press, 1961), p. 291; Victoria Principal, *The Body Principal* (New York: Simon and Schuster, 1983).
3. Advertisement running in *The National Hog Farmer* and *Pork '95*.
4. A surprising example of this dissociation comes in the form of bubble-packed "Pertinent Pig Facts" accompanying a toy Wind-Up Walking Pig. "*Chu,*" it asserts, "is a common Chinese surname

meaning 'pig.' Chinese named themselves this to fool evil spirits into thinking they were animals and therefore not worth bothering" (Seattle: Accouterments, 1995). Note, then, the claim that mere verbal kinship with swine is alone sufficient to define a person out of humanity. It is also worth noting that this "fact" itself is wrong. The common surname and the Chinese name for "pig" may sound similar, but they are different characters, in effect, different words. Hence, if people substitute one for the other (as young people often do), it is a pun, a way of ribbing the person that plays on the obvious difference in the meaning of the two characters. Through this mistake, then, a toy that is made in China bespeaks the culture of its American distributor.

5. George Orwell, *Animal Farm* (New York: Harcourt, Brace and Company, 1946); *Animal Farm*, British film translation by John Halas and Joy Batchelor, 1954; Marilyn Nissenson and Susan Jonas, *The Ubiquitous Pig* (New York: Harry N. Abrams, 1992), p. 122.
6. Steve Davis recently reported to the American Society of Animal Science that the vast majority of Americans whom he has surveyed rate the intelligence of pigs only slightly below their pets and well above that of other commercial livestock. "Should You Mind How Animals Think?," *Pork '98* (April 1998), pp. 14-16.
7. Michael J. Arlen, *Thirty Seconds* (New York: Farrar, Straus and Giroux, 1980), pp. 57-60.
8. "Pork Queens Reign Over Industry: A Look at the Contributions of a Pork Queen," in Don Muhm's *Iowa Pork and People: A History of Iowa's Pork Producers* (Clive, IA: Iowa Pork Foundation, 1995), pp. 99-102.
9. Edited from interview with Paul Reimer, June 15, 1994.
10. On September 3, 1995, in response to low prices in 1994 (and hence reduced check-off receipts and a shortfall of $2.8 million in NPPC funding of the National Pork Board), the check-off rate increased from .35 to .45 percent. "Hog Checkoff Increase Coming," *National Hog Farmer* (March 15, 1995), p. 6; "Iowa Steers Clear of Elite Pork Image," *Iowa City Press-Citizen* (January 18, 1995), p. 10A.
11. For 1996 the NPPC requested about $335 million for the National Research Initiative (38.8 percent), market promotion (32.8 percent), the Emergency Food Assistance Program (19.4 percent), the Water Quality Incentive Program (4.5 percent), pseudorabies monitoring and eradication (2.5 percent), and the National Swine Research Center (1.9 percent). These priorities are consistent with those over the prior two years. *National Hog Farmer* (March 15, 1995), p. 6.

CHAPTER 21

1. Probably the most accessible single survey of pig/person connections is Marilyn Nissenson, *The Ubiquitous Pig* (New York: Abrams, 1992).
2. The prior essay was Richard P. Horwitz, "Multiculturalism and University Lore," in *Multiculturalism and the Canon of American Culture*, European Contributions to American Studies XXIII, ed. Hans Bak (Amsterdam: VU University Press, 1993), pp. 16-26. The subsequent invitation was for the Nordic Association for American Studies Biennial Conference in Oslo, Norway, where I delivered "Husbandry at the Border," August 9, 1995. I took advantage of support for that paper, in the form of travel grants from the Stanley Foundation and the American Council of Learned Societies, to interview Dutch and Danish pig people en route.
3. Ellis Parker Butler, *Pigs is Pigs* (New York: McClure, Phillips and Co., 1906). Biographical information comes from: Clarence Andrews, *A Literary History of Iowa* (Iowa City: University of Iowa Press, 1972), pp. 153-156; Louise Guyol, "Interview with Ellis Parker Butler," *Boston Evening Transcript* (May 1, 1926); Frank Paluka, *Iowa Authors: A Bio-Bibliography of Sixty Native Writers* (Iowa City: Friends of the University of Iowa Library, 1967). An autobiographical sketch appears in Stanley S. Kunitz, *Authors Today and Yesterday: A Companion Volume to Living Authors* (New York: H. W. Wilson Co., 1933), pp. 120-121. A large collection of Butler's papers is available in the Special Collections Department of the University of Iowa Library.
4. Butler, *Pigs is Pigs*, pp. 7-8.
5. Apparently an editor, Ellery Sedgwick, was the one who changed the title to "Pigs is Pigs." Fred W. Lorch, "The 'Pigs is Pigs' Phenomenon," *The Iowan* 12:2 (Winter 1964), p. 52.

6. David Roediger, *Toward the Abolition of Whiteness: Essays on Race, Politics, and Working Class History* (London: Verso, 1994), pp. 181-198, esp. p. 188; *Dictionary of American Regional English*, ed. Federic G. Cassidy (Cambridge: Belknap Press, 1991), vol. 2, pp. 838-840.
7. Butler, *Pigs is Pigs*, pp. 22-23.
8. Ibid., p. 24.
9. Ibid., pp. 35-36.
10. David L. Hull, "Species, Races, and Genders: Differences Are Not Deviations," in *Genes and Human Self-Knowledge: Historical and Philosophical Reflections on Modern Genetics*, eds. Robert F. Weir, Susan Lawrence, and Evan Fales (Iowa City: University of Iowa Press, 1994), pp. 207-219; *Science as a Process: An Evolutionary Account of the Social and Conceptual Development of Science* (Chicago: University of Chicago Press, 1988), p. 79; Strachan Donnelley, Charles R. McCarthy, and Rivers Singleton, Jr., "The Brave New World of Animal Biotechnology," Special Supplement, *Hastings Center Report* 24, No. 1 (February 1994), pp. S8-S11; Ernst Mayr, *The Growth of Biological Thought* (Cambridge: Harvard University Press, 1982), pp. 260-273; Scott H. Podolsky and Alfred I. Tauber, *The Generation of Diversity: Clonal Selection Theory and the Rise of Molecular Immunology* (Cambridge, MA: Harvard University Press, 1997).
11. For a sample of popular (albeit "alternative") musings on the line between humans and non-humans including plants, see: "What Animals Could Tell Us, If We'd Only Listen," Special Issue of *Utne Reader* (March-April 1998), especially "Where Do You Draw the Line?" p. 55.
12. Since foot-and-mouth disease (FMD) broke in the United States six times between the 1880s and 1920s, ranchers' concerns are understandable. The embargo was national by a 1930 act of Congress. Interview with Paul Reimer, June 15, 1994.
13. According to Reimer and many others, the eradication approach was primarily of late nineteenth-century, British origin. Of course, the Peace Corps was not literally "modeled" on the FMD program in Mexico; they just share features (in particular, an approach to staffing sectors) that one successfully employed before the other. Reimer explains:

    > After the five years in Mexico, I was here in Washington, up on the Hill for the testimony. And Sergeant Shriver was there, proposing the Peace Corps. I listened to what he said, and I said, "Hell, we already had that. This is what we were doing in Mexico." What we did was, with the disease in sixteen states, we divided those states up into sectors. A sector was where Mexican and American livestock inspectors could see all those animals—cattle, sheep, goats, and hogs every thirty days. In the meantime, we set up vigilante committees so that, if anything got sick, they'd get in touch. They had to live in that sector, regardless of the terrain. They were allowed out one weekend a month for rehabilitation. Whether it was in the 10,000-foot mountains or the jungle area of southern Veracruz, they had to live there. In many cases, they were the only one who could read and write. So, my feeling was that we were the Peace Corps, you know? We just didn't call it that.

    Reimer interview.
14. In our 1994 discussion, Reimer explained:

    > It turned out that the Mexican entrepreneurs were friends of our entrepreneurs, and they decided they'd take the risks, unbeknown to the governments. They brought these animals into Veracruz. (You know, most of the veterinarians engaged in regulatory work in Mexico were elderly and received their income from issuing health certificates for the movement of animals. In many cases, supposedly, they didn't inspect the animals.) And when the first shipment came into Veracruz, nothing happened. But with the second one, all hell broke loose. Sick animals all over the damned place. . . . So, the two governments decided they were going to work together to eliminate the disease in Mexico.
    >
    > We sent 2,500 U.S. people to Mexico. . . . There were 16 million animals that we had to keep vaccinated every four months, and there was trouble with the vaccine. And we had to go after the mutant strains. We had people killed and died of diseases. One inspector was stoned to death. One was hit by a shotgun. There were all kinds of things that you would not believe until you get into it.
    >
    > But it was eradicated. . . . We spent about $130 million on this thing—peanuts compared to what would have happened if we had the disease in the States.

In subsequent personal correspondence (1997), Reimer added the clarification that he "was also fully aware that Mexico had some very competent veterinarians in their military, since some were my counterparts over the years I was there. There were others in research that helped develop that vaccine that we used that was very effective. There were others in the diagnostic labs, in private practice, and in their veterinary universities. Many of the younger veterinarians in the FMD Program became leaders later in their careers. I feel that they also inspired younger ones under their leadership to become today's leaders in their field not only in Mexico but internationally. I feel that this development was one of the long term benefits of the U.S./Mexican FMD Program."

15. Reimer interview.
16. Interviews with Eric J. Bush, Ken Forsythe and H. Scott Hurd at the Animal and Plant Health Inspection Service of the United States Department of Agriculture in Fort Collins, Colorado, April 7, 1995. During training, Bush also worked on the final stages of the Haitian African swine fever eradication program, helping repopulate Haitian swine with U.S. breeding stock.

    Paul Reimer explains: "I didn't realize it, but in Haiti, their pig was their bank, a piggy bank. When they needed to send a kid to school, they sold a pig. Or when anybody needed medical things or anything, they sold a pig. It's also a part of their voodoo ritual. To get rid of their pigs was a very traumatic situation. But, then again, here's the poorest country in the hemisphere. And the history of this thing is somewhat like cholera. If it exists, you're going to have periodic epidemics that are going to sweep through and kill pigs like flies. So I went down there with the idea of trying to do something about it." Reimer interview, 1994. See also: Kim Taylor, "Stakeholders Add Their Perspectives to Future Search Vision," *Inside APHIS* 15:2 (March 1995), p. 6.
17. In subsequent correspondence (1997), Reimer offered this clarification: "We were aware that when ASF [African swine fever] was first found in Haiti that many hogs died and that 'Boat People' were found with meat when they beached in Florida; so we could assume that some of them left Haiti because of it. . . . The program that we carried out was recommended by the best experts on the disease we could find. The justification for eliminating the swine was to prevent the suffering and deaths of future generations of swine in Haiti and its long term economic effects on education and well being of the owners of swine in Haiti."
18. Reimer interview, 1994.
19. In subsequent personal correspondence (1997), Reimer insisted that well before his retirement, disease eradication programs were conditioned by an expert cost/benefit analysis that included consideration of many factors: "If we thought that we could effectively restrict movements on a regional basis and contain the disease in an area while permitting movements from other parts of a country, we would have done it long ago."
20. Hurd interview, 1995. See also "Regionalization Is Key to Maintaining Exports," *Pork '97* (June 1997), p. 122.

CHAPTER 22

1. Robert Bellah, "Social Science as Public Philosophy," *Habits of the Heart: Individualism and Commitment in American Life* (New York: Harper and Row, 1985), pp. 297-307.
2. This quotation was edited from a tape-recorded conversation with a veterinarian employed by the Ministry of Agriculture at the Institute for Animal Science and Health in Lelystad, the Netherlands, August 13, 1993. When explicitly asked, as the book was going to press, he was reluctant to accept credit for these observations because they were so dated, particularly in the wake of a grave epidemic that broke in February, 1997. So specific plans and practices have changed, and the number of hogs raised in the Netherlands did not significantly decline, 1991-1996. But I think his (and my) more general point—about the distinctive public confidence in state-level planning and the substance of consumer priorities in the Netherlands—remains reliable. Personal correspondence, July 10, 1997. USDA, APHIS, Veterinary Services, "Part III: Changes in the U.S. Pork Industry, 1990-1995," *Swine '95: Grower/Finisher* (Fort Collins, CO: Centers for Epidemiology and Animal Health, NAHMS, 1997), p. 9.
3. See, for example: John Gadd, "European Update," *National Hog Farmer* (August 15, 1995), pp. 44-45.

4. On August 7-8, 1995, my host was Susanne Ammendrup, senior veterinary officer and head of the Department for EU-Trade in the Ministry of Agriculture and Fisheries. In Frederiksberg at the headquarters of the Danish Veterinary Services, she arranged for me to meet Jørgen Flensburg, senior veterinary officer and head of the Department for Zoonoses, Animal Health and Welfare, and his associate, Birte Broberg as well as Finn Sørensen, head of the Veterinary and Meat Hygiene Advisory Section of the Veterinary Department of the Slaughterhouses Union, and Annette Weber, veterinary consultant to the Danish Animal Welfare Society. I also learned from publications of the Danish Animal Welfare Society, such as the *Danish Act on the Protection of Animals* (Act No. 386 of 6 June 1991 as amended by Act No. 183 of 14 April 1993) and those of The Danish Veterinary Service, particularly *The Animal Health and Disease Control Position in Denmark 1994,* and their handouts: "The Danish Plan for Implementation of the Zoonosis Directive" (March 30, 1995), "The Epidemiological Surveillance System in Denmark" (April 25, 1995), "The Eradication of Aujesky's Disease in Denmark" (July 1995), "Porcine Reproductive and Respiratory Syndrome (PRRS)—Status in Denmark" (May 1994), and "Trade with Live Animals and Animal Products in the Single Market of the EEC (September 29, 1992). See also Friland Food A/S, *Code of Practice, Friland Pig Production* (1994); and the series of articles on Denmark's pork industry by Karen McMahon in *National Hog Farmer* (September 15, 1997), pp. 6-12.
5. Interview with Finn Sørensen, the Slaughterhouses Union, Copenhagen, August 7, 1995. For an accessible overview, see: Danske Slagterier [The Federation of Danish Pig Producers and Slaughterhouses], *Statistics 1994* (May 1995).
6. Initial impressions were shaped through e-mail correspondence with Kevin McNamara, Olomouc, Czech Republic, 1995-1996. Refinements came with the help of University of Iowa student Kim Verdeck and her relatives who emigrated from Czechoslovakia to Cedar Rapids, Iowa—Lori Verdeck, Bob Maly and Jerry Stulz. Lori Verdeck told me, "Basically the association of *zabijačka* with what the student said is true. The 'pig slaughter' was not a violent act. The people were very attuned to nature and the cycles of life and death." Personal correspondence, Cedar Rapids, Iowa, April, 1997. See also the 1935 Lada print, "Zabijačka" in Václav Formánek, *Josef Lada* (Praha: Odeon, 1980), p. 27.
7. An author of more than eighty works of fiction and a Nobel Prize nominee, Hrabal was the son of a brewery manager, like the character named Jirí Schmitzer in "Cutting It Short." The imagery evokes provincial experience around the time of the author's birth (1914) on the eve of World War I. Bohumil Hrabal, "Cutting It Short," in *The Little Town Where Time Stood Still,* trans. James Naughton (London: Abacus, 1993), pp. 12-14.
8. Via Internet during the winter of 1995-1996, I solicited hog lore from liberal arts academics around the world. Kevin McNamara responded by asking students in his English class at Palacky University, Olomouc, Czech Republic, to write brief interpretations of a graphic, seasonal postcard, and he sent me a copy. The card featured a particularly graphic image by the so-called primitive artist, Josef Lada (1887-1957). (See page 260.) I here select excerpts from the anonymous papers that McNamara sent to me.
9. Of course, exceptions to these generalization abound. S. Jonathan Bass, for example, finds an important countertradition in the American South. "'How 'bout a Hand for the Hog': The Enduring Nature of the Swine as a Cultural Symbol in the South," *Southern Cultures* 1:3 (Spring 1995), pp. 301-320.
10. See: Stanley E. Curtis, "Animal Welfare Concerns in Modern Pork Production: An Animal Scientist's Analysis," paper presented to the Animal Welfare Committee of the U.S. Animal Health Association at Louisville, KY, November 4, 1980; Stanley E. Curtis and W. Ray Stricklin, "The Importance of Animal Cognition in Agricultural Animal Production Systems: An Overview," *Journal of Animal Science* 69 (1991), pp. 501-507; Hank Davis and Dianne Balfour, eds., *The Inevitable Bond* (New York: Cambridge University Press, 1992); Ian J. H. Duncan, "The Science of Animal Well-being," *Animal Welfare Information Center Newsletter* 4:1 (January-March 1993), pp. 1, 4-7; and Marlene Halverson, "Toward More Humane Husbandry," *ASPCA Animal Watch* (Summer 1995), p. 26. For examples of such work in producers' trade magazines, see: Paul Hemsworth, "Handle with Care," *National Hog Farmer* (November 15, 1993), p. 58; Karen McMahon, "Are You Scaring Your Pigs?" *National Hog Farmer* (October 15, 1995), pp. 48-49; "Research to Prove Happy Pigs are Productive Pigs," *National Hog Farmer* (February 15, 1995), p. 36; and "Sows Respond to TLC," *Pork '94* (January 1994), pp. 26-27.

11. See: Jane Goodall, *In the Shadow of Man* (Boston: Houghton Mifflin, 1971) or "Life and Death at Gombe," *National Geographic* 155:5 (1979), pp. 597-620; Aldo Leopold, *A Sand County Almanac* (New York: Oxford University Press, 1949); Sue Coe, *Dead Meat* (New York: Four Walls Eight Windows, 1995).
12. Interview with Stan Curtis, Pennsylvania State University, State College, June 3, 1994.
13. Temple Grandin speaking at her seminar, "Preserving Pork Quality Through Proper Hog Handling and Loading," at the World Pork Expo, Des Moines, Iowa, June 10, 1995:

> One of the things that's different between the U.S. and other countries is that things get very polarized, very extreme. And one of the problems today is that many children are totally separated from the cycle of nature. You know, for one living thing to survive, another living thing has to die. Many kids are totally separated from that reality. . . . People in Europe and even people in Canada are closer to the cycle, and they are more moderate, have a more sensible view of things like animal welfare.
>
> In Denmark they're very interested in animal welfare, but they're real sensible. In fact, their pork expo, like this, had a big display that shows exactly how a meat packing plant works. Our [U.S.] industry has a tendency to cover up how a meat packing plant works. I think that's a mistake because, I can tell you right now, what people imagine about a meat packing plant is ten times worse than what's actually in there.
>
> I took some of my [animal science] students to a large meat packing plant, and I'm not going to say they loved everything about it, but one thing they said was it wasn't anywhere near as bad as they thought it was going to be. And I asked my students, "What did you think you were going to find there?"
>
> "Oh, buckets of blood dripping down the walls and guts hanging off the ceiling. You know, all the terrible stuff."
>
> And another thing that people forget is that nature is harsh. I mean, if I had a choice: would I rather have a lion or a Stork Automatic Electric Stunner do me in? I think I'd much rather have the Stork Automatic Electric Stunner than a lion dining on my guts. People forget that nature isn't just all Disney World.

Grandin's major publications on hog handling include: "Behavior of Slaughter Plant and Auction Employees toward the Animals," *Anthrozoös: A Multidisciplinary Journal of the Interactions of People, Animals, and Environment* 1:4 (Spring 1988), pp. 203-213; "Canadians Understand Animal Handling," *Meat and Poultry* (September 1993), p. 18; "Environmental and Genetic Factors Which Contribute to Handling Problems at Pork Slaughter Plants," *Livestock Environment IV,* Fourth International Symposium, University of Warwick, Coventry, England, 6-9 July 1993, ed. Eldridge Collins and Chris Boon (St. Joseph, MI: American Society of Agricultural Engineers, 1993), pp. 64-68; "Hog Psychology: An Aid in Handling," *Agri-Practice* 9:4 (Fall 1988), pp. 22-26; "Management Attitude is Vital," *Meat and Poultry* (January 1991), pp. 56-57; "Minimizing Stress in Pig Handling," *Lab Animal: Information, Ideas, Methods and Materials for the Animal Research Professional* 15:3 (April 1986), pp. 1-5; "Nervous Pigs Produce More PSE," *Pork '93* (June 1993), pp. 17-18; *Recommended Animal Handling Guidelines for Meat Packers* (Arlington, VA: American Meat Institute, 1991); "Solving Livestock Handling Problems," *Veterinary Medicine* (October 1994), pp. 989-998; and especially the books that she edited, *Livestock Handling and Transport* (Tucson: University of Arizona Press, 1993) and *Genetics and the Behavior of Domestic Animals* (San Diego, CA: Academic Press, 1997).

14. Grandin, "Management Attitude Is Vital," p. 57.
15. Oliver W. Sacks, "An Anthropologist on Mars," *The New Yorker* (December 27, 1993/January 3, 1994), pp. 106-125; and *An Anthropologist On Mars: Seven Paradoxical Tales* (New York: Alfred A. Knopf, 1995). Grandin is also a vigorous advocate on behalf of people with autism and people who are visually oriented. See, for example: Temple Grandin, *Emergence: Life with Autism* (New York: Warner Books, 1996), and *Thinking in Pictures and Other Reports from My Life with Autism* (New York: Doubleday, 1995).
16. Stephen Budiansky best develops his arguments about the relations between the human and non-human world in: *Man and Beast: The Origins of Domesticated Animals and of Nature* (New York:

W. Morrow, 1992), and *Nature's Keepers: the New Science of Nature Management* (New York: Free Press, 1995).

17. Curtis interview. Stanley E. Curtis, "Ethology: Pigs and People," in *Swine Nutrition,* ed. Elwyn R. Miller, Duane E. Ullrey, and Austin J. Lewis (Boston: Butterworth-Heineman, 1991), p. 31; Tina M. Widowski and Stanley E. Curtis, "The Influence of Straw, Cloth Tassel, or Both on the Prepartum Behavior of Sows," *Applied Animal Behaviour Science* 27 (1990), pp. 53-71; Dale Miller, "A Pig's Intellect: Video Games Help Researchers Peek Inside Pigs' Minds," *National Hog Farmer* (October 15, 1997), pp. 52-55; Paul H. Hemsworth, "Behavioral Problems," in *Diseases of Swine* 7th Ed., ed. Allen D. Leman et al. (Ames: Iowa State University Press, 1992), pp. 653-658.

18. Grandin adds: "The Machinefabriek, G. Nihjuis B.V. in Winterswijk, Holland, named their most highly automated equipment 'Valhalla.' In Nordic mythology, Valhalla is the paradise for warriors who died gloriously in battle." Grandin, "Behavior of Slaughter Plant and Auction Employees," p. 211.

    When I visited Stan Curtis, who has also dedicated much of his professional life to improving animal welfare, the last thing he gave to me was a poem, "Butchering" by Hadley Read. It graphically describes the killing of a hog, the "slashing strokes" and "tubs of intestines," the processing of parts and byproducts, and concludes: "I guess that pig would never know how many things he meant to us." Curtis wistfully advised, "There's a man who really cares about pigs." Curtis Interview; Hadley Read, *Morning Chores and Other Times Remembered* (Urbana: University of Illinois Press, 1977), pp. 173-177.

19. Garrison Keillor, "Hog Slaughter," from original monologues (1983), published in *Meat and Poultry* (August 1989), pp. 23-25. Reprinted by permission of Garrison Keillor.

20. Temple Grandin, "A 'Hog Slaughter' Commentary," *Meat and Poultry* (August 1989), p. 26. Grandin credits the poem to her roommate, Gloria Tester, in 1974. Grandin, "Behavior of Slaughter Plant and Auction Employees." A soul-searching hunter similarly confesses: "I am not a guiltless hunter, but neither do I hunt without joy. What fills me now is an incongruous mix of grief and satisfaction, excitement and calm, humility and pride. And the recognition that death is the rain that fills the river of life inside us all." Richard Nelson, "The Hunt: Wrestling with the Guilty Pleasures of Life in the Wild," *Utne Reader* (March-April 1998), p. 103.

INDEX

Richard P. Horwitz is professor of American studies at the University of Iowa, as well as senior fellow of The Coastal Institute in Narragansett, Rhode Island. Many of the ideas for this book emerged during his part-time labors; he moonlighted for twenty years as a hired hand on a 2,000-acre hog, cattle, and grain farm. His other books include *The American Studies Anthology, Exporting America, The Strip: An American Place,* and *Anthropology toward History.*